Advances in Intelligent and Soft Computing 140

Editor-in-Chief: J. Kacprzyk

Advances in Intelligent and Soft Computing

Editor-in-Chief

Prof. Janusz Kacprzyk
Systems Research Institute
Polish Academy of Sciences
ul. Newelska 6
01-447 Warsaw
Poland
E-mail: kacprzyk@ibspan.waw.pl

Further volumes of this series can be found on our homepage: springer.com

Vol. 124. Yinglin Wang and Tianrui Li (Eds.)
Practical Applications of Intelligent Systems, 2011
ISBN 978-3-642-25657-8

Vol. 125. Tianbiao Zhang (Ed.)
Mechanical Engineering and Technology, 2011
ISBN 978-3-642-27328-5

Vol. 126. Khine Soe Thaung (Ed.)
Advanced Information Technology in Education, 2011
ISBN 978-3-642-25907-4

Vol. 127. Tianbiao Zhang (Ed.)
Instrumentation, Measurement, Circuits and Systems, 2011
ISBN 978-3-642-27333-9

Vol. 128. David Jin and Sally Lin (Eds.)
Advances in Multimedia, Software Engineering and Computing Vol.1, 2011
ISBN 978-3-642-25988-3

Vol. 129. David Jin and Sally Lin (Eds.)
Advances in Multimedia, Software Engineering and Computing Vol.2, 2011
ISBN 978-3-642-25985-2

Vol. 130. Kusum Deep, Atulya Nagar, Millie Pant, and Jagdish Chand Bansal (Eds.)
Proceedings of the International Conference on Soft Computing for Problem Solving (SOCPROS 2011) December 20–22, 2011, 2012
ISBN 978-81-322-0486-2

Vol. 131. Kusum Deep, Atulya Nagar, Millie Pant, and Jagdish Chand Bansal (Eds.)
Proceedings of the International Conference on Soft Computing for Problem Solving (SOCPROS 2011) December 20–22, 2011, 2012
ISBN 978-81-322-0490-9

Vol. 132. Suresh Chandra Satapathy, P.S. Avadhani, and Ajith Abraham (Eds.)
Proceedings of the International Conference on Information Systems Design and Intelligent Applications 2012 (INDIA 2012) held in Visakhapatnam, India, January 2012, 2012
ISBN 978-3-642-27442-8

Vol. 133. Sabo Sambath and Egui Zhu (Eds.)
Frontiers in Computer Education, 2012
ISBN 978-3-642-27551-7

Vol. 134. Egui Zhu and Sabo Sambath (Eds.)
Information Technology and Agricultural Engineering, 2012
ISBN 978-3-642-27536-4

Vol. 135. Honghua Tan (Ed.)
Knowledge Discovery and Data Mining, 2012
ISBN 978-3-642-27707-8

Vol. 136. Honghua Tan (Ed.)
Technology for Education and Learning, 2012
ISBN 978-3-642-27710-8

Vol. 137. Jia Luo (Ed.)
Affective Computing and Intelligent Interaction, 2012
ISBN 978-3-642-27865-5

Vol. 138. Gary Lee (Ed.)
Advances in Intelligent Systems, 2012
ISBN 978-3-642-27868-6

Vol. 139. Anne Xie and Xiong Huang (Eds.)
Advances in Electrical Engineering and Automation, 2012
ISBN 978-3-642-27950-8

Vol. 140. Anne Xie and Xiong Huang (Eds.)
Advances in Computer Science and Education, 2012
ISBN 978-3-642-27944-7

Anne Xie and Xiong Huang (Eds.)

Advances in Computer Science and Education

Springer

Editors
Prof. Dr. Anne Xie
Guangzhou Section of ISER
Association
Guangzhou
China

Prof. Dr. Xiong Huang
Guangzhou Section of ISER
Association
Guangzhou
China

ISSN 1867-5662 e-ISSN 1867-5670
ISBN 978-3-642-27944-7 e-ISBN 978-3-642-27945-4
DOI 10.1007/978-3-642-27945-4
Springer Heidelberg New York Dordrecht London

Library of Congress Control Number: 2011945548

Printed on acid-free paper

Springer is part of Springer Science+Business Media (www.springer.com)

Preface

CSE2011 is an integrated conference concentrating its focus upon Computer Science and Education. In the proceeding, you can learn much more knowledge about Computer Science and Education of researchers all around the world. The main role of the proceeding is to be used as an exchange pillar for researchers who are working in the mentioned field. In order to meet high standard of Springer, AISC series ,the organization committee has made their efforts to do the following things. Firstly, poor quality paper has been refused after reviewing course by anonymous referee experts. Secondly, periodically review meetings have been held around the reviewers about five times for exchanging reviewing suggestions. Finally, the conference organization had several preliminary sessions before the conference. Through efforts of different people and departments, the conference will be successful and fruitful.

CSE2011 is co-sponsored by International Science & Education Researcher Association, Beijing Gireida Education Co.Ltd and Wuhan University of Science and Technology, China. The goal of the conference is to provide researchers from Computer Science and Education based on modern information technology with a free exchanging forum to share the new ideas, new innovation and solutions with each other. In addition, the conference organizer will invite some famous keynote speaker to deliver their speech in the conference. All participants will have chance to discuss with the speakers face to face, which is very helpful for participants.

During the organization course, we have got help from different people, different departments, different institutions. Here, we would like to show our first sincere thanks to publishers of Springer, AISC series for their kind and enthusiastic help and best support for our conference. Secondly, the authors should be thanked too for their enthusiastic writing attitudes toward their papers. Thirdly, all members of program chairs, reviewers and program committees should also be appreciated for their hard work.

In a word, it is the different team efforts that they make our conference be successful on November 26–27, Wuhan, China. We hope that all of participants can give us good suggestions to improve our working efficiency and service in the future. And we also hope to get your supporting all the way. Next year, in 2012, we look forward to seeing all of you at CSE2012.

November 2011 MSEC2011 & CSE2011 Committee

Committee

Honor Chairs

Prof. Chen Bin	Beijing Normal University, China
Prof. Hu Chen	Peking University, China
Chunhua Tan	Beijing Normal University, China
Helen Zhang	University of Munich, China

Program Committee Chairs

Xiong Huang	International Science & Education Researcher Association, China
LiDing	International Science & Education Researcher Association, China
Zhihua Xu	International Science & Education Researcher Association, China

Organizing Chair

ZongMing Tu	Beijing Gireida Education Co.Ltd, China
Jijun Wang	Beijing Spon Technology Research Institution, China
Quanxiang	Beijing Prophet Science and Education Research Center, China

Publication Chair

Song Lin	International Science & Education Researcher Association, China
Xionghuang	International Science & Education Researcher Association, China

International Committees

Sally Wang	Beijing Normal University, China
LiLi	Dongguan University of Technology, China
BingXiao	Anhui University, China
Z.L. Wang	Wuhan University, China
Moon Seho	Hoseo University, Korea
Kongel Arearak	Suranaree University of Technology, Thailand
Zhihua Xu	International Science & Education Researcher Association, China

Co-sponsored by

International Science & Education Researcher Association, China
VIP Information Conference Center, China

Reviewers of CSE2011

Chunlin Xie	Wuhan University of Science and Technology, China
LinQi	Hubei University of Technology, China
Xiong Huang	International Science & Education Researcher Association, China
Gangshen	International Science & Education Researcher Association, China
Xiangrong Jiang	Wuhan University of Technology, China
LiHu	Linguistic and Linguidtic Education Association, China
Moon Hyan	Sungkyunkwan University, Korea
Guangwen	South China University of Technology, China
Jack H. Li	George Mason University, USA
Marry Y. Feng	University of Technology Sydney, Australia
Feng Quan	Zhongnan University of Finance and Economics, China
PengDing	Hubei University, China
Songlin	International Science & Education Researcher Association, China
XiaoLie Nan	International Science & Education Researcher Association, China
ZhiYu	International Science & Education Researcher Association, China
XueJin	International Science & Education Researcher Association, China
Zhihua Xu	International Science & Education Researcher Association, China
WuYang	International Science & Education Researcher Association, China
QinXiao	International Science & Education Researcher Association, China
Weifeng Guo	International Science & Education Researcher Association, China
Li Hu	Wuhan University of Science and Technology, China
ZhongYan	Wuhan University of Science and Technology, China
Haiquan Huang	Hubei University of Technology, China
Xiao Bing	WUhan University, China
Brown Wu	Sun Yat-Sen University, China

Contents

Erratum

Improvement Model of Business Continuity Management on E-Learning

Gang Chen

Shanghai University of Finance and Economic, Shanghai, P.R. China, 200433
School of Economic and Management, Tongji University, Shanghai

Abstract. With the development of IT technology, Business Continuity Management (BCM) should improve to become the most important assurance of E-learning. Improvement model of BCM has seven parts: scope of business continuity management, environmental analysis, risk assessment, business continuity management strategy design, business continuity plan design, training implementation and maintenance, testing and assessment of BCM. Improvement model of business continuity management aims to ensure business continuity and reduce the cost. It can ensure business continuity capabilities and help organizations to confirm key factors affecting business development and the possible threats.

Keywords: Business Continuity Management, Improvement Model, E-learning, Risk analysis

1 Introduction

With globalization of economic and rapid development of science and technology, E-learning survival environment has undergone dramatic changes. Business continuity management is an inevitable outcome of business development in information age. As the developing of E-learning, E-learning management is getting complex and competitive increasingly. With the large number of applications of IT technology, infrastructures which E-learning relies on are increasing complexity. With this complexity increases, business continuity management of E-learning should be improved.

2 Improvement Model of Business Continuity Management on E-Learning

Improvement model of business continuity management on E-learning has seven parts: scope of business continuity management, environmental analysis, risk assessment, business continuity management strategy design, business continuity plan design, training implementation and maintenance, testing and assessment of BCM as shown figure 1.

A. Xie & X. Huang (Eds.): Advances in Computer Science and Education, AISC 140, pp. 1–5.
springerlink.com

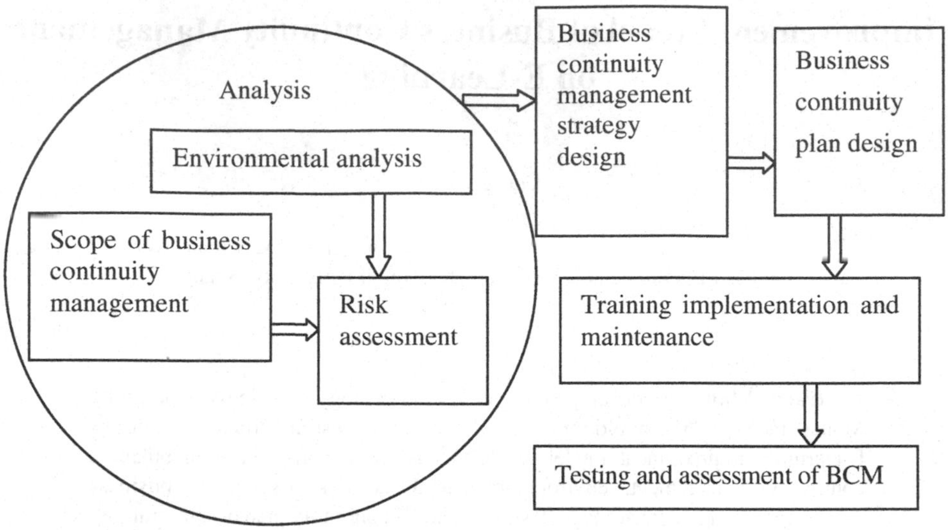

Fig. 1. Improvement model of business continuity management

2.1 Scope of Business Continuity Management

Scope of Business continuity management includes business service grading and sorting. It also should decompose business target into sub-target, grouped the same or similar objects into categories and set management process for each category. Managers should analysis affect of every category which disrupts business activities.

2.2 Environmental Analysis

Environmental of E-learning includes information architecture, application architecture and technical architecture.Information architecture is an information object in order to establish the organization and inter-organizational information model. The purpose of information architecture is to ensure sharing, integration, reuse information and to determine the basic requirements of information management. Application architecture includes information systems which service E-learning such as teaching system, exam system, course resources system, students resign system and teacher service system. Technical architecture includes software, hardware, net work and database center. Software includes operation software and application software. Hardware includes computer system and other equipments in computer center. Net work includes router, switch and other communication software and hardware.

2.3 Risk Assessment

Risk assessment tries to find the problems of E-learning and estimating potential losses of risk. Risk assessment is the tools used by managers to balance risks and security investment. According to organization's situation, managers set up a plan of risk assessment. Managers identify risk according to major elements of risk and their mutual relations. Major elements of risk are assets, threats, weak points and impact risk. The function of Risk (R) has parameters as asset (A), threats (T), weak point (V), impact (I) and security control (C). The function is as following:

R=F (A, T, V, I, C)

The relationship between risk function parameters is shown as figure 2.

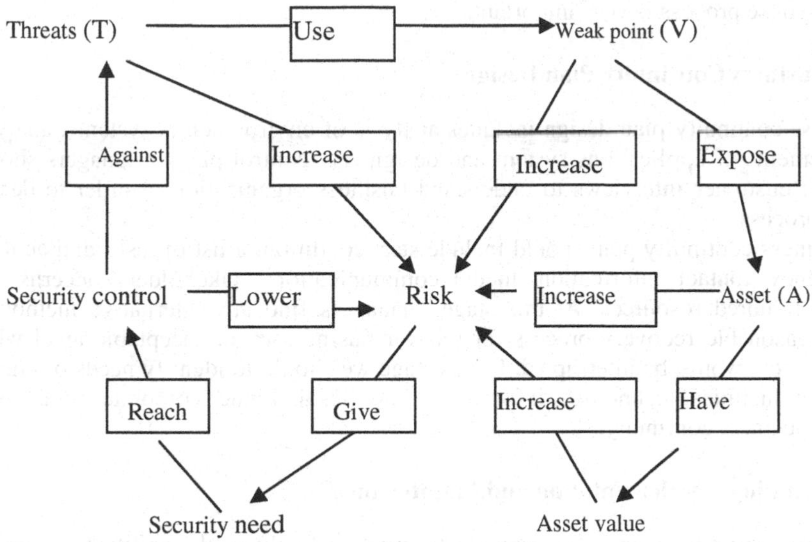

Fig. 2. Relationship between risk function parameters

Assets are anything of value to the organization. Assets have value. Managers should be exhausted less redundant and to identify the organization's important assets as comprehensive as possible.

Weak point is an asset or asset group which can be exploited weaknesses. Weak points of assets exposed to the threat.

Increase of threats, weak points or negative growth of assets value may lead to increase of risk. Managers assess harm caused by risks. According to risk assessment guidelines calculate levels of risk. Managers should confirm right treatment and acceptable of remaining risk if unacceptable risk from previous analysis could not be solved.

2.4 Business Continuity Management Strategy Design

In dynamic environment managers should design business continuity management strategy based on deeply understanding of organization's operational objectives, resources, costs and risk. The strategy should reduce risk and efficient. According to results of risk assessment, business continuity management strategy would be set up with a clear business priorities and identify implementation goals. The main law of determine business continuity management strategy is implement appropriate measures to reduce the risk of accidents and reduce the potential impact of incidents. It should be considered of measures degree, accident and accident protection for the continuity of critical activities, key activities which would not be recognized enough such as human resource, infrastructure and technical facilities.

Business continuity management strategy is a guideline for implementation, so be sure to describe method, class, steps of strategy clearly. A clear response structure and response process is very important.

2.5 Business Continuity Plan Design

Business continuity plan design includes analysis of environmental systems, analysis requirements of applications system and design risk control plans. Managers should conduct customer interviews to understand business organization in order to design BCM process.

Business continuity plan should include start condition, a list of tasks and actions, emergency contact information, media communication, stakeholder concerns and restore required resources. At this stage, managers find any alternative method to make reasonable recovery process for recover business in an acceptable level when key business would be interrupted. In this stage we should to identify needs of human resource, facilities, technology, information, supplies and other corporate resources to ensure business continuity.

2.6 Training, Implementation and Maintenance

Business continuity plan is not enough. In order to facilitate the ability for promote service delivery, testing improvement and maintenance of Business continuity plans is more important. Business continuity management training ensures smooth implementation of plan. The plan also should be improved according to the problems which are found in training.

Training, implementation and maintenance include BCM installation and commissioning, Training staff, preparation of processes and service programs, writing manual and guide. Training, implementation and maintenance are an indispensable stage of business continuity management. This indispensable stage help BCM team make familiar with process and ensure BCM system effectively.

Training and Implementation ensure effectiveness, accuracy and practicality of business continuity management. According to Business continuity plan, training BCM team members ensure business continuity management plan can be regularly verified and continuous improvement. Maintenance refers regularly revised and updated procedures to ensure that the plan would not be out of time.

2.7 Testing and Assessment of BCM

Testing and assessment of business continuity management ensure that business continuity management strategies could be adjust and adapt to changing environmental. It is the most important stage in business continuity management life cycle. Testing and assessment of business continuity management include project preparation and formulation, incident response. When business is interrupted, it should be maintained though step by step recovery. At this stage, senior manager should check plan of business continuity management completely in order to make sure whether the plane is adequate, effective and meet the business continuity wants.

If it is found that the process of plan is unreasonable and insufficient, we must take corrective to prevent the business continuity management in an unacceptable state. Managers should examine business continuity management to ensure BCM is running well, and monitor performance.

3 Conclusion

In order to adapt dynamic environment, improvement of business continuity management can be defined as a way to deal with most crises and to adapt changes in the environment. Thus it can ensure business continuity capabilities and help organizations to confirm key factors affecting business development and the possible threats. Improvement model of business continuity management aims to ensure business continuity and reduce the cost.

Acknowledgment. This work is supported by Leading Academic Discipline Program 211 Project for Shanghai University of Finance and Economics (the 3rd phase).

References

1. Chen, G.: Design service level agreement based on dynamic IT. In: 2009 Pacific-Asia Conference on Knowledge Software Engineering, KESE 2009, pp. 21–24 (2009)
2. Chen, G.: BCM mechanism based on infinite-horizon growth model in E-commerce. In: 2nd International Symposium on Electronic ISECS 2009, vol. 1, pp. 435–438 (2009)
3. Chen, G.: Design of Digital Educational Resource Platforms in University. In: Proceedings - International Symposium on Computer Science and Computational, pp. 142–145 (2008)
4. Chen, G.: IT service system design and management of educational information system. In: 1st International Conference on Information Science and Engineering, ICISE 2009, pp. 307–310 (2009)

Using Computer Simulation to Improve Teaching and Learning Effectiveness on DC/DC Converter

LiPing Fan*

College of Information Engineering, Shenyang University of Chemical Technology, Shenyang, 110142, P.R China
flpsd@163.com

Abstract. DC/DC converter is an important component element of the course of Power Electronics for electrical engineering and automation specialties. The complex topological structure and the inenarrable working processes make it difficult to finish teaching and learning smoothly. Engineering simulation technology is integrated into the teaching process to improve the teaching and learning effectiveness on DC/DC converter. Practice results show the validity of this teaching method combining computer simulation.

Keywords: DC/DC converter, computer simulation, teaching effectiveness.

1 Introduction

Power Electronics is an important area within electrical engineering, which includes advances in semiconductor technology and new applications in industrial power electronic equipment, such as power converter, variable-speed motor drivers, and others. Nowadays, more than 75% of all generated electric power was refined by power electronic converters into various forms of power sources, which are optimally satisfied with the requirements of the different user loads [1]. Power electronics is a very difficult and complex technical area for students. The complex topological structure and the inenarrable working processes make it difficult to finish teaching smoothly. Education in Power Electronics is enough daunting for students due to difficulties at proper description complexity of physical phenomena, understanding the circuit operation and complicated mathematical description [2]. Traditional learning and training methods are not able to cope with demand of easy understanding the matter. In education of power electronics, a proper balance between different teaching methods is requested, with final goal – development of student skills. There are many interesting and good papers describing current state and trend in education of power electronics [3-5]. To make teaching highly efficient and successful, it is required to combine traditional teaching approaches with new learning techniques.

The advances in information technology have reflected themselves in development of modern education methodologies. They are also changing the traditional teaching

* This work was supported by the Project of Teaching Reform in Higher Education of Liao Ning Province, China (No.2010-2-3-53), and the Project of Teaching Reform in Education of Shenyang University of Chemical Technology (No. 2009A006).

styles in high education. When digital computers or microprocessors as well as various software and hardware are used in classrooms and/or laboratories, it can provide a much more effective and efficient way in teaching and learning, and make a lot of mathematics related to engineering problems much easier to understand. This is also true for teaching the course of Power Electronics [6].

Engineering simulation is one of the most important tools in engineering education [7]. Simulation and animation often enrich modern education in the field of power electronics. MATLAB is a widely used tool in electrical engineering community. It is now accepted that MATLAB and its numerous toolboxes can replace and/or enhance the usage of traditional simulation tools for advanced engineering applications [8].

Within the context of teaching power electronic technology, MATLAB/Simulink is used as a tool to augment effectiveness of teaching and learning. This paper describes using simulation in teaching DC/DC converter.

2 Problem Description

DC-DC converters have been widely used in most of the industrial applications such as DC motor drives, computer systems and communication equipments. It is an important component element of the course of Power Electronics for electrical engineering and automation specialties. The principle of power conversion is the main content of learning converter. Wave analysis is an important part of explaining the principle of various power converters. Usually, in order to explain these waves to students, some skilled teacher like drawing these waves in blackboard, but it will spend much time, and it is difficult to draw all waves because of the complexity of most waves. For some other teachers, they would rather let students watch pictures in textbook, and in the meantime explain these waves. Such teaching method usually can not attract student and can not receive good teaching effect. More effective teaching methods are needed.

3 Simulation Example

A basic DC-DC converter circuit known as the Buck converter is illustrated in Fig. 1, consisting of one switch VT, a fast diode VD and RLC components. The switch can be implemented by one of three-terminal semiconductor switches, such as IGBT or MOSFET. Such a buck converter is the most typical structure that must be introduced in the course of Power Electronics.

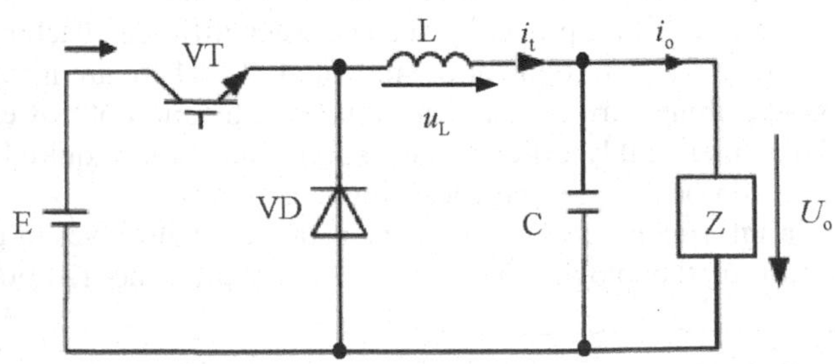

Fig. 1. Buck DC/DC converter

Use the MATLAB SimPowerSystem blockset can set up the Simulink simulation model of the DC/DC converter circuit. The MATLAB simulation model is shown in Fig. 2. The simulation results are shown in Fig. 3 and Fig. 4.

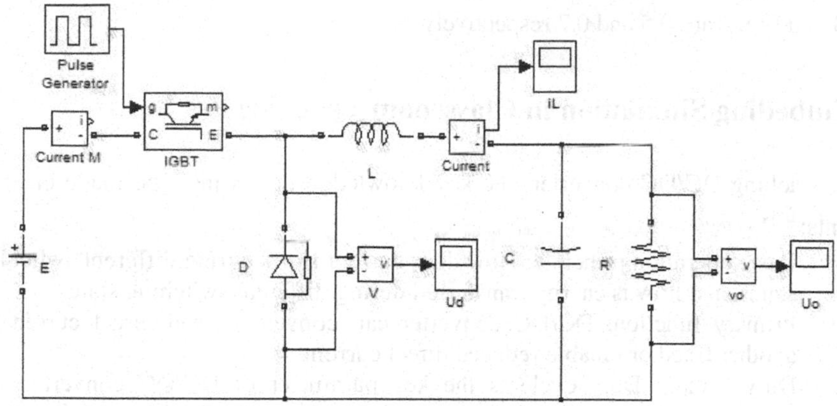

Fig. 2. Simulation model of Buck DC/DC converter

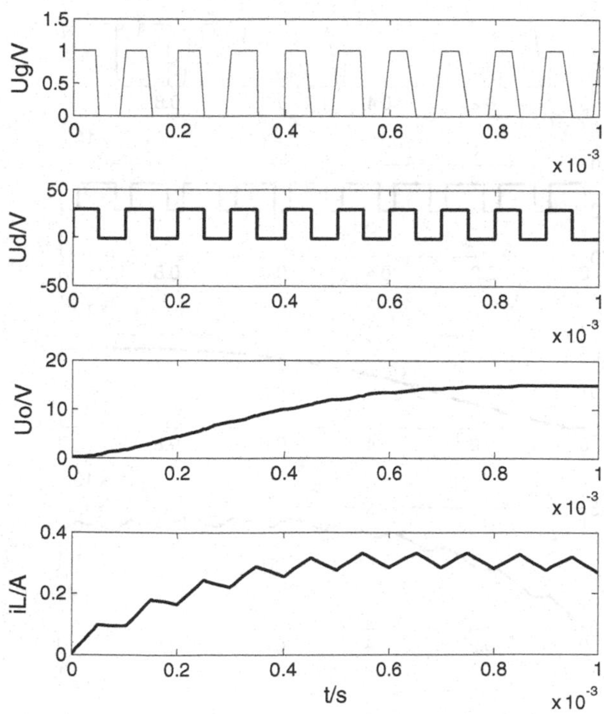

Fig. 3. Simulation operation waves of Buck DC/DC converter (D=0.5)

The designing requirement of the converter is: input voltage E=30V, the switching frequency is 10 kHz, the output voltage is 15~25V. According to the designing requirements, the main parameters used in the simulation are designed as follows: L is chosen as 15mH, C is chosen as 3.5 μ F. The load resistance is 50 Ω . The duty cycle of Fig. 3 and Fig. 4 are 0.5 and 0.7 respectively.

4 Embeding Simulation in Classroom Teaching

When teaching DC/DC converter, the key knowledge points must be made clear to students:

- Basic operation principle: How does current flows during different switching states, and how is energy transferred during different switching states
- Primary function: DC/DC converter can convert a fixed direct current to another fixed or tunable reduced direct current.
- Duty cycle: Duty cycle is the key parameter of DC/DC converter. By regulating duty cycle, the output direct current which satisfying load requirement can appear.

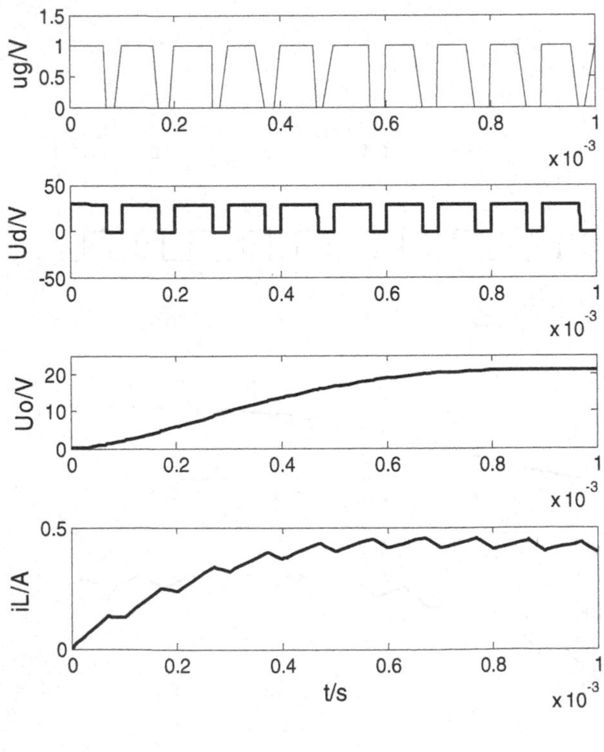

Fig. 4. Simulation operation waves of Buck DC/DC converter (D=0.7)

To explain these knowledge points clearly to students, the waves for input and output are needed to shown to the students. Here we can switch the scene from blackboard to the MATLAB/Simulink simulation platform, carry out a simulation running of the buck DC/DC converter circuit shown in Fig.1, and then show the simulation running results shown in Fig. 3 and Fig. 4 to students, explain how the DC/DC converter works contrasting with these simulation waves. By observing the simulation running process from beginning to end, students can find the change as time goes on, Meanwhile, by comparing the respective waves corresponding to different duty cycle, such as shown in Fig. 3 and Fig. 4, students can understand the important parameter of duty cycle D, and experience the varying process of output waves along with D. Students usually feel as though suddenly seeing the light.

5 Conclusions

Integrating computer simulation technology into the teaching process of DC/DC converter can make the classroom teaching vivid and valid, and stimulate the students' learning enthusiasm on the course of power electronics. Meanwhile, by using computer simulation in learning power electronics, students can grasp more knowledge about computer software and hardware simultaneously.

References

1. Wang, S.C., Chen, Y.C., Su, J.H.: Teaching Switching Converter Design Using Problem-Based Learning with Simulation of Characterization Modeling. Journal of Power Electronics 10(6), 595–603 (2010)
2. Bauer, P., Fedak, V.: Philosophy of Interactive e-Learning for Power Electronics and Electrical Drives: a Way from Ideas to Realization. Journal of Power Electronics 10(6), 587–594 (2010)
3. Hurley, W.G., Lee, C.K.: Development, Implementation and Assessment of a Web-based Power Electronics Laboratory. IEEE Transactions on Education 48(4), 567–573 (2005)
4. Williams, J.M., Cale, J.L., Benavides, N.D., et al.: Versatile Hardware and Software Tools for Educating Students in Power Electronics. IEEE Transactions on Education 47(4), 436–445 (2004)
5. Dudrik, J.: New Methods in Teaching of Power Electronics Devices. Iranian Journal of Electrical and Computer Engineering 4(2), 117–120 (2005)
6. Jakopovic, Z., Sunde, V., Bencic, Z.: Undergraduate Power Electronics Laboratory–Applying TSMST Method. Journal of Power Electronics 10(6), 621–628 (2010)
7. Ibrahim, D.: Engineering Simulation with MATLAB: Improving Teaching and Learning Effectiveness. Procedia Computer Science 3, 853–858 (2011)
8. Varadarajan, M., Valsan, P.: MatPECS-A MATLAB-based Power Electronic Circuit Simulation Package with GUI for Effective Classroom Teaching. International Journal of Engineering Education 21(4), 606–611 (2005)

[illegible] effectively [illegible]

[illegible]

The Improved Diagonal-Matrix-Weight IMM Algorithm Based on LDU Factorization

Liang Gao[1], Jianping Xing[1,*], Fan Shi[2], Zhenliang Ma[1], Junchen Sha[1], and Juan Sun[1]

[1] School of Information Science and Engineering, Shandong University, 250100 Jinan, China
[2] Dept. of Network Engineering Electronic Engineering Institude 230037 Hefei, China
gaoliang2004319@gmail.com, xingjp@sdu.edu.cn

Abstract. The diagonal-matrix-weight IMM (DIMM) algorithm can solve the IMM algorithm confusions of probability density functions (PDFs) and probability masses of stochastic process. However, the DIMM algorithm may generate calculating divergence, caused by the fact that the state error covariance matrix is repeatedly applied to calculate the inverse matrix. A new method, LDU factorization diagonal-matrix-weight IMM (LDU-DIMM) algorithm, is proposed in this paper to solve the factorization of asymmetric matrix error covariance of DIMM algorithm. Experiment results verify the effectiveness of the proposed algorithm.

Keywords: LDU factorization, IMM, Diagonal-matrix-weight.

1 Introduction

The state estimation problem of discrete-time stochastic systems with Markov switching parameters is always the focus of interest in the community of maneuvering target tracking. The IMM algorithm has confusions of probability density functions (PDFs) and probability masses that denote the continuous-valued and discrete-valued parameter of stochastic process respectively, which usually has been neglected [1]. In the process of IMM algorithm, updating weights of models are derived from the mixture of PDFs and probability masses, however, probability mass must be a value in the interval [0, 1], and any PDF has no such restriction. Thus, the two kinds of values are at different levels that lead to a certain error in many cases.

There have been many related researches to improve the IMM algorithm. For example, Johnstona.L et.al [2] derived a weighted IMM with a recursive implementation of maximum posteriori (MAP) state sequence estimator; Genovese et.al [3] proposed an algorithm of accurate state estimation of targets with changing dynamics can be achieved through the use of multiple filter models. Dahmani Mohammed et.al [4] gives a new approach based on two filters $\alpha\beta$ and $\alpha\beta\gamma$ using IMM design instead of a Kalman filter second and third order for the tracking maneuver target. But these improvements can't be devoted to the problem that the confusion of probability density functions (PDFS) and

* Corresponding author

A. Xie & X. Huang (Eds.): Advances in Computer Science and Education, AISC 140, pp. 13–21.
springerlink.com

probability masses in IMM algorithm. An improved IMM algorithm [1], using multi-model fusion criterions weighted by diagonal matrices, respectively, together with the step of calculating the mixed initial model weights, state and corresponding covariance, named diagonal-matrix-weight IMM (DIMM), was proposed to solve the above problem.

However, there are some shortcomings of the DIMM, for example, calculating divergence, leading to the deteriorated filter performance, which the round-off error is caused by the fact that the state error covariance matrix was repeatedly applied to calculate the inverse matrix in the updating diagonal matrix weight step of DIMM. System parameters setting up unsuitably will create calculating divergence. The U-D factorization for the error covariance is used to improve the curacy and calculating divergence of Kalman filter in [5]. U-D factorization is applied to symmetric matrix of Kalman filter, so we proposed the LDU factorization to solve the factorization of asymmetric matrix error covariance of DIMM. In this paper, an improved IMM algorithm is proposed: LDU-DIMM. The simulation example shows that the proposed algorithm is effective solving the problem of calculating divergence. In particular, the LDU-DIMM algorithm is superior to the IMM algorithm in accuracy estimation.

2 The LDU-DIMM Algorithm

In this section, we describe the Markov jump linear systems target tracking equation, then illustrating the parameters of diagonal-matrix-weight IMM algorithm based on LDU factorization (LDU-DIMM), and the last is providing the steps of LDU-DIMM algorithm.

2.1 Markov Jump Linear Systems Tracking Equation

Consider the following Markov jump linear system [1]

$$x(k+1)=F_j(k)+T_j\omega_j(k) \tag{1}$$

$$z(k+1)=H_j(k)+v_j(k) \tag{2}$$

Where the state vector $x(k)$ is an n-dimensional vector, the observation process $z(k)$ is an m-dimensional vector and the subscript $j\in S=\{1,2,\cdots s\}$ denotes the model. The matrix functions, F_i, T_i and H_i are known. The model dependent process noise is assumed to be a Gaussian random process with

$$E\left[\omega_j(k)\right]=0,\quad E\left[\omega_j(k)\omega_j(k)^T\right]=Q_j \tag{3}$$

The model dependent measurement noise term is also assumed to be a Gaussian random process with

$$E\left[v_j(k)\right]=0,\quad E\left[v_j(k)v_j(k)^T\right]=R_j \tag{4}$$

Let M_j^k denote the model j at time k. The model dynamics are model led as a finite Markov chain with known model-transition probabilities from model i at time $k-1$ to model j at time k.

$$\pi_{ij} \triangleq prob\left\{M_j^k \mid M_j^{k-1}\right\} = P\left\{M_j^k \mid M_j^{k-1}\right\} \tag{5}$$

$$0 \leq \pi_{ij} \leq 1, \quad \sum_{j=1}^{s} \pi_{ij} = 1, \quad i, j \in S \tag{6}$$

The initial state distribution of the Markov chain is $\varphi = \left[\varphi_1, \ldots, \varphi_s\right]$, where

$$0 \leq \varphi_j \leq 1, \quad \sum_{j=1}^{s} \varphi_j = 1, \quad j \in S \tag{7}$$

This Markov chain description of the target's models is used to model the unknown inputs.

2.2 The Steps of LDU-DIMM Algorithm

In order to solve the problem of calculating divergence which caused by the state error covariance matrix was repeatedly applied to calculate the inverse matrix in the step of updating diagonal-matrix-weight, the method of LDU factorization used to factorize the asymmetric matrix error covariance of DIMM algorithm. The steps of algorithms are as follows:

Step1: Calculating the mixed initial diagonal-matrix-weight for the filter matched to model $M_j^k \left(j \in S\right)$

$$B_{i|j}\left(k \mid k\right) \triangleq P\left\{M_i^{k-1} \mid M_j^k, Z^{k-1}\right\} = \frac{\pi_{ij} B_i^{k-1}}{\sum_{i=1}^{s} \pi_{ij} B_i^{k-1}}$$

$$= \begin{pmatrix} \dfrac{\pi_{ij} b_{i1}}{\sum_{i=1}^{s} \pi_{ij} \pi_{ij} b_{i1}} & \cdots & 0 \\ \vdots & \ddots & \vdots \\ 0 & \ddots & \dfrac{\pi_{ij} b_{in}}{\sum_{i=1}^{s} \pi_{ij} \pi_{ij} b_{in}} \end{pmatrix}, \quad i, j \in S \tag{8}$$

$$\text{where } B_i^{k-1} = diag\left(b_{i1}, b_{i2}, \cdots, b_{in}\right) \triangleq P\left\{M_i^{k-1} \mid Z^{k-1}\right\} \tag{9}$$

Step2: Calculating the mixed initial state and corresponding covariance for the filter matched to model $M_j^k\left(j\in S\right)$

$$\hat{x}_{0j}\left(k\mid k\right)=\sum_{i=1}^{s}B_{i|j}\left(k\mid k\right)\hat{x}_j^{k-1} \tag{10}$$

$$\begin{aligned}P_{0j}\left(k\mid k\right)=\sum_{l-1}^{s}B_{i|j}\left(k\mid k\right)\{P_i^{k-1}+\left[\hat{x}_j^{k-1}-\hat{x}_{0j}\left(k\mid k\right)\right]\\ \times\left[\hat{x}_j^{k-1}-\hat{x}_{0j}\left(k\mid k\right)\right]^T\}\end{aligned} \tag{11}$$

Step3: Kalman filtering and factorize the asymmetric matrix error covariance

$$\hat{x}_j\left(k\mid k-1\right)=F_j\hat{x}_{0j}\left(k\mid k\right) \tag{12}$$

$$\begin{aligned}P_j\left(k\mid k-1\right)&=F_jP_{0j}\left(k\mid k\right)F_j^T+T_jQ_jT_j^T\\ &=L_j\left(k\right)U_j\left(k\right)D_j\left(k\right)\end{aligned} \tag{13}$$

Where L_j is a lower triangular matrix, U_j is a diagonal matrix, D_j is a upper triangular matrix

$$E_j\left(k\right)=D_j\left(k\right)U_j\left(k\right)H_j^T \tag{14}$$

$$G_j\left(k\right)=L_j\left(k\right)A_j\left(k\right) \tag{15}$$

$$S_j\left(k\right)=H_j^TG_j\left(k\right)+R_j \tag{16}$$

$$K_j\left(k\right)=P_j\left(k\mid k-1\right)H_j^T\left(H_jP_j\left(k\mid k-1\right)H_j^T+R_j\right)^{-1}=G_j\left(k\right)S_j^{-1}\left(k\right) \tag{17}$$

$$\begin{aligned}\hat{x}_j&=\hat{x}_j\left(k\mid k-1\right)+K_j\left(k\right)\left(z\left(k\right)-H_j\hat{x}_j\left(k\mid k-1\right)\right)\\ &=\hat{x}_j\left(k\mid k-1\right)+G_j\left(k\right)S_j\left(k\right)^{-1}\left(z\left(k\right)-H_j\hat{x}_j\left(k\mid k-1\right)\right)\end{aligned} \tag{18}$$

$$\begin{aligned}P_j^k&=\left(I-K_j\left(k\right)H_j\right)P_j\left(k\mid k-1\right)\\ &=L_j\left(k\right)\left(D_j\left(k\right)-E_j\left(k\right)S_j\left(k\right)^{-1}E_j\left(k\right)^T\right)U_j\left(k\right)\end{aligned} \tag{19}$$

Step4: Combining of the state estimations and corresponding covariance according to the updated diagonal-matrix-weight

$$\hat{x}\left(k\right)=\sum_{j=1}^{s}B_j^k\hat{x}_j^k \tag{20}$$

Updated diagonal-matrix-weight of model M_j^k is

$$B_j^k = diag\left(b_{j1}, b_{j2} \cdots b_{jn}\right) \tag{21}$$

which is derived from vector $b_i \ (i = 1, \cdots, n)$

$$b_i = \left[b_{1i}, b_{2i} \cdots b_{si}\right] = \frac{e^T \left(\mathrm{P}^i\right)^{-1}}{e^T \left(\mathrm{P}^i\right)^{-1} e} \tag{22}$$

with
$$e = \begin{bmatrix} 1 \\ \vdots \\ 1 \end{bmatrix}_{s\times 1}, \quad \mathrm{P}^i = \begin{bmatrix} P_1^{(ii)} & \cdots & 0 \\ \vdots & \ddots & \vdots \\ 0 & \cdots & P_s^{(ii)} \end{bmatrix}, \tag{23}$$

and $P_j^{(ii)}$ is the ith diagonal element of matrix P_j^k

The error variance matrix of the optimal fusion estimation is

$$P(k) = diag\left[P_1, P_2 \cdots, P_n\right] \tag{24}$$

$$P_i = \left[e^T \left(\mathrm{P}^i\right)^{-1} e\right]^{-1} \tag{25}$$

3 Simulation and Comparison Results

In this section, some experiments are conducted to evaluate the performance of IMM, DIMM, LDU-DIMM algorithm. Two goodness-of-fit are used to assess the tracking efficiency of the results: running time and Root Mean Square Error (RMSE). The running time can be calculated by the software and the RMSE is calculated as

$$RMSE_k = \sqrt{\frac{1}{M} \sum_{i=1}^{M} \left(\hat{x}_k^i - x_k^i\right)^2} \tag{26}$$

Where x_k^i and $\hat{x}_k^i$ denote the true and estimated state at the ith Monte Carlo run in time k , respectively, M is the total number of independent Monte Carlo runs.

3.1 Simulation Settings

The target trajectory is generated as follows: The target moves in different state for four periods. First, it moves with constant velocity $v_x = 0, v_y = -1$ in a straight line from 0s to 50s. Then it maneuvers and turns right with $\omega = \pi / 20$ from 51s to 70s. From 71s to 120s it moves with constant velocity $v_x = 1, v_y = 0$. For the last 30s, it maneuvers and turns right with $\omega = \pi / 20$. The position trajectory of the target is shown in Figure1, and the velocity trajectory of the target is shown in Figure2.

For the tracking of the target, two models are employed: constant-turn (CT) model with constant angular rate and constant-acceleration (CA) model, the state representation for each model refers to [6].

The simulation parameters setting are as follows:

Sampling interval $T = 1s$, Noise covariances are

$$R_i = diag\left([\sigma_1^2, \sigma_2^2]\right) \quad \sigma_1 = \sigma_2 = 0.5$$

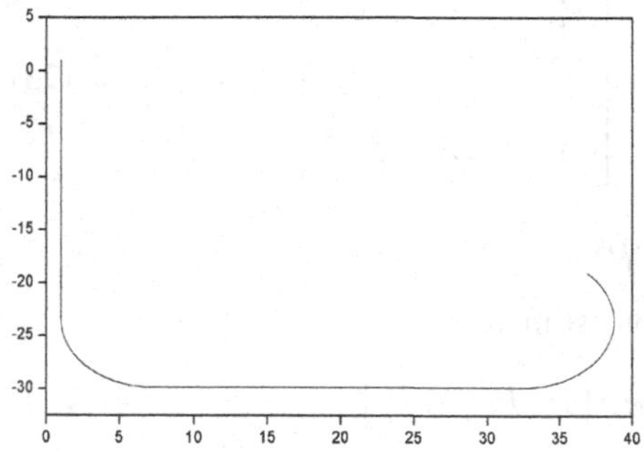

Fig. 1. True position trajectory

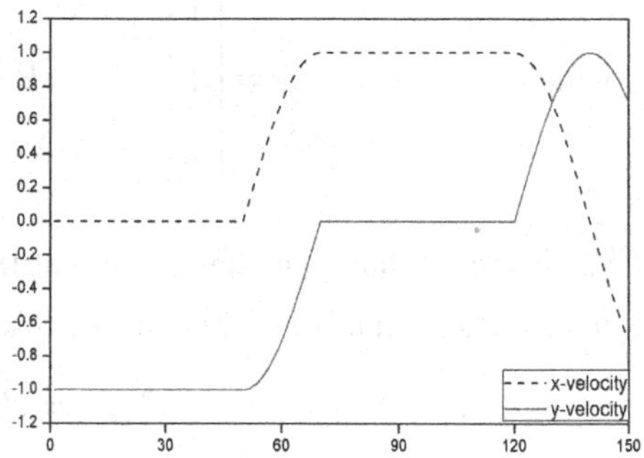

Fig. 2. True Velocity trajectory

Initial state and error covariance for the target are

$$x(0) = [1\ 1\ 0\ \text{-}1\ 0\ 0]' \quad P(0) = diag\left([0.1, 0.1, 0.1, 0.1, 0.1, 0.1]\right)$$

The model transition matrix and initial distributions are

$$\pi_{ij} = \begin{bmatrix} 0.95 & 0.05 \\ 0.05 & 0.95 \end{bmatrix}, \quad \phi_1 = 0.9, \phi_2 = 0.1$$

$$B_1 = diag\left([0.9,\ 0.8,\ 0.2,\ 0.5,\ 0.6,\ 0.1]\right)$$

$$B_2 = diag\left([0.1,\ 0.2,\ 0.8,\ 0.5,\ 0.4,\ 0.9]\right)$$

3.2 Result Analysis

The results are obtained from 100 Monte Carlo simulation runs. Figure 3 shows the IMM position RMSE in x direction, Figure 4 shows the DIMM position RMSE in x direction with calculating divergence from step 126(In order to facilitate drawing, the RMSE value of divergence represented by 1 in y-axis), and Figure 5 shows the LDU-DIMM position RMSE in x direction. From the figures, it can be obviously seen that the tracking performance of IMM algorithm in linear motion is relatively worse than that in circular motion. The DIMM algorithm (before calculating divergence) and LDU-DIMM algorithm have almost the same tracking performance and they both

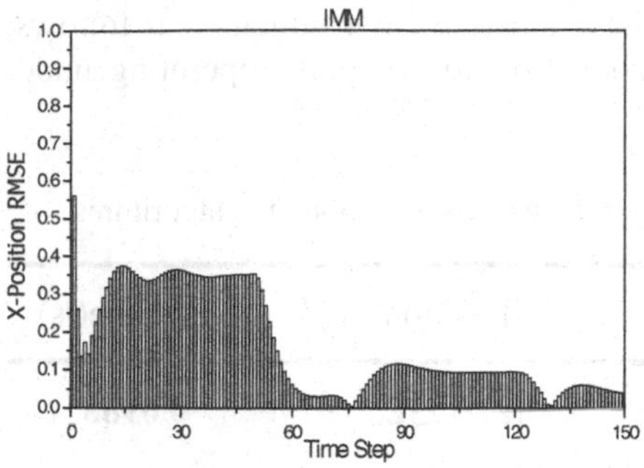

Fig. 3. The IMM position RMSE in x direction

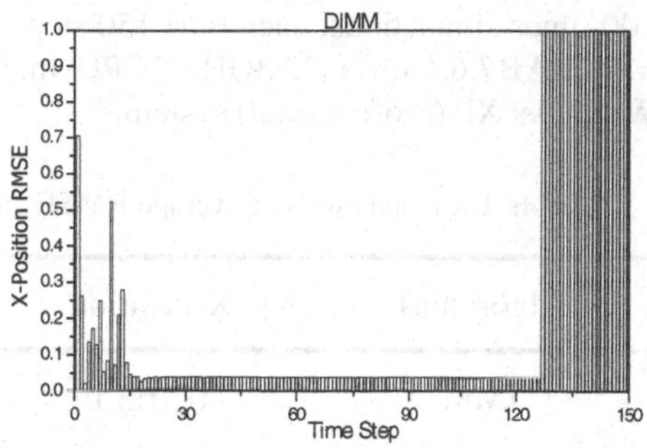

Fig. 4. The DIMM position RMSE in x direction

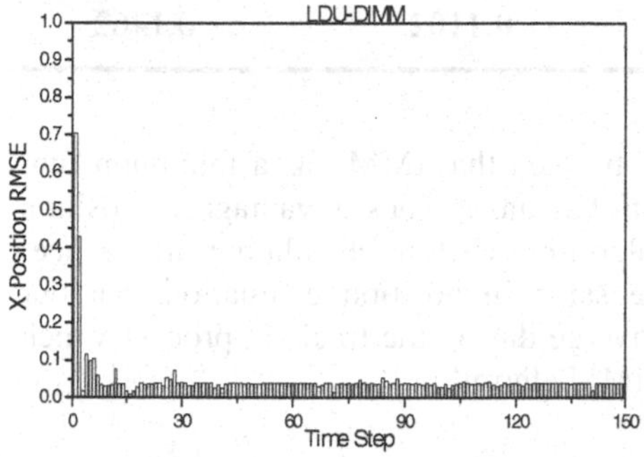

Fig. 5. The LDU-DIMM position RMSE in x direction

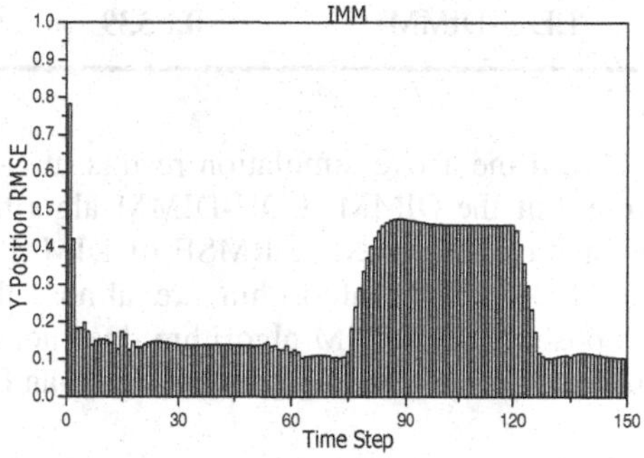

Fig. 6. The IMM position RMSE in y direction

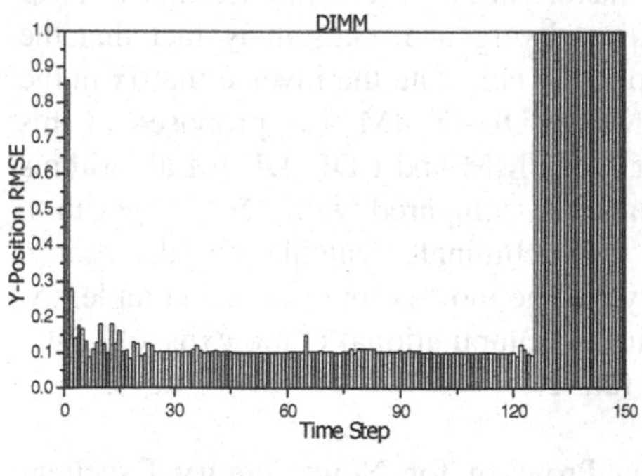

Fig. 7. The DIMM position RMSE in y direction

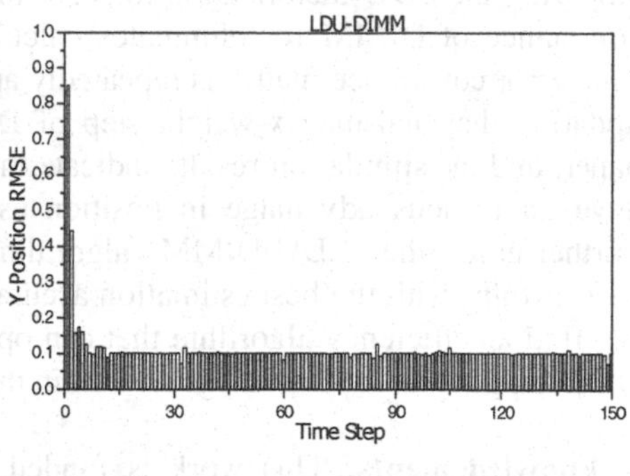

Fig. 8. The LDU-DIMM position RMSE in y direction

have a better tracking accuracy than IMM algorithm in x direction. What is the most Important is that our proposed LDU-DIMM algorithm eliminates the calculating divergence as can be seen from Figure 3.

Figure 6, Figure 7 and Figure 8 shows the IMM, DIMM and LDU-DIMM position RMSE in y direction. Table 1 shows the comparisons of average RMSE and run-time statistics among the algorithms. The running time is taken average after summing for

100 times simulations, each time 150 steps. The simulation environment is as follows: MATLAB7.6.4 on a 2.93GHz 2CPU Intel core 2-based computer operating under Windows XP (Professional) system.

Table 1. Comparisons of average RMSE and run-time statistics among the algorithms

Algorithms	X-Position	Y-Position	Run-time(/s)
IMM	0.1635	0.236532	**0.0983**
DIMM	0.0542 (before calculating divergence)	0.1113 (before calculating divergence)	0.1346
LDU-DIMM	**0.0539**	**0.1102**	0.1462

From the above simulation results, it can be seen that IMM has a minimum run-time, but the DIMM, LDU-DIMM algorithm has an obvious advantage in position estimation. The average RMSE of DIMM algorithm (before calculating divergence) and LDU-DIMM algorithm are almost the same in position estimation, but our proposed LDU-DIMM algorithm does not diverge during the tracking process which indicates better tracking performance than DIMM algorithm.

4 Conclusion

Applying the LDU factorization to solve the factorization of asymmetric matrix error covariance of DIMM for eliminates calculating divergence, caused by fact that the state error covariance matrix is repeatedly applied to calculate the inverse matrix in the updating diagonal-matrix-weight step of DIMM. LDU-DIMM was proposed in this paper, and the simulation results indicate that the DIMM and LDU-DIMM algorithms have an obvious advantage in position estimation compared with IMM algorithm. Furthermore, the LDU-DIMM algorithm can eliminate calculating divergence successfully, with the best estimation accuracy but the most computational complexity. So, find an efficiency algorithm that can optimize computational complexity of LDU-DIMM algorithm is a meaningful work in the future.

Acknowledgments. This work is funded by Program for New Century Excellent Talents in University (Grant No.NCET-08-0333) and Independent Innovation Foundation of Shandong University (Grant No.2010JC015).

References

1. Fu, X., Jia, Y., Du, J., Yu, F.: New interacting multiple model algorithms for the tracking of the manoeuvring target. IET Control Theory and Applications 4(10), 2184–2194 (2010)
2. Johnston, L.A., Krishnamurthy, V.: An improvement to the interacting multiple model (IMM) algorithm. IEEE Transactions on Signal Processing 49(12), 2909–2923 (2001)

3. Genovese, A.F.: The interacting multiple model algorithm for accurate state estimation of maneuvering targets. Johns Hopkins APL Technical Digest 22(4), 614–623 (2001)
4. Mohammed, D., Mokhtar, K., Abdelaziz, O., Abdelkrim, M.: A new IMM algorithm using fixed coefficients filters (fastIMM). AEU - International Journal of Electronics and Communications 64(12), 1123–1127 (2010)
5. Bierman, G.J.: Measurement Updating Using the U-D Factorization. In: Proceedings of the IEEE Conference on Decision and Control, pp. 337–346 (1975)
6. Julier, S., Uhlmann, J., Durrant-Whyte, H.F.: New method for the nonlinear transformation of means and covariances in filters and estimators. IEEE Transactions on Automatic Control 45(3), 477–482 (2000)

Performance Analysis of VCN and IVCN Localization Algorithms in WSN

Yang Liu[1], Jianping Xing[1,*], Hua Wu[2], Xiaoming Wu[2], and Yingbing Zhou[2]

[1] School of Information Science and Engineering, Shandong University
[2] Department of Information Engineering, Shandong Jiaotong University
250100 Jinan, China
{ly0314,wuhua1982111}@126.com, sduxingjp@163.com

Abstract. In this paper, we made comprehensive comparisons of two localization algorithms in wireless sensor network (WSN): A localization algorithm based on virtual central node (VCN) and an improved 3D node localization algorithm based on virtual central node (IVCN). VCN and IVCN algorithms are both adapted to the wireless sensor network (WSN) that anchor nodes present a uniform distribution in three dimensional sensor spaces. During the localization process, by deducing a 3D special node, which is called the virtual central node, unknown nodes can compute their own positions automatically. However in IVCN algorithm, localization problem is solved by deducing a 3D special node that is called virtual central node (VCN) from three different anchor nodes and the deducing process is more simplified than that of our previous VCN. This IVCN algorithm overcomes the defects of VCN algorithm which is pointed out in our previous work. In order to explore to what extent the performances are improved, explicit analysis of differences are made in this paper. From the simulation graphs the localization error of IVCN is even better than CVN in some situations. Further the proposed IVCN algorithm costs the least localization time of the two and less communication overhead.

Keywords: WSN, performance analysis, VCN, IVCN.

1 Introduction

Wireless sensor network (WSN) has become a hot research field due to its all kinds of applications, such as environment, medical care, military fields and industry [1]. In all these applications, node location is important because the information of node physical location is useful for coverage, routing, target tracking and rescues [2]. Especially the 3D node localization problem in wireless sensor networks (WSNs) is more critically important with the WSN used broader and broader in each corner of our life.

There are many localization algorithms for WSNs have been proposed. Almost all localization methods take use of a part of nodes with prior knowledge of their absolute physical positions called anchor nodes, and inter-sensor measurements, to

* Corresponding author

A. Xie & X. Huang (Eds.): Advances in Computer Science and Education, AISC 140, pp. 23–29.
springerlink.com

obtain the practical location of nodes with unknown position information called unknown nodes. All these methods can be mainly classified as range-based algorithm and range-free algorithm. The former approaches compute the node position fully based on distance or angular information acquired by using the Time of Arrival (ToA), Angle of Arrival (AoA), Time Difference of Arrival (TDoA), or Received Signal Strength Indicator (RSSI) techniques[3][4][5][6][7]. In this method node localization accuracy is high but it also calls for more expensive hardware consumption and the whole spending of the network is high. On the contrary, range-free localization schemes merely rely on the existence of radio connectivity to a neighbor instead of measuring distance or angle to that, which further decrease the consumption power and hardware requirements. Range-free schemes mainly explore the local network topology and the coordinate computation is derived from the locations of the surrounding anchor node position coordinate information [8][9][10][11][12].

In [13] and [14] we propose two range-free localization algorithms in wireless sensor networks called VCN and IVCN for short respectively. The two algorithms both achieve good performances. In this paper, explicit comparisons are given to illustrate how the differences happen and explore the different performances between the two.

The rest of this paper is organized as follows. In Section 2, the descriptions of the two algorithms are given respectively. In section 3, the concrete simulation analysis is described. Conclusions are listed in Section 4.

2 Descriptions of VCN and IVCN Algorithms

This section describes the two localization algorithms mentioned above. The basic principles are given first and then algorithm processes can be found.

2.1 Algorithm Description Based on Virtual Central Node (VCN)

In our proposed algorithm, all anchor nodes are supposed to present a uniform distribution in the special 3D sensing space. In this algorithm each sensor node estimates its position solely based on the information gathered directly from the anchor nodes. Since it does not depend on neighboring sensor node communication, it is independent of network connectivity and it is more suitable for all kinds of complicated applications.

The algorithm first begins from flooding data packages from anchor nodes to the whole sensor network. Each anchor node is able to broadcast information packages periodically. This time slice can be set manually. The data information includes anchor node ID, and coordinates of corresponding anchor nodes. Unknown nodes' task is easy and energy efficient because they are only in charge of listening to these packages in the time slice T. Unknown nodes can memory how many packages have received from different anchors. Then they judge whether the time slice T is arrived. If so, the information can be recorded, or go on waiting. As we said in the last part, once the package enters the communication range unknown nodes can detect it immediately and record the information data contained in the corresponding packages. At last by the information provided in these packages, virtual central node is formed and can be computed. Finally unknown position can be derived using the virtual sensor node above. By using the center of the square, virtual central node can be computed through adding half of communication range on one of three ordinate directions. But there is a problem in this algorithm. As shown in Fig.1, after the fourth node is determined, we cannot make sure the virtual central node is on which side of the plane. It may be on the same side with unknown node which means low estimation error. However if is on the opposite side of the unknown node, it will produce a lot of localization error. During our localization algorithm, the position of virtual central node on which side is decided randomly and of course it is not a perfect solution which can induce lots of uncertainty. Of course how to solve this problem completely is also a research direction in the future to make higher and better localization accuracy.

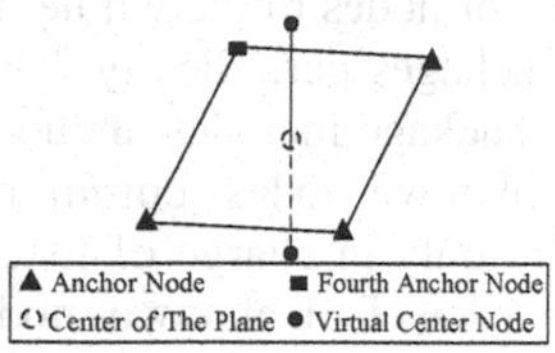

Fig. 1. Computation of VCN

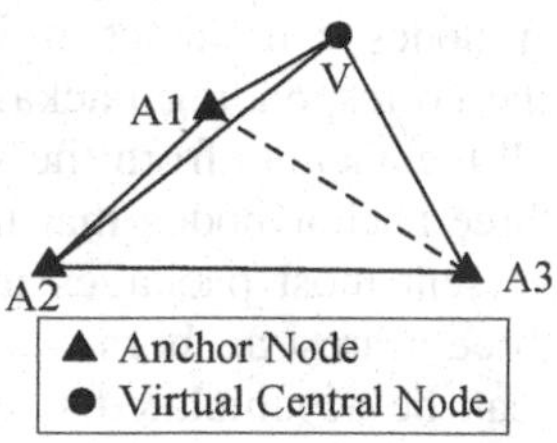

Fig. 2. Derivation of unknown node position

When the virtual central node is found out, unknown nodes can finish the localization process. As shown in Fig.2, three anchor nodes (A1, A2, and A3) and virtual central node V forms a tetrahedron. Of course this tetrahedron is anomalous. The virtual central node could be either side of the plane determined by node A1, A2, and A3. Then we can use the similar way of Centroid algorithm. That is to say the center of the four special nodes is the estimated position of the unknown node. It is necessary to illustrate the feasibility of this method. As we known, virtual central node is important in this algorithm. Based on this node unknown node could compute the center of a 3D graph.

2.2 Description of Improved VCN Localization Algorithm (IVCN)

As shown in Fig.3, there are eight anchors which form a cube. A_i (i =1, 2… 8) is anchor node and V is virtual central node.

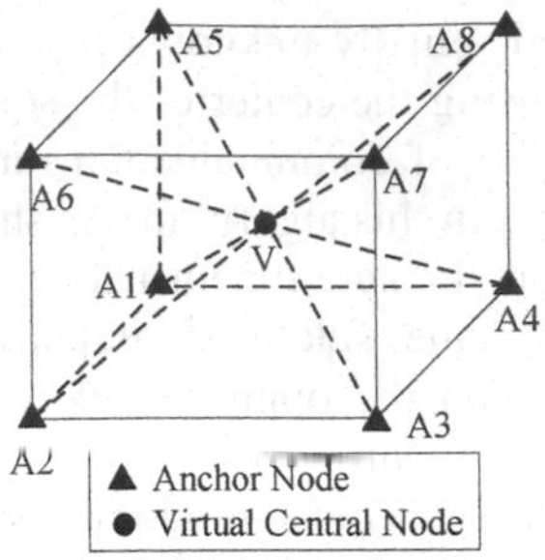

Fig. 3. Formation of VCN

First a random fixed time slice T is produced then the IVCN algorithm begins from flooding data packages from anchor nodes to the whole network. Each anchor node is able to broadcast information packages periodically. This time slice can also be set manually. The data information package includes anchor node ID, and coordinates of corresponding anchor nodes. Unknown nodes' communication function is easy and energy efficient because they are only in charge of listening to these packages from anchors in the presetting time slice T. Unknown nodes can memory how many packages have received from different anchors and record the amount of packages. Then they judge whether the time slice T is arrived. If so, the information can be recorded, or go on waiting. As we said in the last part, once the package enters the communication range unknown nodes can detect it immediately and record the information data contained in the corresponding packages. All unknown nodes can equally receive packages from all the anchors from the whole network and they only record the full information of three anchor nodes that have sent the most packages. Because three anchors that have sent most packages mean the most nearest to the unknown nodes. Choosing the three nearest anchors stand for little localization error. That is because distance error can be accumulated from node to node. Once three anchors are recorded the next key step of IVCN localization algorithm is how to deduce virtual central node.

In IVCN algorithm we don't need to compute a fourth node that is on the same plane with three anchors as in our precious VCN. In this way the computation time is diminished sharply and their localization accuracy is more or less the same. In some situations IVCN get better results than VCN.

The centroid of the four nodes (three anchors and virtual central node) with determined position is used as the estimated coordinates of unknown nodes.

3 Simulation Analysis

In this section, we are going to study the performance of VCN and IVCN in MATLAB software. The related assumptions and network parameters are made as follows. The localization space is supposed to be a cube with the side length 100m. That is to say the whole volume of the 3D localization space is 100×100×100m^3. The anchor nodes are in a uniform distribution on 3D grid with side length 20m in this sensing cube which is to say there are 216 anchor nodes in all in the localization space. Every eight anchor nodes form a cube also with the side length 20m in the space. Unknown nodes are randomly deployed in WSNs and the number can be changed artificially to evaluate the localization performance of our proposed IVCN. The communication range of each unknown node is a changeable parameter. Once the packages, which are sent by anchor nodes, enter the communication radius, unknown nodes can detect them and record the corresponding anchor information. Package information from anchor nodes includes the anchor node ID and the source anchor's corresponding coordinate.

In the simulation the estimated error is defined as the difference between estimated position and real position to different communication range of sensor nodes which can be defined by the following equation.

$$Estimation\ Error = \frac{\sum_{i=1}^{n} \sqrt{(\hat{x} - x_i)^2 + (\hat{y} - y_i)^2 + (\hat{z} - z_i)^2}}{n} \quad (1)$$

In the equation described above $(\hat{x}, \hat{y}, \hat{z})$ and (x_i, y_i, z_i) denote estimated and true position respectively and n is the number of nodes needs to be localized.

3.1 Localization Error Comparisons of VCN and IVCN

In this part the comparisons of VCN, IVCN and other two classic localization algorithms are given in the following figures.

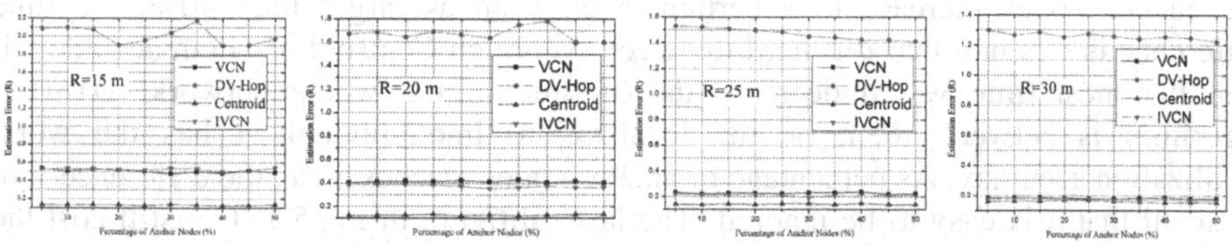

Fig. 4. Error comparisons of IVCN, VCN, and other two with different communication ranges

We vary the number of deployed unknown sensors to get different node density and connectivity with R=15m, 20m, 25m, and 30m respectively. Here percentages of anchor nodes are altered from 5% to 50% and estimation error is record in each situation with other network settings the same. The four sub-graphs in Fig.4 describe the estimation error under different percentages of anchors of IVCN, VCN, DV-Hop, and Centroid with varied communication range R. DV-Hop and Centroid algorithms are two classic localization algorithms which are admitted by most researchers. All the four localization algorithms take advantage of the distance estimation to anchors for estimating sensors' positions in WSNs. There is no doubt that more accurate distance measurement leads to better position estimation. The four curves preserve approximate straight line. DV-Hop algorithm is the worst of the four no matter how the environmental parameters change. Its estimation error always stays at a high level. Centroid algorithm performs the best when communication range is no larger than 35m. From Fig.4 we can conclude that communication range is the main factor of IVCN, VCN, and Centroid which affect the localization accuracy.

3.2 Localization Time Comparisons of VCN and IVCN

Localization time here is defined from the beginning of whole algorithm to the localization accomplishment of all unknown nodes in WSNs. Localization time reflects the operating efficiency of an algorithm. Little time cost means high efficiency of energy consumption and lifetime of network. We make records of the localization time under different communication range in MATLAB software which is given in the following figure.

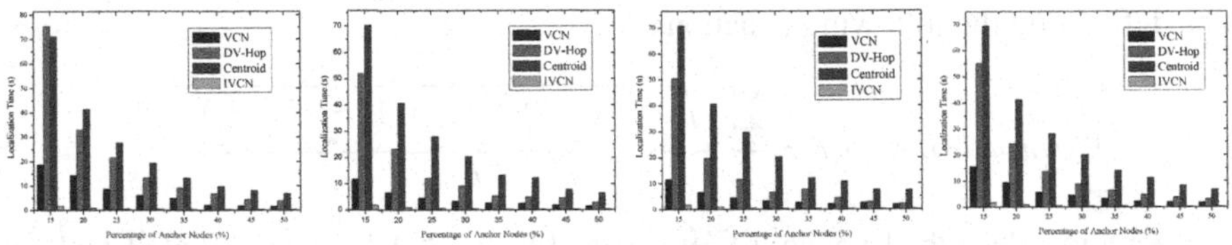

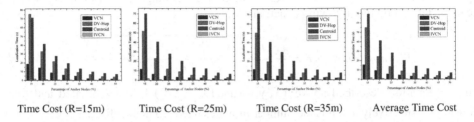

Time Cost (R=15m) Time Cost (R=25m) Time Cost (R=35m) Average Time Cost

Fig. 5. Comparisons of time cost

From Fig.5, the histograms all present a downtrend. No matter how the percentage of anchors changes IVCN needs the least localization time in the four figures. On the contrary Centroid algorithm needs the most which means it cost most energy. When the anchor nodes increase to a certain level, such as larger than 40%, the time difference is so small that it is hard to detect. Although Centroid has high accuracy, it cost the most time which decrease the lifetime of the whole network. DV-Hop algorithm is between them on the localization time although it has the worst localization accuracy. As percentage of anchor nodes reaches 50%, the differences are so small that it is easy to be ignored. The last subfigure in Fig.5 IVCN still cost the least average localization time. So in some situations that need quick localization IVCN is definitely the best choice and it has the highest energy efficient.

4 Conclusions

In the simulation graphs provided in Section 3, IVCN overcomes the defects of VCN. Also it retains the advantages of VCN. IVCN is not only the improvement of VCN but also extends the application fields of the algorithm. So it is suitable for the situation needs quick localization. Based on these properties we are thinking how to use it in mobile node localization using IVCN in mobile wireless sensor network environment.

Through analysis we can find that VCN and IVCN algorithms are both two high efficient which used in 3D wireless sensor networks. They can both realize localization with high accuracy. But to some extent IVCN is better than VCN.

Acknowledgement. This paper is supported by New Century Talents Supportive Program of China (No. NCET-08-0333), Independent Innovation Foundation of Shandong University (No. 2010JC015) and Shandong Province Natural Science Foundation (Grant No. ZR2011FM039).

References

1. Mao, G., Fidan, B., Anderson, B.D.O.: Wireless sensor network localization techniques. Computer Networks 51, 2529–2553 (2007)
2. Gezici, S.: A survey on wireless position estimation. Wireless Personal Communications 44, 263–282 (2008)
3. Kannan, A.A., Mao, G., Vucetic, B.: Simulated annealing based localization in wireless sensor network. In: Proceedings of the IEEE Conference on Local Computer Networks, pp. 513–514 (2005)
4. Caffery Jr., J., Stuer, G.L.: Subscriber Location in CDMA Cellular Networks. IEEE Trans. Vehicular Technology 47(2), 406–416 (1998)
5. Klukas, R., Fattouche, M.: Line-of-Sight Angle of Arrival Estimation in the Outdoor Multipath Environment. IEEE Trans. Vehicular Technology 47(1), 342–351 (1998)
6. Cong, L., Zhuang, W.: Hybrid TDOA/AOA Mobile User Location for Wideband CDMA Cellular Systems. IEEE Trans. Wireless Comm. 1(3), 439–447 (2002)
7. Aspnes, J., Eren, T., Goldenberg, D., Morse, A.S., Whiteley, W., Yang, Y., Anderson, B.D.O., Belhumeur, P.: A theory of network localization. IEEE Transactions on Mobile Computing 5(12), 1663–1678 (2006)
8. Savvides, A., Garber, W.L., Moses, R.L., Srivastava, M.B.: An analysis of error inducing parameters in multihop sensor node localization. IEEE Transactions on Mobile Computing 4(6), 567–577 (2005)
9. Chan, F.K.W., So, H.C.: Efficient weighted multidimensional scaling for wireless sensor network localization. IEEE Transactions on Signal Processing 57(11), 4548–4553 (2009)
10. Costa, J.A., Patwari, N., Hero, A.O.: Distributed weighted-multidimensional scaling for node localization in sensor networks. ACM Transactions on Sensor Networks 2(1), 39–64 (2006)
11. Shang, Y., Ruml, W., Zhang, Y., Fromherz, M.: Localization from mere connectivity. In: Proc. of ACM MobiHoc, Annapolis, MD (June 2003)
12. Langendoen, K., Reijers, N.: Distributed localization in wireless sensor networks: A quantitative comparison. Computer Networks 43 (2003)
13. A 3D Node Localization Algorithm Based on Virtual Central Node (VCN) in Wireless Sensor Networks. Submitted to Journal of Networks (under review)
14. Liu, Y., Xing, J., Zhou, Y., Wu, H.: IVCN: An Improved 3D Node Localization Algorithm Based on Virtual Central Node (VCN) in Wireless Sensor Networks. Source: Journal of Information and Computational Science 8(8), 1395–1403 (2011)

References

1. [illegible] Wireless sensor network localization techniques. [illegible] Networks 51, 2529–2553 (2007)
2. [illegible] based position estimation [illegible] Personal [illegible] 2008)
3. [illegible] Wireless [illegible]
4. [illegible] AOA [illegible]
5. Kaplan, E.D., Hegarty, C.J.: Understanding GPS [illegible]
6. [illegible] AOA [illegible] Systems [illegible] Wireless Comm. [illegible] 2003)
7. [illegible] B.D.O., [illegible] localization [illegible] Transactions on Mobile Computing [illegible] 2006)
8. [illegible] An analysis of error inducing [illegible] Transactions on Mobile [illegible]
9. [illegible] W., [illegible] IEEE [illegible] Processing [illegible] (2005)
10. [illegible] Networks [illegible] (2007)
11. [illegible] W., [illegible]
12. [illegible] Networks [illegible]
13. [illegible] IVCN) in Wireless Sensor Networks [illegible] Journal of [illegible]
14. [illegible] Node Localization Algorithm Based on [illegible] VCN [illegible] Networks. Journal of [illegible]

Impact Analysis of Environment Factors on Iterative Calculation of Secondary Grid Division (ICSGD) Localization Scheme for Wireless Sensor Networks

Yuxin Ren, Jianping Xing*, Yang Liu, and Can Sun

School of Information Science and Engineering, Shandong University, 250100 Jinan, China
{renyuxin180708,sduxingjp}@163.com, ly0314@126.com, suncan0203@gmail.com

Abstract. Localization scheme is a fundamental and essential issue for wireless sensor networks (WSNs). Iterative calculation of secondary grid division (ICSGD) localization scheme could solve the inconsistency between calculation amount and location accuracy. The performance of the localization scheme was evaluated in a series of simulations performed using MATLAB software and was compared to the grid division localization scheme (GDLS). The simulation results demonstrated that the scheme outperformed the GDLS in terms of higher location accuracy, and lower location amount.

Keywords: Wireless Sensor Networks, Localization, Grid Division, Iterative Calculation, Radio Irregularity.

1 Introduction

Recent years, a rapid growth of research into wireless sensor networks has been achieved as a means of developing advanced monitoring, data acquisition and control applications [1]. In many applications, sensor nodes transmit not only real-time data but also their current position. Therefore, how to locate sensor node effectively is becoming a critical issue in the development of wireless sensor networks.

In the literature, a number of localization schemes have been proposed. These schemes take different approaches to solve the localization problem based on different assumptions. According to these assumptions, the available localization schemes could be divided into two categories: range-based localization schemes[2],[3],[4],[5],[6],[7] and range-free localization schemes [8],[9],[10],[11]. The range-based localization schemes get the sensor position based on distance or angular information acquired. Although range-based schemes are reasonable accurate(less than 2m [12],[5]),each sensor node must be equipped complex hardware. Conversely, range-free schemes achieve sensors' position information without the need of ranging measurements. The minimum location error in range-free schemes could reach about 10 percent of the radio range [8].

* Corresponding author.

A. Xie & X. Huang (Eds.): Advances in Computer Science and Education, AISC 140, pp. 31–38.
springerlink.com

Wireless sensor networks may often be deployed over three-dimensional areas in real applications, such as military sensing devices deployed in a nonplanar battlefield, a sensor network floating in the air for tracking chemical plumes. Therefore, sensor localization for the three-dimensional space is needed. A few schemes are proposed in [13], [14], and [15]. Chia-Ho Ou and Kuo-Feng Ssu present a scheme [16] to locate unknown nodes using flying anchors in three-dimensional space. The scheme could acquire good performance, but had to rely on infrastructure. Our group has developed a grid division localization scheme (GDLS) for three-dimensional WSNs; however, we meet a problem, which is the inconsistency between localization accuracy and amount of calculation. As shown in Fig.1, the cube was divided into many cubes in the form of grid first, and we could estimate the cube in which an unknown node was located. Then, the first estimated cube was divided into smaller cubes in the form of grid again and we could estimate the smaller cube in which the unknown node was located. At last, we assumed the centroid of the smaller cube as the true position of the unknown node.

In this paper, simulation showed the impact of four factors on the performance of ICSGD localization scheme. As the power range of beacons increased, the location accuracy increased, but also the location time was longer. Space size would make little impact on the scheme.

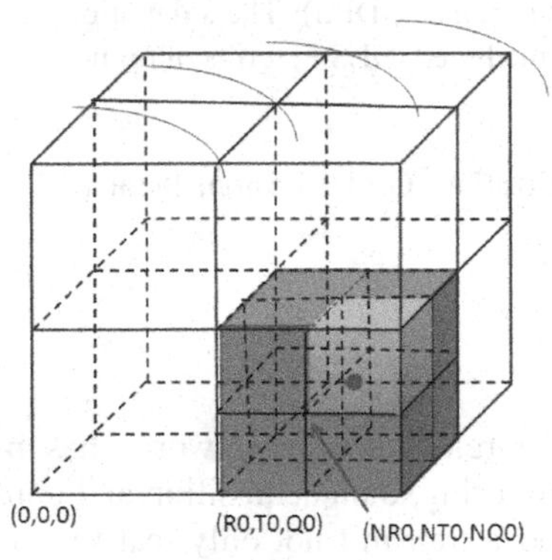

Fig. 1. Illustration of secondary grid division

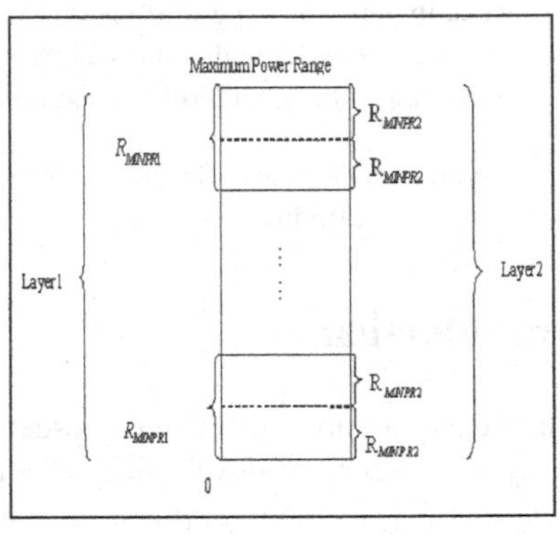

Fig. 2. Two layers controlling power of beacon

2 Scheme Analysis

2.1 Beacon Node and Unknown Node

In this scheme, nodes were divided into two categories: beacon node and unknown node. Beacon nodes could get their accurate position information with the help of GPS receivers, while unknown nodes had to calculate their position according to the position of beacon nodes.

The beacon nodes could locate themselves accurately by GPS receivers, and they could control transmitting power. The power was divided into two layers, and had two minimum power ranges R_{MINPR}, which were intervals of transmitting power as shown in Figure 2. Take an example, if a beacon transmitted information at level N, the maximum transmitting distance of information was $N \times R_{MINPR}$. The R_{MINPR} of layer 1, R_{MINPR1}, was larger than the R_{MINPR} of layer 2, R_{MINPR2}. In this way, sensor nodes could quickly calculate their probable position at a lower cost. Furthermore, a smaller R_{MINPR2} could increase the locating accuracy. The beacon information included beacon ID, accurate coordinates (X, Y, Z), flag of layer, level N and minimum power ranges R_{MINPR}, as shown in Table 1.

When an unknown node received the information of beacon nodes, it stored the information into its local information table, which was shown in Table 2. The table included sensor ID, beacon ID and beacon coordinates. After calculation, it also stored the upper limit of layer 1 and layer 2, lower limit of layer 1 and layer 2, first locating coordinates and second locating coordinates into its table.

Table 1. Illustration of beacon information

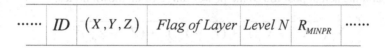

……	*ID*	(X,Y,Z)	*Flag of Layer*	*Level N*	R_{MINPR}	……

Table 2. Information table of unknown node

……	*Sensor ID*	*Anchor ID*	*Anchor Coordinates*	*Upper Limit of Layer 1*	*Lower Limit of Layer 1*	*Upper Limit of Layer 2*	*Lower Limit of Layer 2*	*First Locating Coordinates*	*Second Locating Coordinates*	……

2.2 System Environment and Assumptions

The system environment included forty beacon nodes and five hundred unknown nodes. All nodes are randomly deployed in a three-dimensional cube.

The localization scheme required some fundamental assumptions as follows:

(1)All nodes were static. Once nodes were randomly deployed, the position of each node was fixed.

(2)Each beacon node was equipped with a GPS receiver or some other forms of localization device to get accurate position information.

(3) Each node was equipped with omni-directional antenna, which could receive the signal of all directions.

2.3 System Environment and Assumptions

All simulations were completed on PC, whose CPU was Celeron M 1.73GHz, and RAM was DDR II 1GHz.

Simulation Parameters:

There are four simulation parameters, whose settings were summarized in Table 3.

Power range. This parameter was the maximal power range of beacon nodes.

Space size. Space size was represented by the cube side length in simulation.

Division way. The ICSGD(X+Y) divided the cube into X3 smaller cubes first, and then divided the new small cube into Y3 smaller cubes.

Radio irregularity. DOI model [20] was used to simulate the radio irregularity in this paper.

Table 3. Simulation parameters

Parameter	Value(s)
Power range	40,45,50,55,60,65,70,75,80
Space size	20,100,200,300
Division way	(20+5),(10+10),(10+5),(5+5),(5+10),(5+20)
Radio irregularity	0,0.02,0.04,0.06,0.08,0.1

Performance Index:

There were two main performance indexes to evaluate the scheme.

Average location error A_e. The ratio of the average distance between the estimated position and the actual position to the power range of unknown nodes.

$$Ae = \frac{\sum_{i=1}^{K}\sum_{j=1}^{n}\sqrt{(X_{ej}-X_{aj})^2+(Y_{ej}-Y_{aj})^2+(Z_{ej}-Z_{aj})^2}}{n'RK}, 0 \le n' \le n \quad (1)$$

(*Xej*, *Yej*, *Zej*) was the estimated position of unknown node *j*, (*Xaj*, *Yaj*, *Zaj*) was the actual position of node *j*. *K* was the number of simulations. *n* was the number of all unknown nodes, and *n'* was the number of located unknown nodes. If an unknown node couldn't be located, its location error was 0. *R* was the power range of unknown nodes.

Average location time A_t. The total time taken for all unknown nodes to finished their localizations.

2.4 Grid Division Location Scheme

In GDLS, each unknown node established its own 3-dimensional coordinate system. The origin point *O* was the centroid of beacon nodes from which information could be received. The *X*-axis was the latitude crossing the origin, and *Y*-axis was the longitude. The *Z*-axis was the line crossing origin and perpendicular to the Plane *XOY*. Then, the unknown node divided the whole cube into many smaller cubes. After that, each beacon node began to vote. If the smaller cube was in the power range of beacon node, the beacon node voted for this cube. At last, the smaller cube which received the largest number of votes was the most probable cube, in which the unknown node was. And the centroid of the smaller cube was regarded as the position of the unknown node. In the scheme, we set a threshold value. If the largest number of votes was larger than this value, it was supposed that the unknown node could be located. Otherwise, the unknown node couldn't be located, and started locating again.

2.5 Flow of ICSGD Localization Scheme

Each beacon node broadcast its beacon information in the whole network. At first, it increased its power level according to layer 1, after certain time, it increased power according to layer 2. When an unknown node received its beacon information, stored the information into local information table, and calculated the upper and lower limit of layer1 and layer2, which was shown in Fig.3. With the help of upper and lower limit of layer 1, it could estimate the smaller new cube in which the unknown sensor node was the most probable using GDLS. Then, it could calculate the estimating coordinates according to the upper and lower limit of layer 2 using GDLS again.

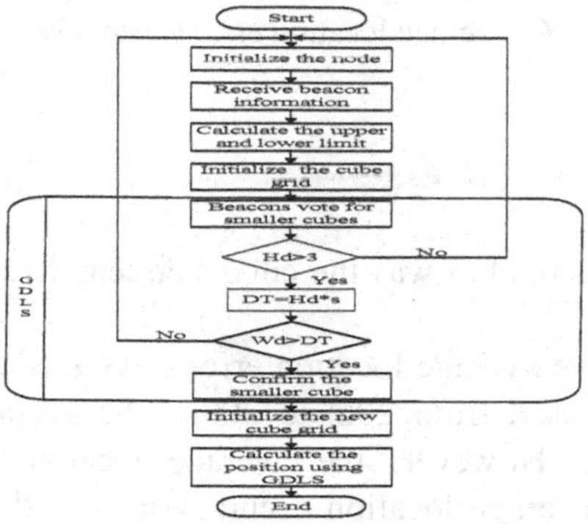

Fig. 3. Flow of an unknown node localization

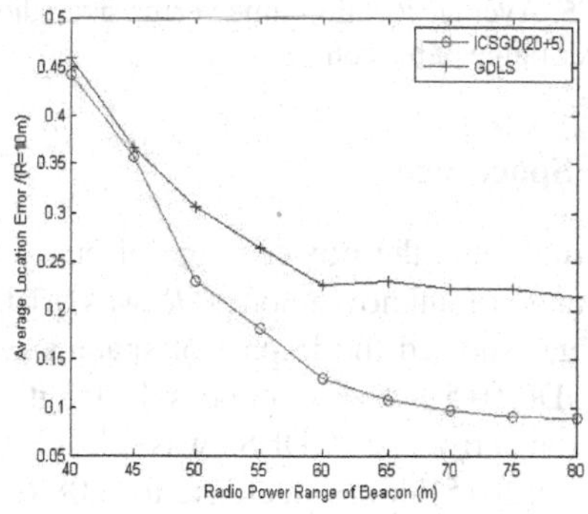

Fig. 4. Average location error vs radio power

3 Result and Analysis

In the paper, we analyze the impact of four factors on the performance.

3.1 Radio Power Range of Beacons

In simulations, All nodes were deployed in a 100*100*100m^3 cube, and the power range of unknown node, *R*, was 10m.

Fig.4 showed the impact of radio power range of beacon on average location error of GDLS and ICSGD(20+5). The average location error of GDLS fell from 0.4579*R* to 0.2137*R* as the radio power range of beacon increased from 40m to 80m. However, the average location error of ICSGD(20+5) decreased from 0.4417R to 0.0886*R*.

In the whole network, beacons broadcast their information. Relatively speaking, an unknown node could receive information from more and more beacons as the radio power range of beacon increased step by step. So, the localization of unknown nodes was more accurate. Hence, the average location error was lower. However, when the radio power range of beacon was equal in both schemes the smallest cube of ICSGD(20+5) was much smaller than that of GDLS after secondary division. So, the average location error of ICSGD(20+5) was obviously lower than that of GDLS.

In Fig.5, it could be seen that the average location time increased as the radio power range of beacon increased. As power range increasing, the number of broadcasting beacon information increased.

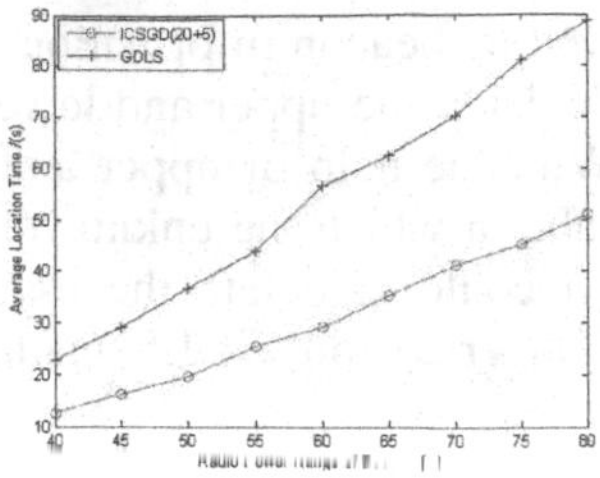

Fig. 5. Average location time versus sizeradio power range of beacon

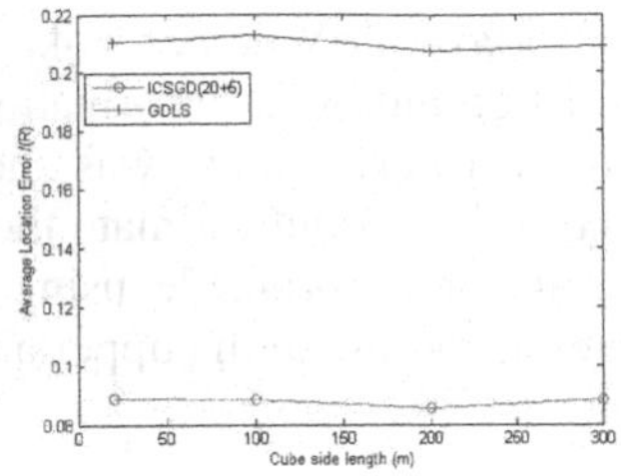

Fig. 6. Average location error versus space

3.2 Space Size

In simulation, the power range of beacons was 0.8d, d was the cube side length, and the range of unknown nodes, *R*, was 0.1d.

Fig.6 showed the impact of space size on the average location error of GDLS and ICSGD(20+5). As the cube side length increased from 20m to 300m, the average location error of GDLS was about 0.21R. However, the average location of ICSGD(20+5) could decrease to 0.08R. The average location accuracy increased by 162.5%, and the average location error could remain unchanged as space size varied.

As shown in Fig.7, the average location time to finish locating was about 88.5s in GDLS, but it just need about 52s to finish location in ICSGD(20+5), and the calculation amount decreased by 70.19%. Also, the average location time in ICSGD(20+5) could remain unchanged no matter how the space size varied.

During locating, GDLS divided the cube into 65525 smaller cubes, but ICSGD (20+5) divided the cube into 8125 smaller cubes in total. Therefore, the average location time decreased greatly. Though the calculation amount decreased, the shortest cube side length of ICSGD (20+5) was shorter than that of GDLS. The shortest cube side length of GDLS was 0.04d, while that of ICSGD (20+5) was just 0.01d. As a result, location accuracy could increase greatly. To sum up the points, ICSGD could achieve better results, and the performance could hold the line as space size varied.

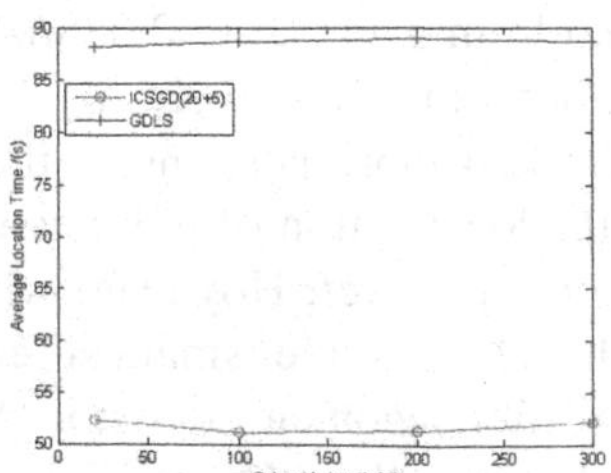

Fig. 7. Average location time versus space size

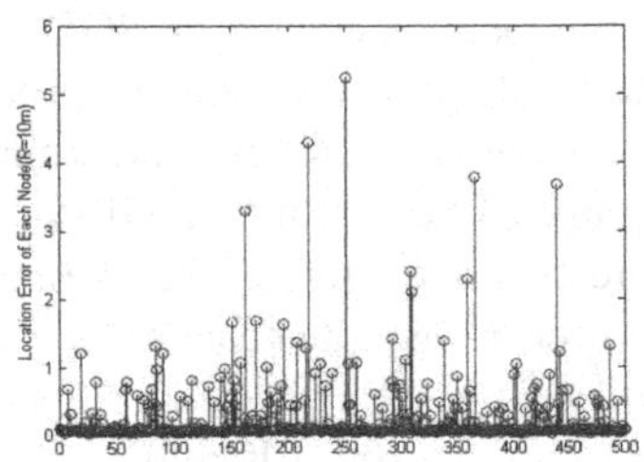

Fig. 8. Location error in ICSGD (5+20)

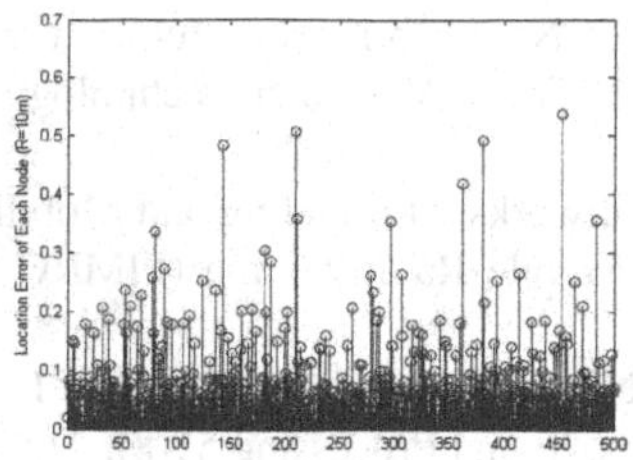

Fig. 9. Location error in ICSGD (20+5)

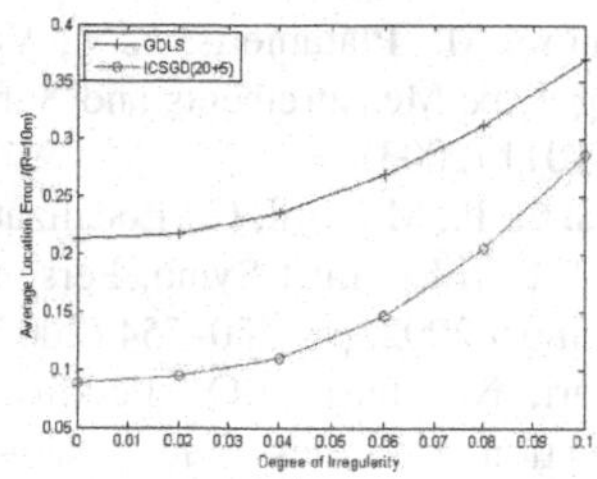

Fig. 10. Average location error versus radio irregularity

As shown in Figure.8, location error of 15% of unknown nodes was higher than 0.5R, and several were even higher than 3R. However, there were 73% of nodes whose location error was lower than 0.1R in Figure.11, and 93.8% of nodes were lower than 0.2R. The location error of those nodes increased the average location error greatly on the whole. Therefore, error during the first localization caused the disparity between ICSGD (5+20) and ICSGD (20+5).

To sum up, when the cube was divided in different way, the average location error was different. On the whole, the divided cube was smaller, the performance of scheme was better, also the calculation time was longer. In actual applications, we could select correct ICSGD location scheme according to conditions.

4 Conclusion

In this paper, performance of ICSGD localization scheme was analyzed in different environments, and compared with that of GDLS. Simulation had shown that space size made little impact on ICSGD localization scheme. However, power range of beacon and division way could make great impact on the performance of ICSGD. Also, the radio irregularity would decrease the location accuracy. In the same conditions, performance of ICSGD was much better than that of GDLS. Therefore, ICSGD localization scheme was distributed and accurate, and suitable for different scale network.

Acknowledgement. The paper is supported by New Century Talents Supportive Program of China (No. NCET-08-0333), Independent Innovation Foundation of Shandong University (No. 2010JC015) and Shandong Province Natural Science Foundation (Grant No. ZR2011FM039).

References

1. Afyildiz, I.F., Su, W., Sankarasubramaniam, Y., Cayirci, E.: A Survey on Sensor Networks. IEEE Comm. Magazine 40(8), 102–114 (2002)
2. Cong, L., Zhuang, W.: Hybrid TDOA/AOA Mobile User Location for Wideband CDMA Cellular Systems. IEEE Trans. Wireless Comm. 1(1), 439–447 (2002)

3. McGuire, M., Plataniotis, K.N., Venetsanopoulos, A.N.: Location of Mobile Terminals Using Time Measurements and Survey Points. IEEE Trans. Vehicular Technology 52(4), 999–1011 (2003)
4. Bergamo, P., Mazzini, G.: Localization in Sensor Networks with Fading and Mobility. In: Proc.13th IEEE Int'l Symp. Personal, Indoor and Mobile Radio Comm (PIMRC 2002), September 2002, pp. 750–754 (2002)
5. Patwari, N., Hero, A.O., Perkins, M., Correal, N.S., O'Dea, R.J.: Relative Location Estimation in Wireless Sensor Networks. IEEE Trans. Signal Processing 51(8), 2137–2148 (2003)
6. Niculescu, D., Nath, B.: Ad Hoc Positioning System (APS) Using AOA. In: Proc. IEEE INFOCOM 2003, March 2003, pp. 1734–1743 (2003)
7. Nasipuri, A., Li, K.: A Directionality Based Location Discovery Scheme for Wireless Sensor Networks. In: Proc. First ACM Int'l Workshop Wireless Sensor Networks and Applications (WSNA 2002), September 2002, pp. 105–111 (2002)
8. He, T., Huang, C., Blum, B., Stankovic, J., Abdlzaher, T.: Range-Free Localization Schemes for Large Scale Sensor Networks. In: Proc. ACM MobiCom 2003, September 2003, pp. 81–95 (2003)
9. Liu, C., Wu, K., He, T.: Sensor Localization with Ring Overlapping Based on Comparison of Received Signal Strength Indicator. In: Proc. ACM MobiCom 2003, September 2004, pp. 81–95 (2004)
10. Savarese, C., Rabaey, J., Langendoen, K.: Robust Positioning Algorithms for Distributed Ad-Hoc Wireless Sensor Networks. In: Proc. USENIX Technical Ann. Conf., June 2002, pp. 317–327 (2002)
11. Niculescu, D., Nath, B.: DV Based Positioning in Ad Hoc Networks. Kluwer J. Telecommunication Systems 22(1), 267–280 (2003)
12. Savvides, A., Han, C.C., Srivastava, M.B.: Dynamic Fine-Grained Localization in Ad-Hoc Networks of Sensors. In: Proc. First IEEE Int'l Conf. Mobile Ad-Hoc and Sensor Systems (MASS 2004), October 2004, pp. 516–518 (2004)
13. Zhang, L., Zhou, X., Cheng, Q.: Landscape-3D:A Robust Localization Scheme for Sensor Networks over Complex 3D Terrains. In: Proc. 31st IEEE Int'l Conf. Local Computer Networks (LCN 2006), November 2006, pp. 239–246 (2006)
14. Liang, J., Shao, J., Xu, Y., Tan, J., Davis, B.T., Bergstrom, P.L.: Sensor Network Localization in Constrained 3-D Spaces. In: Proc. IEEE Int'l Conf. Mechatronics and Automation (ICMA 2006), June 2006, pp. 49–54 (2006)
15. Kushwaha, M., Molnar, K., Sallai, J., Volgyesi, P., Matoti, M., Ledeczi, A.: Sensor Node Localization Using Mobile Acoustic Beacons. In: Proc. Second IEEE Int'l Conf. Mobile Ad-Hoc and Sensor Systems (MASS 2005) (November 2005)
16. Ou, C.-H., Ssu, K.-F.: Sensor Position Determination with Flying Anchors in Three-Dimensional Wireless Sensor Networks. IEEE Transactions on Mobile Computing 7(8) (August 2008)

Research on Data-Driven in Business Workflow System Design

ChuanSheng Zhou

Software College, Shenyang Normal University, Shenyang, Liaoning 110034, China
jasoncs@126.com

Abstract. Today, the workflow technologies have been used in quite a lot of web based business system development. However, in most of these type systems, even they ware designed and developed with component and MVC technologies to improve the system flexibilities and scalability, but the workflow procedures are still not easily changed and are not satisfied to the enterprise requirements for the business and market changing. Here by research on the XML and MVC, it uses XML based data-driven idea to design workflow based business web systems and provide its design model. The practice indicates this design model can meet enterprise flexible business workflow requirements.

Keywords: XML, Data-Driven, Workflow, MVC.

1 Introduction

Since 1993 the WFMC (Workflow Management Coalition) was setup and released a series of models and specifications, the workflow technologies have been applied into many areas and enterprises, like ERP, CRM, SCM, OA etc. and even in normal MIS and web based business systems. However, alone with market and enterprise business grows up, the workflow system need be easily changed and upgraded to satisfy to the enterprise requirements. But look at the current work-flow systems, most of them exits following problems: (1) the business workflow not easily changed. Most of current system fixed the business workflow in their source code; some systems provided mechanisms for flexible workflow, but enterprise need go back to the system provider to change it; (2) the business workflow execution rules and policies are not easily changed. Most of workflow systems used predefined rules and policies (specifications) to route workflow and once the system is done, all these are hardcode in workflow engine; (3) the business data are not easily changed. In most of current systems, once the system is running, the business dada (like purchase order, etc.) are not easily changed. Some systems used component and MVC (Model-View-Controller) technologies to separate the data and display, but they are coupled so closed. As even in both side uses same data objects and accessing methods internally to do data binding. This not only limits the system scalability and flexibility, but also limits the business data changeable to meet enterprise market changes and business changes.

As discussed above, it can be found that the reason of these problems occurred are because of focusing on workflow execution and not on the business data itself.

A. Xie & X. Huang (Eds.): Advances in Computer Science and Education, AISC 140, pp. 39–45.
springerlink.com

The workflow procedure in fact is business data flow and not the control flow. The business data moves from one people (role) to another people (role) is because of the data status changed, it is that data drives itself moving around in system. The designers of workflow system need turn back to business data and use data to design the system. Comparing with program logics, people are better at data processing.

2 Data Driven and Business Workflow

2.1 The Data-Driven Concept

From computer architecture, the computers can be categorized into control-flow computers and data-flow computers. In control-flow computers, the code and program executed sequences are pre-programmed by programmer; it exactly follows the programmer control (program code) to execute one by one. This is control-driven. Nowadays, most computers are control-flow computers. The data-flow computers are base on data and data relationship, its execution sequence is not fixed and data will drive the computer and system running [1].

Data-Driven is a control theory and methodology which using online and offline data and knowledge from data processing to design controllers, and in some assumptions, it has astringency, stability guaranty and robust results. Generally it is a theory and methodology to design system directly from data to system [2]. The root of design and programming is data representation. The core of data-driven is to separate the business definition and business implementation, which means using data to define business and according to these data to design and implement the system. The system is responsible to use data, explain data, manage data and transfer data and it doesn't care about whether the data will be changed or not. Data-driven includes three major steps, which is data and data set definition, using data, and data set maintenance.

2.2 Business Workflow Specific Requirements

Workflow is the movement of documents, tasks and information among people in certain rules and policies throughout an organization in-order to make decisions and complete any required processing that they create. It is a totally or partially auto-execution process [4]. Business workflow is that enterprises uses workflow technologies to process their huge business data to improve their work performance and effectiveness. But when analyzing enterprise business, it can be found that there are some special requirements: (1) there are many types business data need to be processed, and alone with enterprise and market changes, some new types of business data need be easily appended and some old types of data need be removed; (2) each type of business data has its own work flow and the flow need be changeable and configurable to meet enterprise new requirements which means the flow control rules and policies can be changeable; (3) each type of business data needs multiple presentation patterns. As alone with the data moving in the flow, different people (role) has different permission to read data and different right to deal with the data and make decisions. Meanwhile, even to same people (role), the display style and appears of data may be changeable also, for example personnel customized user interface; (4) during one business data workflow upgrading, the other workflows shouldn't be interrupted that means the enterprise internal changes, market changes and system upgrading should not affect the online system running.

3 Data-Driven Business Workflow Design

From discussion above, it is obvious that the control-driven workflow design methodology is not very good at business data workflow. The data-driven workflow means the workflow runs dependent on the data status and its situation, its running path cannot be fixed or predefined in the system source code. Using data-driven methodology to design and implement business workflow is more close to human's thinking habits. There are four steps to use data-driven business work-flow design:

1. Define business data and their usage (usage data), define rules data to control the workflow movement, which means to build data model first.
2. According to the data model to design the workflow model.
3. Using data to drive the workflow model running, and keep the model as simple as possible, and make sure each component in the model are driven by data.
4. Keep the data (including business data, usage data and control data) as flexible as possible, like use XML format.

3.1 Design of Business Workflow Data Model

All the business workflow requirements are that the workflow design and implementation must be flexible enough to allow enterprise changes. XML is eXtensible Markup Language; it is extensible, across platforms and internet and programming language independent. It's very suitable to describe business data and its workflow. In order to build a data-driven business workflow, based MVC model concept, the data model needs to be created first. By study business workflow, the following data needs to be defined:

(1) Define business data (like purchase order, etc) in XML format, and also including attributes data, like the current business data status, etc.
(2) Define data flow rules and policies in XML.
(3) Define components connection in XML.

As illustrated in Fig.1, it is designed to isolate business data layer which can make sure data access layer independent on the database information, like database name, connection and table filed name, etc.

Data Mapping: It is XML format description which describes the mapping between data specified in Business Data Patterns and table/fields in Business Data Layer.

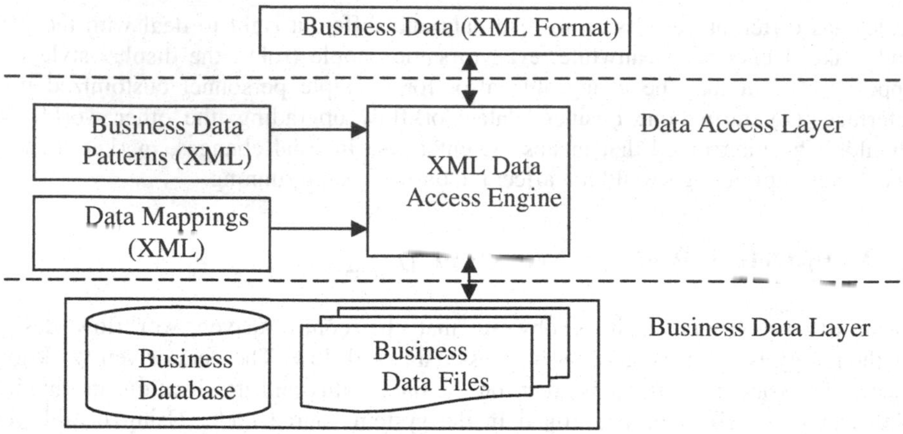

Fig. 1. Business workflow Data Model

Business Data Pattern: An XML description which describes the business data used in whole system. It is template specifying how to build Business Data in XML format.

XML Data Access Engine: It has two works to be done. (1) According to request and Data Mapping to fetch data from Business Data Layer and based on the Business Data Patterns to package retrieved data into specified XML; (2) with input of XML data to decode it with Data Mapping and Business Data Patterns and store them into Business Data Layer.

3.2 Design of Business Workflow View

In order to design a generic view with reasonable structure used to present business data, the view should only concern about the display style and data itself; it shouldn't care about the business logics and should be as simple as possible of the connection with data model and controller. As data model is only output the XML information, so during the view design, we should think about how to display the business data (XML format) correctly and don't care about what content is inside. That means the view's responsibility is only to serve the business data.

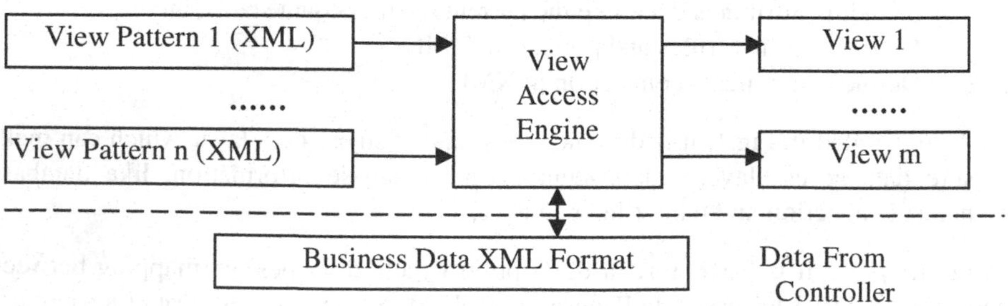

Fig. 2. Design of Business workflow View

For same business data, during it moves around in the system, different people requires system to display it in different way and even for some security reason etc. the content to be displayed may be different also. In traditional mode, software engineers will design and implement different pages, this will cost many work efforts for engineers. In order to save engineers time and even release engineers from this repeatable workloads, it uses XML format view pattern to define the display pages with different styles. It has following advantages: (1) XML format, extendable and flexible, user can easily add or remove some information to change display style. (2) Text format, user can easily edit it and no need programming skills. (3) User can add any more patterns for their business changes. Or even define customized display style personally.

View Access Engine: It has two major functions. (1) It interprets business data (XML) with selected view pattern and generate web page. (2) It captures the business data changes and package it XML format and pass back to controller.

3.3 Design of Business Workflow Controller

As specified in the diagram above, the business workflow is described in XML file. This file contains all the status of business data during its movement and the status change rules and policies. All the status are based on the business data movement and their policies and are not fixed alone with its moving. The flow status will be changed, and during this period, the user just need uses text editor to change the rules and policies, the workflow changes will be done. The workflow engine works based on the business data itself and its rules and policies to determine which the next status is and then tell the data model to update it and dispatch it.

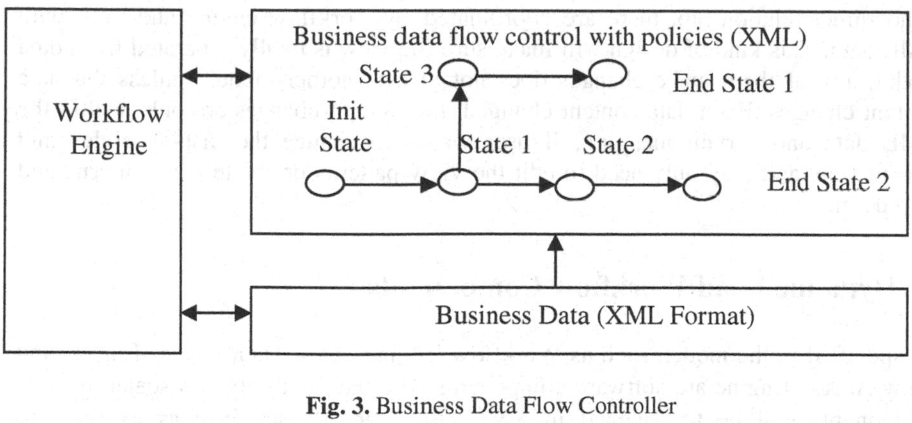

Fig. 3. Business Data Flow Controller

3.4 Design of Business Workflow System Model

For any enterprise, there are a lot of businesses to be processed; each business data has its own workflow. In order to manage and control all these flows, it uses XML to manage them. Each flow has an identifier and corresponding to a business data. This allows users and enterprise easily to add and remove certain workflows to start a new business or stop some useless workflow.

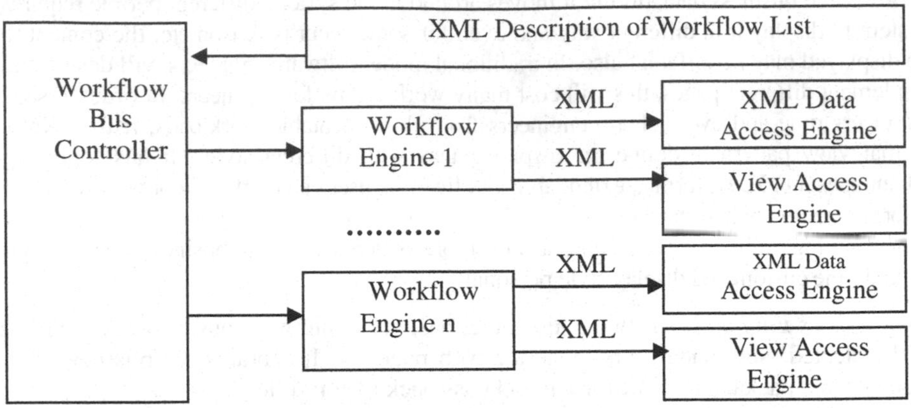

Fig. 4. Design of Business workflow Working Model

Workflow Bus Controller: a software component which used to manage the workflow list and coordinate each workflow engine. Because Workflow Engine, Data Access Engine and View Access Engine are only relate to XML data, but there are not dedicate to any specific business data and its flow, so for all business data, they can use same engines, of course in multiple instances to control all business data and workflows. Of course, the designer can design dedicate engines for some specific business data.

In this system model, between Data Access Engine and View Access Engine, there is no direct relationship, there are coordinated by workflow engine and only with XML data. This kind of design will make sure the view is totally separated from data model, any of them make changes does not affect another module unless the data content changes. Even data content changed, the system changes are only within the XML data and certain mapping; if user wants to change the display styles and information, and they only need to edit the view patterns or create new patterns and map them.

4 Dynamic Load Workflow Components

As specified in the model, such as Workflow Engine, XML Data Access Engine and View Access Engine are software components. For the flexibility and scalability the components will be text named in XML file. Here it uses Java as example to demonstrate dynamically loading software components and execution of it methods.

Java Classloader[5], package mechanism provides a facility to dynamically load java based component. It uses Classloader to load component and uses reflect method to call and execute component methods dynamically. For example, wfEngine.class is business data workflow engine and the class name is wfEngine, one method name is xmlFetch(String dataID), the implementation detail is listed Fig.5 below:

```
public class EngineLoader  {
......
   // Using ClassLoader to load Java component wfEngine
   ClassLoader loader;
   loader = new ClassLoader ("C:\\wfEngine.class");
    Class objClass = loader.loadClass("wfEngine");
    Object obj = objClass.newInstance();

    // Using reflect mechanism to get the xmlFetch method of wfEngine
    Method x1 = objClass.getMethod("xmlFetch", new Class[ ]{String.class});
    Object objS[] = { new String ("dataID") , null};
    // Start the Java Component wfEngine
    x1.invoke(obj,objS);
......
}
```

Fig. 5. Dynamic Load Workflow Component

5 Summary

Comparing to control-driven, data-driven is another methodology to solve computer problems. Here by study data-driven methodology and its application, it illustrates a new method to design and implement a business workflow; with XML and MVC technologies, it uses XML data as media to simplify the couples of the components connections. This kind of workflow design and application and will hugely improve the workflow suitable to the business changes or upgrades and will enhance its flexibility and configurability. Of course using XML data as the communication among components will drop down the system and workflow work performance, but it can memory cache certain stable XML to improve it.

References

1. Shu Xiao-Hua. "Data-Driven Calculation and Its Applications" [J]. Design & Appliations, 2009.3: 82-84.
2. Hou Zhong-Sheng, Xu JIan-Xin. "On Data-driven Control Theory: the State of the Art and Perspective" [J]. ACTA AUTOMATICA SINICA, 2009.35(6): 650-666.
3. Zhang Jie, Zhao Wen-Yun, Ni Xiao-Feng, "Workflow Based Changes Management Tools" [J]. Computer Engineering, 2005.31(11):50-51.
4. Zhang Zhu-Hong, Wang Qian, Zhou Xiao-Ping. "Research and Development of Distributed Workflow based ERP System" [J]. Industrial Control Computer. 2005.18(2): 5-7.
5. Liu Xiao-Dong, Cao Yun-Fei, Hu Zhao. "Dynamic Component based Compose Component ". Mucroelectronics & Computer. 2005.22(2): 100-102.
6. ibm/developerWorks. Understanding the Java ClassLoader [DB/OL]. http://www.panix.com/~mito/articles/articles/classloader/j-classloader-ltr.pdf.
7. WfMC.Workflow Management Coalition Terminology and Glossary[R]. Brussels: Workflow Management Coalition, 1996.
8. Xu Jia-Xin, Hou Zhong-Sheng. "Notes on Data-Driven System Approaches" [J]. ACTA AUTOMATICA SINICA. 2009.35(6): 668-674.

Research on XML Data-Driven Based Web Page Design

ChuanSheng Zhou

Software College Shenyang Normal University, Shenyang, Liaoning 110034, China
jasoncs@126.com

Abstract. Alone with Internet appeared, the web applications have been used in many business areas. But the flexibility, changeful and reuse of web page is always one of core difficult problems to many web designers and developers. Here start at introduction of data-driven method and analysis problems of web pages design and implementations, by research and study on the XML, XSL and MVC technologies, it introduces an XML based data-driven dynamic web page design method and model to provide a dynamic web page generation and then to improve web page reusability, scalability, flexibility, and extensibility. At the end, it demonstrates the implementation with XSL templates.

Keywords: XML, Data-Driven, Web Page, XSL.

1 Introduction

Alone with web application quickly development and spread out, there are a lot of web based software appeared in market, like web based MIS, ERP, OA, etc. Today, almost all enterprises need these systems to help them deal with and complete their business jobs. But in traditional web systems, all of the web pages are designed and developed in control-driven, which means the web pages and work flow are designed from user actions point of view, and are customized and hard coded in the system design and development time, and cannot be easily changed. While to modern enterprises, their business and process for management are always changing, and the user's actions cannot be exactly predefined or predicted earlier, therefore their web system and web pages need to be easily upgraded to adapt to these changes. In another aspect, to web system designer and developers, even in the same system, web pages design and development normally cost them much effort. Here, in order to resolve these problems, it first introduces data-driven method and analyzes the web pages, by study on XML and MVC model, to illustrate a dynamic web page design model with XML based data-driven method and at the end by using XSL to demonstrate its implementation.

2 The Web Page and Its Problems

Generally a web page has two responsibilities. One is to display data and another one is to accept user inputs. To display data, the first is that there are some data in some format to be displayed which are defined the content in a web page (it is called

A. Xie & X. Huang (Eds.): Advances in Computer Science and Education, AISC 140, pp. 47–53.
springerlink.com

content data here); the second is that there is some display templates used to define the web page appearance (it is called web page template here); the third is to bind the data to the display template and then generate the final web page. For instance, there is one web page used to display customer information in CRM system. First there is customer detail information data which can be retrieved from CRM system; Second need to define a web page display template which specifies how to arrange and display the customer information data; and at the end binding the customer detail information data to the display template and generate the finial web page to output customer information. Regarding to the web page accepts user inputs, actually it depends on the display template, as the template can specify if the display items in a web page can be changeable. If it is then the web page can accept user inputs, otherwise it cannot accept user inputs.

In many web systems, the web page is becoming a headache problem to web system designers as there are two major reasons. (1) The web page display style and format is predefined in design and implement time, and fixed in the source code. (2) The binding of business data to display format is done in the source code. So when enterprise need upgrade their system and web pages to meet the business and market changes, they have to look for the system provider to redesign and re-code the system. This is not only cost enterprise more money, but also interrupts their working procedure.

3 Data-Driven Web Page Model

According to MVC mode, content data can be processed by Model and also can be displayed in web pages to users by View. Data is core and the web page is method which can be changed. With data-driven method, to design a generic web page generation model, the following steps can be followed:

(1) Define content data to be displayed in web page;
(2) For each content data, design display templates;
(3) For same data, design multiple display templates if the content data has different status (like edit).
(4) Design and implement web page engine to bind the content data to templates and generate web page.

3.1 Data-Driven Web Page Generation Model

From Fig.1, it is clear that the data-driven web page generation model is based on MVC model (Model-View-Controller). The dash line means that any content data in any status will map to one specified web page template, and content data in Model can be changed from one status to another status. Although this model is MVC, but it changed from traditional MVC. In this design it removed the direct connection between Model and View and uses the Controller to coordinate the Model and View communications in XML format data. That means the communication between Model and View is through Controller in XML format data.

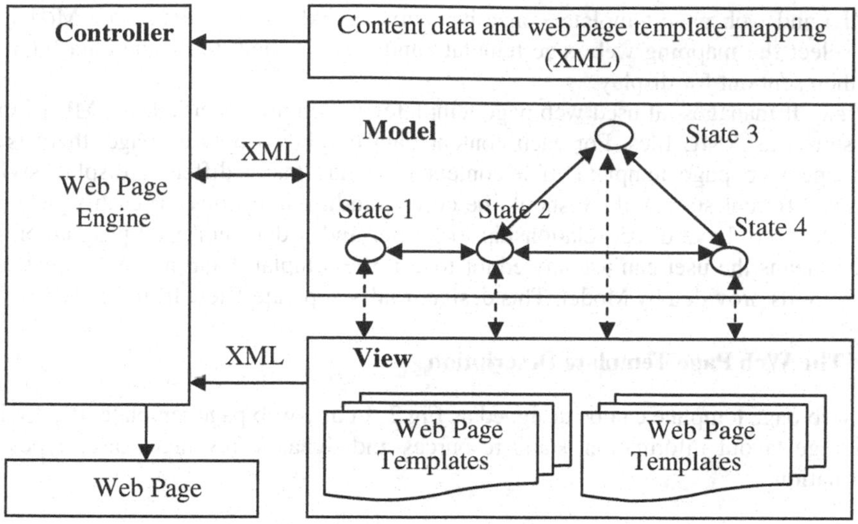

Fig. 1. Data-Driven Web Page Generation Model

3.2 The Component Descriptions in MVC Model

XML (eXtensible Markup Language) is extensible language, which is across platforms and internet and it is programming language independent. Using XML to package content data (processed by Model and to be display by View) and define web page templates can resolve the close coupling relationship between View and Model in traditional MVC. At same time, once the display style need to be upgraded, the user only need using any XML editors or text editors to change the web page templates and no need go back to the system providers.

Model: It manages all types of content data used in the system including the data computation. Normally it has two responsibilities; one is to receive request data (XML format) from Controller, understand the request and deal with related operations (like status movement, database query, transaction process, etc.); the second one is to package the results of the request into content data (XML format) and pass back to Controller. For each type of content data, its format is predefined in XML template. If some content data format changed or need add new types of data, the Model only needs replace one data processing component or add new data processing component. This can keep the Model changeable and scalable, and meanwhile make sure Model is focusing on data processing and no need care about how to display it. The communication between Model and Controller is XML information (XML request and content data in XML format).

Controller (Web Page Engine): It has two jobs to be done also. The first one it to resolve the HTTP request from web page and package it as XML request based on predefined request type and its XML template and then sending to Model. As in XML template, it can specify the mapping between request data and HTTP request parameters, so it is easily to resolve the HTTP request. The user can manually add some new XML template to scale up the user request processing. The second one is to generate a web page. According to the predefined mapping between content data

(XML) and web page templates, once it receives XML content data from Model, it will select the mapping web page template and resolve them to do the data binding and then sent out for display.

View: It manages all used web page templates which are specified in XML format and stored as XML files. For each content data displayed by web page, there is at least one web page template. If a context data has many different display styles (means different status), there should be corresponding templates for each style. The View does not have direct relationship to Model and it does not need programming, which means the user can use any editor to edit the templates and no need direct call the methods provided by Model. This design totally separate View from Model.

3.3 The Web Page Template Description

The web page template can be designed as Fig.2. In any web page template, it contains web page layout information, static resources and dynamic resources three types of information.

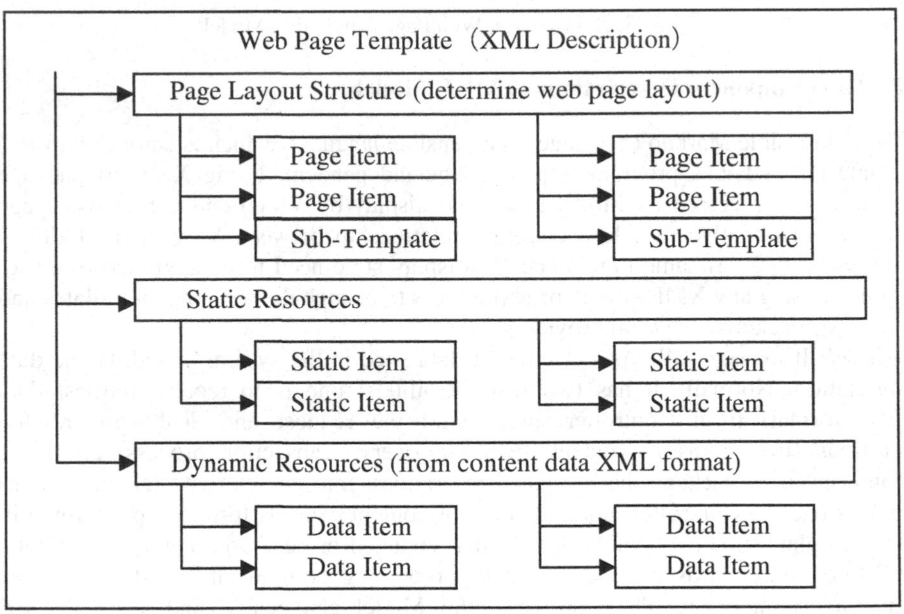

Fig. 2. XML Description of Web Page Template

Page Layout Structure: It defines a web page layout and its page items information. The page item can be any one, like background, font size, position, etc. A web page can be embedded into another web page (here it use sub-template to describe it).

Static Resources: Some static resources used in the web page, like icon file, image files, CSS, etc. One static resource can be used and reused in any page items.

Dynamic Resources: It means the data used in page item which extracted from content data. Because in a web page, maybe not all of data in context data are displayed, it only specifies the used portion. And also for each data item, it needs to specify its attributes (like changeable, etc).

3.4 Content Data and Web Page Template Mapping

As an example in table 1, in this XML mapping, the tag <DataType> used to specify what kind of content data to be displayed. This is first level element in the mapping file, as the web page template selection is based on content data types and its status; putting it as first level element can improve the system performance. The content data type can be predefined and can be added later, each data type has a system level unique identification, For example, "CustomerList" indicates the list data of all customers in the system. If there are other types of customer list (like only list customers' name), user can assign another unique data type to it, or user changes the web page template to realize it. The later method performance is lower as there are much more data (e.g. customer's other information) will be processed and packaged in Model.

Table 1. XML Mapping Between Data and Template

Tags in XML file	Meaning of Tag
< DataType="CustomerList">	Data Type
<Status_1="T_CL_1.xml">	the data in Status 1, using template T_CL_1.xml
<Status_2="T_CL_2.xml">	the data in Status 2, using template T_CL_2.xml
<Status_3="T_CL_3.xml">	the data in Status 3, using template T_CL_3.xml
</DataType>	
< DataType="Login">	Data Type
<Status="T_Login.xml">	In status, the Login Template using T_Login.xml
</DataType>	
……	
……	
……	
……	
< DataType="Order_DVD">	Data Type
<Status1="T_DVD1.xml">	the order in status1, using T_DVD1.xml template
<Status2="T_DVD2.xml">	the order in status2, using T_DVD2.xml template
</DataType>	

4 Example of Web Page Generation

Here we use customer list as example and working in IIS and programming in ASP to demonstrate how to dynamically generate the web page.

(1) Define customer-list.xml (e.g. in Fig. 3) contains a customer list; each customer row will contain his/her details. It example assumes the Model creates a XML file and pass to Controller (web page engine). Of course in practice, the Model may create an XML string or XML object.

(2) According to customer-list.xml (e.g. in Fig. 4), define its web page template customer-list.xsl, in this example; we use XSLT to define it.
(3) Design the web page engine in ASP and do the binding of customer-list.xml and customer-list.xsl to generate the web page (e.g. in Fig. 5).

```
//Customer-list.xml
<?xml version="1.0"
encoding="ISO-8859_1>
<customer-list>
   <customer>
      <name>Tony Yong</name>
      <title>Manager</title>
      <company>IBM</company>
      <phone>12345678</phone>
   </customer>
   ……
</customer-list>
```

Fig. 3. Example of Customer-list XML file

```
//Customer-list.xsl
<?xml version="1.0" encoding="ISO-
8859_1>
<xsl:stylesheet version="1.0">
xmlns:xsl=http://www.w3.org/1999/xsl/transfor
m>
<xsl:template match='/'>
<html><body>
<table border="1">
   <tr bgcolor="blue">
   <th>Name</th>
   <th>Telephone</th>
   </tr>
  <xsl:for-each select ="customer-
     list/customer">
  <tr bgcolor="yellow">
   <td><xsl:value-of select="name"></td>
   <td><xsl:value-of select="phone"></td>
  </tr>
</table>
</html>
```

Fig. 4. Example of Customer-List XSLfile

In this example, the content data is stored in XML file, but in actual implementation, it can be an XML string or XML object, but any XML programming package can parse it and load it. Regarding to server side data binding, in practice, the designer can develop a software component, which can accept the XML data and XSL template as parameters to implement it. In fact, this is just an example; the web page templates may not be defined in XSL but need in XML format and can be understood by web page engine. Another point is this example uses the Microsoft component "Microsoft.XMLDOM" to do the data binding, but in practice, the designer can design and implement their own engine to do the data binding.

5 Summary

Anyway, data-driven is a methodology for problem solving and system design compared with control-driven. As web page is used to present data information to users for actions in a web based system, data-driven is very suitable to do web page dynamic generation and which can help designers totally separate the data processing from view. Here it bases on XML and MVC technology to design a data-driven web

```
// binding customer-list.xml to customer-list.xsl on server in ASP language,
// similar as web page engine
<%
    ' load customer-list.xml
    set xml=Server.CreateObject("Microsoft.XMLDOM");
    xml.async = false;
    xml.load(Server.MapPath("customer-list.xml");

    'load customer-list.xsl
    set xsl=Server.CreateObject("Microsoft.XMLDOM");
    xsl.async = false;
    xsl.load(Server.MapPath("customer-list.xsl");
    Response.Write(xml.transformNode(xsl));
%>
```

Fig. 5. Example of Data binding of XML and XSL in ASP

page generation model, and practice indicates it is successful. Meanwhile it also indicates more flexibility, scalability and more convenient to users for system upgrade.

References

1. Hou, Z.-S., Xu, J.-X.: On Data-driven Control Theory: the State of the Art and Perspective. Acta Automatica Sinica 35(6), 650–666 (2009)
2. Xu, J.-X., Hou, Z.-S.: Notes on Data-Driven System Approaches. Acta Automatica Sinica 35(6), 668–674 (2009)
3. Zhang, C.: Research on Component Service Framework Base on Data-Driven. Journal of Computer Engineering and Application (18), 39–41 (2005)
4. Wang, L., Zhang, G.-H., Liu, L.: Research on Data-Driven Based Object Model. Journal of Yunnan University 31(S1), 48–51 (2009)
5. Zhang, Z.-H., Wang, Q., Zhou, X.-P.: Research and Development of Distributed Workflow based ERP System. Industrial Control Computer 18(2), 5–7 (2005)
6. Liu, X.-D., Cao, Y.-F., Hu, Z.: Dynamic Component based Compose Component. Mucroelectronics & Computer 22(2), 100–102 (2005)

```
'populate customer list and to each customer list of orders ... [illegible]
'Generate the requested data
...
[illegible]
Set ... = Server.CreateObject("Microsoft.XMLDOM") [illegible]
[illegible]
[illegible]

[illegible]
Set ... = Server.CreateObject("Microsoft.XMLDOM")
[illegible] = False
[illegible] Server.MapPath(...) [illegible]
Response.Write ... [illegible]
```

Fig. 3 [illegible] XML and XSL [illegible]

[illegible]

References

1. [illegible]
2. [illegible]
5. [illegible]
6. [illegible]

Research and Application of Data-Driven in Web System Design

ChuanSheng Zhou

Software College, Shenyang Normal University, Shenyang, Liaoning 110034, China
jasoncs@126.com

Abstract. The system flexibility and scalability is always the problem to web system designers and developers. During Web based system development and design, how to make the system flexible and scalable is a headache issue to software architects and designers. Here, it analyzes the problems of current web system design and researches on XML, MVC and data-driven technologies, to introduce a data-driven based web system design and describe its model.

Keywords: XML, Data-Driven, Web System, MVC.

1 Introduction

Alone with web technologies developed, web based systems have been used into many areas and enterprises, like e-Commerce, e-Government, ERP, CRM, OA etc. During the application of these systems in practice, the enterprise and end users found that there are following problems: (1) once the system is deployed, the business functions and related work flow are fixed; the users and even enterprise cannot change them; (2) the system cannot be easily upgrade. For example, if enterprise needs add some new business or change some old business procedure, the current system is very difficult to do these changes; (3) the system web page including display styles, content, reports, etc. cannot be easily customized without further development; (4) the business data cannot be easily changed and maintained to meet market and enterprise management changes. The reason of these problems happened is because of that the web system designer tried to understand all business procedures and related business data process and tried to use web system to package them together for users and enterprises. But in fact, it is impossible as different enterprise has different business, and even to same enterprise, alone with their business movement and market, they will change or upgrade their business procedures and business data also. In another side, the working model of web system is "request/response" which means the system receives request data and then sends back the response data to users. Is there a way to save web system designers from complicated business and leave the business procedure and data to enterprise? Here, by research on XML and MVC technologies, it illustrates a data-driven web system design model and try to solve the problems stated above.

A. Xie & X. Huang (Eds.): Advances in Computer Science and Education, AISC 140, pp. 55–61.
springerlink.com

2 Data Driven and Web Systems

2.1 The Data-Driven Concept

Data-Driven is a control theory and methodology which using online and offline data and knowledge from data processing to design controllers, and in some assumptions, it has astringency, stability guaranty and robust results. Generally it is a theory and methodology to design system directly from data to system [1].

Data-Driven is a control theory and methodology which using online and offline data and knowledge from data processing to design controllers, and in some assumptions, it has astringency, stability guaranty and robust results. Generally it is a theory and methodology to design system directly from data to system [2]. The core of data-driven is to separate business definition from business implementation.

2.2 Web Systems

Web system uses "request/response" as its standard operation. There are following characteristics (using MVC as example):

(1) The request data are sent to back end in HTTP;
(2) The request data are accepted by some module in back end (as Controller, etc);
(3) The request data are processed by some module in back end (as Model, etc);
(4) The response data are generated by some module in back end (as Model, etc.);
(5) The response data are bind in some format by some module in back end (like View in MVC, etc.);
(6) The response data are sent back to browser in HTTP;

As discussed above, all of web system processes request and generates the response. If using data-driven method to design a web system, it can follow the next steps:

(1) Define data set to be processed by web system;
(2) For each data, define its operations (including the request and response data);
(3) Using XML to describe these data (or templates) to keep the data and their operation are flexible and scalable.
(4) Using data and data operation to drive the web system model design.

3 Data-Driven Web System Design

3.1 Design Data-Driven Web System Model

According to the discussion above and MVC model, a data-driven web system model can be designed as follow:

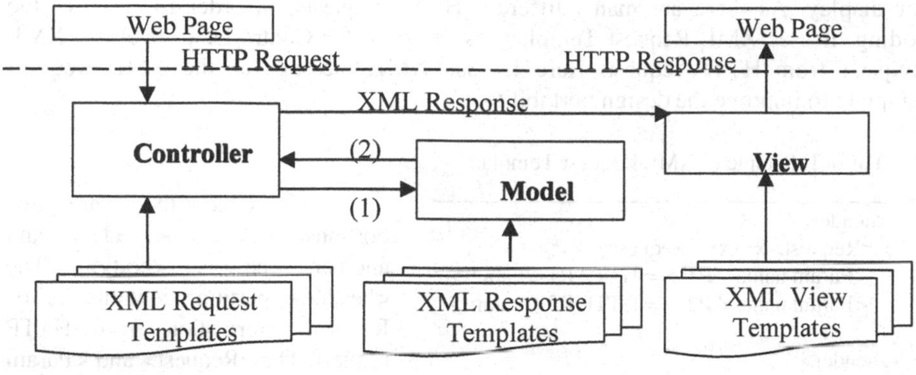

Fig. 1. Data-Driven Web System Design Model (1) XML Request (2) XML Response

As in Fig.1, it is modified MVC, as it removes the direct communication between Model and View and uses Controller as data interchange. In the meantime, the data used for communication between Controller and Model or View is only based on XML messages or XML format data. This is not only separate Model from View, but also quickly drops down the coupling relationship between Controller and Model or View. And finally it improves the web system scalability and upgradable functionalities.

3.2 The General Working Procedure of the Model

As web system normally uses "request/response" as its standard working mode. Its standard work procedure is below:

(1) the Controller receives the HTTP request from web page;
(2) the Controller bases on the predefined XML Request templates to parse and package the HTTP request to XML Request (XML format request) and pass the XML Request to Model;
(3) the Model parse the XML Request and according to the pre-designed XML Response templates for the request, do certain computing and data accessing and transforming (like access database, transaction, etc.) and then package the results data in XML format (XML Data) and pass back to Controller;
(4) the Controller receives the XML Response from Model and pass to View;
(5) once the view receives XML Response and based on pre-designed view templates for that data to do the data binding and then generate the new web page and send out as HTTP response.

3.3 The Design of Controller

In this model, the controller has 4 responsibilities: (1) accept the HTTP request and parse it; (2) according to predefined XML Request template, it transfer HTTP request to XML Request; (3) pass the XML Request to predefine de data Model for data further processing; (4) accepts the XML Response from data Model and pass to View for display. As there are many different HTTP requests, in order to simplify the coding, it uses XML Request Template as the base for Controller to package XML Request from HTTP request. Here it uses XML message as the XML Request template to improve the design portability.

Table 1. Sample of XML Request Template

```
<header>
  <Request> xxxxx</Request>
   <Param name="P1"> = HTTP R1</Param>
   <Param name="P2"> = HTTP R2</Param>
   ……
</header>

<body>
<data_field_1>"HTTP parameter 1"</field_1>
< data_field_2>"HTTP parameter 2"</field_2>
< data_field_3>"HTTP parameter 3"</field_3>
< data_field_4>"HTTP parameter 4"</field_4>
< data_field_5>"HTTP parameter 5"</field_5>

   ……
   ……
< data_field_n>"HTTP parameter n"</field_n>
</body>
```

In this table, there are two portions. One is <header> and another one is <body>. The <header> portion contains XML Request information from HTTP request. The <Request> and <Param name=""> are predefined during the system design and directly map to HTTP request parameters. The <header> can be understood by both Controller and Model. The <body> contains all needed information if the <Request> need to update database or storage information, and the <data_field> should be same as defined in XML Response Template and their values mapping to HTTP request parameters. All the XML Request and related parameters should be predefined; of course if there is some new request, the predefined list can be easily updated.

3.4 The Design of Model

The model has two jobs. One is to understand the XML Request and do some processing; another one is to package the response data in XML Response based on the XML Response Template and pass back to Controller.

Table 2. Sample XML Response Template

```
<header>
<Request>xxxxxx</Request>
<Response>xxxxx</Response>
 <DataStatus> ReadOnly</DataStatus>
  ……
</header>
<body>                     // View data
<data_field_1>"value 1"</field_1>
< data_field_2>"value 2"</field_2>
< data_field_3>"value 3"</field_3>
< data_field_4>"value 4"</field_4>
< data_field_5>"value 5"</field_5>
  ……
< data_field_n>"value n"</field_n>
</body>
```

Similar to XML Request, the XML Response should be predefined also and the Model should follow this XML Response Template to package the response of Model to Controller. As shown in the sample, the XML Response is response to the XML Request;

During some data processing, the response data may be in different status and need different display style. Because data actual process may be very complex and may be associated to many tables, databases and files, and even related to a lot of dedicate algorithms. So in order to simplify the complexity, it uses a software bus to manage all the data process components and uses a table to map them which can be easily changed with text editors if some data or data process components needs to be upgraded without interrupt the system running, as shown in next table of the mapping predefined and used by Model.

Table 3. Mapping of XML Request/Data Component and XML Response

XML Request	Data Component	XML Response
xmlRequest1	DataModel_1	xmlResponse1
xmlRequest2	DataModel_1	xmlResponse2
xmlRequest3	DataModel_3	xmlResponse3
xmlRequest4	DataModel_4	xmlResponse4
……		
……		
xmlRequest-n	DataModel_n	xmlResponse_n

As shown in table 3, each XML Request will map to a data process component and map to a predefined XML Response template. Of course user can map more than one XML Requests to one data component, but this will increase the data component complexity and drop its performance. In the table, we use *xmlRequest* to represent XML request, *DataMode* to represent data process component and *xmlResponse* to represent XML response template.

3.5 The Design of View

In order to design a generic view with reasonable structure used to present business data, the view should only concern about the display style and data itself; it shouldn't care about the business logics and should as simple as possible of the connection with data model and controller. As data model is only output the XML information, so during the view design, we should think about how to display the business data (XML Response) correctly and don't care about what content is inside. That means the view's responsibility is only to serve the business data.

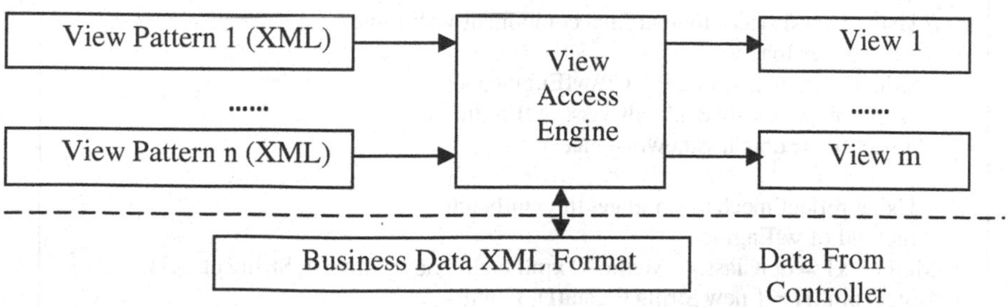

Fig. 2. Design of Business Data View

As shown in table 4, some XML Response has more than one status, if the display need different, then it cam map to different view patterns (which is based on XML Response template to pre-define in XML format). Some XML Response has no status, which means all the response uses same view pattern to display. During system running, if the enterprise need change their display, they just need change the view pattern. If the business data changed, besides the view pattern need change, there are including XML Request (templates), XML Response (templates) and the data process component need to be changed. For the second situation, the user can use any text editor to update the XML Request (templates) and XML Response (templates), but for the data process component need to be re-programming and then replace old one.

Table 4. Mapping of XML Response/Data Status and XML View Pattern

XML Response	Data Status	View Pattern
xmlResponse1	Status_1	xmlView_1
	Status_2	xmlView_2
xmlResponse2		xmlView_P
xmlResponse3	Status_1	xmlView_P1
	Status_2	xmlView_P2
……		
……		
xmlResponse-n		xmlView_n

4 Dynamic Access Software Components

As specified in the model, such as Workflow Engine, XML Data Access Engine and View Access Engine are software components. For the flexibility and scalability purpose to allow user easily name them and change some flows and patterns, the components will be text named in XML file. This needs the system can dynamic load the components and run them. Here it uses Java as example to demonstrate dynamically loading software components and execution of it methods.

```
public class EngineLoader  {
……
// Using ClassLoader to load Java component wfEngine
 ClassLoader loader;
 loader = new ClassLoader ("C:\\wfEngine.class");
  Class objClass = loader.loadClass("wfEngine");
  Object obj = objClass.newInstance();

// Using reflect mechanism to get the xmlFetch
// method of wfEngine
Method x1 = objClass.getMethod("xmlFetch", new Class[ ]{String.class});
Object objS[] = { new String ("dataID") , null};

// Start the Java Component wfEngine
x1.invoke(obj,objS);
……
}
```

Fig. 3. Dynamic Access Software Component

Java Classloader[5], package mechanism provides a facility to dynamically load java based component. As Classloader appointment mechanism specified, it can appoint the load task to upper layer loaders unless the class is loaded. And also, Java provides reflect method to call and execute component methods dynamically. For example, wfEngine.class is business data workflow engine and the class name is wfEngine, one method name is xmlFetch(String dataID), the implementation detail is listed in Fig. 3.

5 Summary

The Data-driven is a methodology to use data driving system design and problems solving. Here by study on data-driven methodology, MVC and XML technologies, it illustrates a new method to design and implement a web based systems. As XML is flexible and web system "request/response" as its standard operation, it uses XML to define a set of XML Requests data and related XML Responses data, by binding to different view patterns (XML view template data) to improve the whole system flexibility and scalability; in the meantime, the end user or enterprise just need use any text editor to achieve their system, display information changed unless the changes related to XML Response template.

Acknowledgement. This project is sponsored by:

Science & Technology Department of Liaoning Province (Project Number: LiaoKeHan [2010] 36)

References

1. Shu, X.-H.: Data-Driven Calculation and Its Applications. Design & Appliations 3, 82–84 (2009)
2. Hou, Z.-S., Xu, J.-X.: On Data-driven Control Theory: the State of the Art and Perspective. Acta Automatica Sinica 35(6), 650–666 (2009)
3. Zhang, J., Zhao, W.-Y., Ni, X.-F.: Workflow Based Changes Management Tools. Computer Engineering 31(11), 50–51 (2005)
4. Zhang, Z.-H., Wang, Q., Zhou, X.-P.: Research and Development of Distributed Workflow based ERP System. Industrial Control Computer 18(2), 5–7 (2005)
5. Liu, X.-D., Cao, Y.-F., Hu, Z.: Dynamic Component based Compose Component. Mucroelectronics & Computer 22(2), 100–102 (2005)
6. ibm/developerWorks. Understanding the Java ClassLoader (DB/OL), `http://www.panix.com/~mito/articles/articles/classloader/j- classloader-ltr.pdf`
7. WfMC. Workflow Management Coalition Terminology and Glossary. Workflow Management Coalition, Brussels (1996)
8. Xu, J.-X., Hou, Z.-S.: Notes on Data-Driven System Approaches". Acta Automatica Sinica 35(6), 668–674 (2009)

[illegible] ClassLoader [illegible] mechanism provides a facility to dynamically load [illegible] based component [illegible] ClassLoader appointment mechanism specifies [illegible] point to load [illegible] class is loaded [illegible] also Java [illegible] method to call [illegible] methods dynamically [illegible] class [illegible] and the class name [illegible] (engine [illegible]) [illegible]

[illegible] Summary

[illegible] data [illegible] methodology [illegible] XML [illegible] based systems. A[illegible] its standard operation, it uses [illegible] XML Response data [illegible] XML view [illegible] whole system [illegible] just need use [illegible] changed only [illegible] XML Response [illegible]

Acknowledgement. [illegible]

[illegible] Foundation [illegible] Project [illegible]

References

[illegible] Automation [illegible]

2. [illegible] Perspective [illegible]

[illegible]

[illegible]

[illegible] Automation [illegible]

Cogitation of Information Technology Construction in University Library

Ning Xu

Library, Jilin Agricultural Science and Technology University,
Jilin 132109, Jilin Province, China
420421354@qq.com

Abstract. With the rapid development of computer network information technology, libraries of colleges and universities have already entered an internet era. Digitization, networking, intellectualization have been the signs of modern library. In the network environment, the value of the university library is not only reflected in its collection, but also reflected on if it can satisfy users' requirements on knowledge information. The author expounded the present situation and problems of the construction of universities library information technology and the concrete measures that should be taken, and proposed two important problems which should be paid attention to in the construction of universities library information technology.

Keywords: Colleges and universities, library, information technology.

University library is the information center, along with the computer network information technology flying development, libraries of colleges and universities have already entered an internet era. Digitization, networking, intellectualization have been the signs of modern library.

1 The Overview of University Library Information Technology [1]

In the process of the information construction, Library has formed a complete service system in using scientific technology principle, computer equipment, network resources, and software and hardware tools. Generally, the Library Information Technology includes traditional service technology, modern information technology and humanized management technology; they constitute the whole of technology of the library service developed in an all around way. Among them, the library management technology is an important part of modern library and even the core. It covers the scientific goal management, division of labor management, process management, and decision-making management, information management and social system management, etc.

Modern library management provides a good platform for the fusion of its science technology and humanity spirit, it is a good regulator of them. At the present, the

A. Xie & X. Huang (Eds.): Advances in Computer Science and Education, AISC 140, pp. 63–69.
springerlink.com

humanized information technology based on the modern management idea has become the main stream of the integrated development of library, including automation technology, retrieval technology, network technology and the WEB management technology, etc. In library management practice, the modern information technology becomes the core pillar of the library management. They influence each other, promote each other, and dependent on each other, and collaboratively construct comprehensive modern system. From the point of view of integrated concept, the library automation technology realizes the management that is a collection of characteristic services including purchasing& cataloguing, retrieval, circulation, and reservation etc. with advanced language programming. Network technology is a medium providing highly sharing platform for all types of library information resources. It makes the heterogeneous database of every large library to realize a solid unity, and uses the scientific ways such as cross-database retrieval, full-text retrieval, and navigation index proved the tremendous power of information sharing once again. Library WEB2.0 technology realizes reasonable changes from deeply core to practice in library. WEB2.0 formed a strong impact towards the library's information construction by its idea of unique internet, sharing and building together, and fully reflects the decisive advantage of characteristics service of library.

2 The Present Situation of the Construction and Problems of Information Technology in University Library [2]

2.1 The traditional ideas restricted the development of information technology construction. Firstly, there is a popular bias on acknowledge to library. Most people think of the role of library are collection and preservation, this concept to the can far not adapt to the development of contemporary library, and not further understand the service functions of library. Secondly, the staff is not conscious of the information. Many college school library services obviously has not kept up with the development of The Times, the development of information resources is not enough.

2.2 The funds of universities can't support the high cost of information technology construction. The present situation which most universities face is the shortage of funds and it can't enrich the information technology construction. To digitize books gets the cost on the high side, which seriously hindered the development of the construction of information technology. After investigation, when network information in a period of high tide, most electronic books nets show quite practical problem. If we want to produce an electronic book, it costs nearly 1000 yuan to type into electronic text. In this way, we need about 100 million yuan to construct a small library with 100000 collections.

2.3 The database has not been perfectly established. In the information technology construction, the establishment of database is not perfect, and both of quality and quantity can't meet the requirements. Lots of university libraries didn't take the subject indexing in completed database and the machine-readable format is not very standard. The present database is primarily written one; other types of database or the storage quantity are not very much. The retrieval functions of journal articles' category-based searching are a little weak, which is not only unfavorable for readers to retrieve, but also goes against the construction and management of library information technology.

2.4 The personal professional level of librarians limits the library informatization. The construction of the library information technology requires the librarians must have perfect information knowledge. While, at present, the university libraries still use the traditional library collecting and borrowing working patterns. The librarians who are majored in library information technology are few and most of them are not good at information technology. Therefore, university library should implement digitization construction. First of all, the prominent problem that human resources shortage should be solved. And we should pay attention to the information technology quality training of the existing librarians. Pay attention to introducing professional and technical personnel, including computer exploitation, operation management and maintenance professionals that are essential.

3 Concrete Measures for Library Construction of Information Technology [3]

3.1 To renew the idea and get away from the traditional constraints. To carry out the construction of information technology, the most urgent thing that university libraries need to do is to get away from the traditional constraints, emancipate the mind, and abolish the old management mode and obsolete concepts. The service idea of innovation, information, and use should be set up. Mutual complementary advantages should be practice between the literature material and instrument equipment. To enforce corporation, perfect resources sharing mechanism, renew the idea, completely change inherent management ideas, and strengthen the consciousness of the information service.

3.2 Improving hardware, accumulating construction fund. Universities library digital information construction is in transition development. In construction of information, hardware is the basis.Hardware construction should be forward-looking. A set of advanced computer network system and flexible structural system should be established. In the choice of technology and products, we should choose products that are more advanced, real-time expanding and easy maintenance; keep for future digitization construction the development room.

3.3 The maintenance of the traditional library and digital library of fusion technology. The traditional library and digital library are different just in the stage of development. Because the massive popularization of computer and freely flowing network provide the premise for the development of the digital library, While the traditional university library collection, classification, cataloging, inquisition etc. make solid foundation for the generation and development of digital library.

3.4 To establish and perfect the mechanism of talent management.The development of information technology inevitably impact the original library management system, change the library borrowing environment and information provide ways. The traditional library's social position has been moved. The construction of the digital library needs professional staff, management staff and subsequent reserve talents. The traditional library times work pattern of librarians is the combination of the books and people or the combination of the readers and them by book, also is the combination of the people.

4 The Two Important Problems That Should Be Paid Attention to in University Library Information Technology Construction

4.1 The copyright problems and avoidances in the construction of information technology [3].

Digital library operation comes down to the right of communication in information network which should be according to "the protection regulations of information network transmission right". After 2002, in which the first case that digital library infringe upon the author's information network transmission right happened, domestic digital libraries involved in more and more cases which are about dispute of information network transmission right. Therefore, how does digital library do to reasonably avoid infringing upon the holder's right of information network transmission relates to the sustainable development of the digital library. The paper focuses on this problem, and then puts forward some principles and measures that can reasonably avoid infringing upon the holder's right of information network transmission.

4.1.1 The problems of information network transmission right in digital library

Digital library offers digital works to readers as a kind of information service by the way of network, in order to realize the resources sharing in the corresponding users group. While digital library operation mainly includes the following three segments: information resources construction, the information resources management and information resources service. All the three segments are closely related to network transmission of works.

The problems of information network transmission right in library information resources construction. Information resource construction which mainly deals with the library collection resources construction, construction of the digital resources, online information resources collection and integration is the base and the core of digital library.

The problems of the right of communication through information network in digital library information management. Digital library information management, that is, in the process of information resources construction, the storage, organization and management and maintenance of the flood of digital information resources have been done, so that make it more orderly and normally, and then release to the network to facilitate readers' retrieval use.

The problems of the right of communication through information network in digital library information service. Information service is the ultimate purpose of the construction of digital library. It includes document delivery service, heterogeneous resources retrieval service, digital reference service, personalized service, video on demand service, information push service and interlibrary loan service, etc, a series of service projects for users.

4.1.2 Reasonable principles of avoidance

Legal principles. Legal principle is the basis of legal standard or a relatively secure principle and criteria in law. The law is scared and can't be offended. "The protection regulations of information network transmission right" as the legal basis which standardize information communication activities in network environment, has the legal effect.

The oblige priority principles. The holder of the right of communication through information network is not only the owner of the rights, but also the creator of the rights object. The holder's spirit rights and economic rights are protected by law, so that keep holder's continuous creation. When digital library reasonably avoid the interests conflict with the holder's right of communication through information network, it should be according to the holder's priority principles, in order to avoid the disputes of right of communication through information network with the holder.

The interests balance principles. The balance principles of interests are important basis of copyright protection. Digital library operation touches upon multipartite interests, including obliges, users, digital library and the press and other related interests institutions. In the case of abiding by the legal principles and according to the holder right, weighing multipartite interests can not only ensure digital library interests, but also create a harmonious network environment which is good for its own development.

4.1.3 The reasonable measures to circumvent the problems of information network transmission right in digital library

Authorize the mass works. In the construction of digital library information resources, the greatest puzzle is authorization of mass works in network dissemination. Most digital libraries are in the embarrassing situation because they didn't get the author's authorization. The holder's authorization can be obtained by the following four means: contacting with the author directly, "authorized offer", cooperating with the press and acquiring through the collective management institutions of copyright.

Effective use of the terms of "reasonable use"

The institution of reasonable use is an effective form to balance the right of information network transmission and the interests of digital library. In "The protection regulations of information network transmission right", there are provisions on the institution of reasonable use. "Citing reasonably is required in introducing works, reporting news and executing official business; the little use of school teaching research; with the purpose of offering service to the blind; offering minority language edition works to domestic users, and so on. Secondly, the library, memorial, museum and gallery, etc, can provide digital works which are legally collected and published and those are copied in digital form for preserving the edition to the service object, but directly or indirectly benefit is forbidden". So, those nonprofit university libraries and public libraries, as far as possible, can use "fair use" system to avoid problems related to the right of communication through information network in the process of digital library service, while digital libraries for the purpose of profit such as the star digital library, the intellectual scholar at digital library are beyond privilege.

Make full use of social public information

Strengthen the technical measures. One of the problems related to the right of communication through information network in the process of digital library service is that some of the users would intend to avoid or destroy technical measures devices to get resources. Protection only taken by user terminals is not enough, the digital library as a network information content provider and network service provider, was asked to protect the holder of the right of communication through information network by technical measures. The so-called technical measures means that the transmitter of network works limits access to the works of others or prevents another person to exercise the right of the holder, such as user registration, software encryption technology, access controlling technology, digital watermarking technology and so on, are all the effective way to enhance protection of the technical measures.

4.2 Note the combination of advanced technology and individualized service of digital library [4]
How to introduce the information technology on the basis of humanized service into contemporary digital library, thus to offer the readers more, better, and faster service? This is an urgent question which needs to be solved in the process of construction of university digital library. In my opinion, modern information technology with the combination of personalized service shall include the following aspects: friendly service interface, database construction and convenient search service, daily readers training, good personality service.

4.2.1 The friendly service interface. Service interface is a terminal where a digital library provides information to the reader groups; it is also a necessary road where the readers query information. A friendly interface helps readers not only more convenient, quicker, and easier to find information, at the same time the designing of the interface provide users feelings like at home, so as to better involve in information query and interactive experience.

4.2.2 Database construction. Digital library construction shall take the database construction as the core, the information resources sharing and network information service as the purpose, cover information collection, information management, information service the whole process on the function. To build database by classify and index different types of information resources (collection and network resources) with unified format and standard. By full text search, information dynamic release, the individualized information service, to release the information in the database.

4.2.3 Convenient search services. Digital library has a world of information resources, to search for information from this ocean quickly and easily, the efficient retrieval tool is absolutely necessary. A digital library can not reflect his essence without good retrieval tools. At present, each of the digital library searching services has certain problem. The more remarkable is in university digital library, the retrieval service has very big difference from users' daily used network searching engines.

4.2.4 Regular readers training. As digital library has greatly enriched resources, although homepage interface can be as friendly as possible, it will be difficult to satisfy all the users' hobbies. Even some users will lose the direction in the boundless sea of information, and ultimately fail to the information. So readers training are necessary. It includes regular library information, database introduction, and information retrieval, download etc.

4.2.5 Thoughtful personality service. Besides of information services, in order to achieve the best of personal and humanistic goal, digital library should also provide a series of thoughtful personality service. Some scholars have begun to realize the importance of this problem.

Based on the function orientation of university library, university libraries must strengthen the construction of information technology, providing better services for the colleges and social development.

References

[1] Feng, J.: Fusion of Library Information Technology and the Humanities Spirit, Science and Technology Innovation Herald (2011)

[2] Li, Y.: An Analysis of Importance of Library Informatization Construction, Brilliant Work (2011)

[3] Lin, Y., She, J.: On the Right of Communication through Information Network in the process and the Avoiding Measures. Zhejiang University Library and Information Work (2009)

[4] Bao, C.: An Analysis of Digital Library Information Technology and Individual Character Service, Library and Information (2009)

Study on E-Commerce Application in Manufacturing Enterprises: A Case Study of Ruijie Network Co., Ltd.

Ling Tian, Xiaohong Wang, and Liwei Li

School of Management, Beijing Union University,
Beijing, 100101, China
{glttianling,gltxiaohong,gltliwei}@buu.edu.cn

Abstract. Ruijie Network Co., Ltd., a Chinese network equipment manufacturing enterprise to build e-commerce website actively for improving competitiveness, was built in Jan 2000. Now, Ruijie has been one of the most famous brands of network equipment manufacturing enterprise in China. This paper surveys the status quo of e-commerce application comprehensively from the perspective of technology, management, operation and cooperation in Ruijie Network Co., Ltd. Based on the analysis and current development trend of e-commerce application in manufacturing enterprises, this paper puts forward with some suggestions.

Keywords: Ruijie Network Co., Ltd. (Ruijie Networks), e-commerce (EC), manufacturing enterprises, application.

1 Introduction

With the rapid development of China's internet and e-commerce, many manufacturing enterprise had applied e-commerce for improving competitiveness. It is true that e-commerce application can expansion of corporate, enhance the corporate image, optimize the structure of them, reduce propaganda costs, and simplify communication procedures. In the late 1990s, when e-commerce emerged as a new mode of expanding markets, it was popularly received by manufacturing enterprises. Many strive for registration of their own websites and start e-commerce activities.

In the past some studies on e-commerce application of many enterprises in different geographical and cultural environments and work settings have been conducted, unfortunately scant information is available about e-commerce application in manufacturing industry, especially in network equipment manufacturing enterprise. Therefore, there is a need to make a case study about the e-commerce application in manufacturing enterprises.

Ruijie Network Co., Ltd. is a Chinese leading IP solution and product supplier, it is called Ruijie Networks for short in this paper. Its product line covers routers, switches, data center, wireless, security gateways and application software, which are widely deployed in government, financial service, education, healthcare, enterprises and telecom carriers. Ruijie.com.cn, founded in Jan 2000, it focuses on items that have

A. Xie & X. Huang (Eds.): Advances in Computer Science and Education, AISC 140, pp. 71–75.
springerlink.com

basic enterprise-related information in Chinese and English to build internet website actively for produces, solutions, support, partners, about Ruijie Networks, and so on. By now, Ruijie has been one of the most famous brands in China, which is obviously way ahead other manufacturing enterprise. Although providing comprehensive information and services, Ruijie.com.cn also has some shortcomings in website features, website information update, and website ranking.

2 The Survey of the Status Quo of E-Commerce Application in Ruijie Networks

In the face of new economic format and information technology, what is the status quo of e-commerce application in Ruijie Networks? To answer this question, this survey selects the more than 120 persons covering managers, staff, registered users, page visitors, and partners as the investigating targets and issues the questionnaires to them about the status quo of e-commerce application in Ruijie Networks comprehensively from the perspective of technology, management, operation and cooperation. One hundred and three effective questionnaires are returned and analyzed in combination with visiting them in order to understand the status quo.

2.1 Advantages of Ruijie Networks

The result indicates that, 83% of the investigated enterprise staff pays more attention to e-commerce application in Ruijie Networks; they think technology model of Ruijie Networks is special. With the support of advanced IT technology, such as OA system, CRM system, ERP system, and internet website, Business process and communication is easy, efficiency, convenient and security. So, more staff thinks that e-commerce application brought lower marketing costs, the expansion of the scope of the enterprise market, and exaltation of brand awareness. More than 70% of the investigated enterprise managers pay more attention to e-commerce application; they usually increase annual investment on the website, management information system and training about information technology for all formal staffs. So, Although Ruijie Networks has above 37 branches over the world with 3,000 employees, covering Beijing, Tianjin, Sichuan, Shandong, and other provinces, Malaysia, Hong Kong and abroad regions, management model of Ruijie Networks is efficient.

About Ruijie.com.cn, 58% of the investigated person pays less complaint to website, very few of them put forward a few requirements about enterprise's website including features, color, framework, information update, and website ranking; only a few give some advice to improve the website. 82% of users expressed satisfaction with the features, 78% of users were satisfied of range of produce, 66% of users were pleased of the services for registered users, convenience for online, and authenticity website information. Above all, there are many advantages in Ruijie.com.cn:

Consulting in Ruijie.com.cn is convenient. Ruijie.com.cn has offered maximal convenience and best communication services to customers by cooperating with the traditional producers and using advanced IT technology, customers contact with the advisory services for 24 hours every day, both online consulting services and their call center.

Framework of Ruijie.com.cn is clear. Ruijie.com.cn has six columns, such as produces, solutions, support, partners, about Ruijie Networks, education and certification, and so on.

The number of produce is large. Ruijie.com.cn offers more than six series of products, such as switches, routers, wireless, data center, security gateways, application software, and so on.

2.2 Shortcomings of Ruijie Networks

The result also indicates that, very few of the investigated enterprise staff put forward a few requirements about enterprise's website including features, color, information update, and website ranking; only a few give some advice to improve the website. But, most of the investigated registered users, page visitors, and partners think some problems exist in the process of doing e-commerce, all most of the investigated customers hope that enterprise should improve the website and add a series of reasonable services as soon as possible.

First, personalized service should be improved. For example, by now, Ruijie.com.cn hasn't provided any shopping service in the website, so a lot of registered users hope that the website should provide them with more network-related produces for online shopping in the future, customers can browse, search, select, reserve, review, and confirm it.

Second, Ruijie.com.cn hasn't provided any payment option in the website, so all most of the users hope that the website should provide them with more choice for online payment than many other network-related companies in the future, as soon as the development of IT technology.

Third, 68% of the investigated person pays some complaint to website updating and information. Judging from the updating and maintenance, Ruijie.com.cn don't have their own specialists, the updating cycle is very different. Sometime, website keep updating once two weeks, sometime, over one month.

3 The Analysis of the Status Quo of E-Commerce Application in Ruijie Networks

3.1 The Awareness of E-Commerce Has Been Built Up

Just like above, managers of Ruijie Networks have realized the revolutionary changes that e-commerce brings. So, they usually increase annual investment on the website, management information system and training about information technology for all formal staffs. But there are still some shortcomings in website including online searching, online booking, online paying management system, and so on.

In one word, they already have formed real idea of e-commerce, but not perfect.

3.2 Website Is Imperfect

In Jan 2000, when e-commerce emerged as a new mode of expanding markets, it was popularly received by Ruijie Network Co., Lt. timely. Not only Ruijie Networks set up its own websites to adapt to the development of the internet in new era, but also put management information system and information technology into company. Now, Ruijie.com.cn has been one of the most famous brands of IP solution and product supplier in China, which is obviously way ahead other manufacturing enterprise.

Unfortunately, owing to market situation and other factors, Ruijie Networks bids farewell to its e-commerce practice far more. That is to say, Ruijie Networks usually use website as the window of propaganda, instead for business purposes.

4 Suggestion

4.1 The Improvement of Website Functionality Is a Key Factor

To win in the more competitive world, Ruijie Networks must fully adopt information technology and optimize the use of resources to improve website functionality. It must make efforts to make up shortcomings in website. In detail, Ruijie.com.cn should develop the registration, searching and selecting, submitting applications, offering solutions, signing a reservation, and paying Management System, based on its practice on network-related services.

In addition, in order to attract more customers, website promotion should be paid more attention to enhance website ranking and attract more customers.

4.2 A Team of Skillful Talents Should Be Constructed

The application of e-commerce doesn't necessarily mean that it is ok to buy some computers and software, or to promulgate some standards, or to set up some communication networks. For Ruijie Networks, skillful talents are inadequate.

First, actively introduce talents. Ruijie Networks should carry out a series of preferential policies, which can construct a more relaxing environment to introduce talents from other cities or enterprises. This conduct will produce a big impact on introducing skillful talents.

Second, engage in vocational training of the new working staff. This is also a means to solve the talent shortage problem in Ruijie Networks. Many universities have taken much collaboration with Department of Laboring Force to train primary and intermediate e-commerce practicing and achieved good results. Only after a team of talents with reasonable layout is set up, could e-commerce in Ruijie Networks maintain a sustainable development.

5 Conclusion

Based on the above analysis, the status quo of e-commerce application in Ruijie Networks is imperfect, though which is obviously way ahead other manufacturing enterprises. As the macro-environment for e-commerce all over the country gets updated thoroughly, we can believe that Ruijie.com.cn will have more achievements and contributions to e-commerce application in manufacturing industry.

Acknowledgments. This work reported in this paper is supported by the Funding Project for Academic Human Resources Development in Institutions of Higher Learning under the Jurisdiction of Beijing Municipality, which is from "Study of application model in Beijing's Small and Medium Sized E-commerce Enterprises" (PHR201108387). Please allow me to express my gratitude here.

References

1. http://www.ruijie.com.cn
2. Tian, L., Wang, X., Li, L.: Exhibition E-commerce Enterprises Analysis: a Case Study of Expo-china.com. In: 2009 IEEE International Conference on Web Services (IITAW 2009) (November 2009)
3. Wang, X., Li, L., Tian, L.: Business to Consumer E-commerce Enterprises Analysis: a Case Study of Eguo.com. In: 2008 International Seminar on Future Information Technology and Management Engineering, November 20, pp. 98–101 (2008)
4. Tian, L., Wang, X., Xue, W.: Analysis of E-Commerce Application in Pharmaceutical Enterprises: a Case Study of Yaofang.cn. In: 2010 Future Information Technology and Management Engineering (FITME 2010) (October 2010)
5. Wang, X., Li, L., Tian, L.: Business to Consumer E-commerce Enterprises Analysis: a Case Study of zgpgc.com. In: ICCTD 2009, November 2009, pp. 194–197 (2009)
6. Tian, L., Wang, X., Li, L.: Research on Pharmaceutical E-commerce Enterprises: a Case Study of Jxdyf.com. In: 2010 Second International Conference on E-Learning, E-Business, Enterprise Information Systems, and E-Government (EEEE 2010) (December 2010)

Discussion on Occupation Moral Education in Higher Vocational Colleges

Jushun Li

Hainan College of Software Technology, Qionghai City,
Hainan Province, 571400, China
ashunzi@163.com

Abstract. Occupation moral education is becoming an increasingly popular subject areas in the higher occupation education. The following discussion in this paper will address those aspects of what has been learned in the areas of developmental and educational psychology that can help educators engage in meaningful moral and character education.

Keywords: Occupation moral, education, moral development.

1 Introduction

Nowadays, there has been much debate about the main purpose of the University, particularly in Higher Vocational colleges. Some suggest it is the dissemination of knowledge, others the innovation of knowledge, some emphasis on basic research, applied research, some highlight academic or general education, others like the vocational or professional education. After 20 years of teaching practice, President of Harvard University Derek Bok thinks, in addition to the students in a competitive economy, the modern university should be to instill a sense of civic responsibility and concern for others, but in the present higher vocational education system emphasis has given to transmission of knowledge and cultivation of occupational skills. I believe that a university education, especially in the occupation school, should be, first of all, a moral education. The purpose of university education, then, is produced in the students' moral consciousness and a strong sense of civic responsibility. Some scholars and education experts point out that we should modify the current method and order in which moral education is imparted. But this should be a process in which the elementary ideals enlarge themselves to include the lofty ideals.

The goal of moral education is to encourage individuals to develop to the next stage of moral reasoning. Moral development, is not merely the result of gaining more knowledge, but rather consists of a sequence of qualitative changes in the way an individual thinks. Within any stage of development, thought is organized according to the constraints of that stage. An individual then interacts with the environment according to their basic understandings of the environment. Individuals at the conventional level of reasoning, however, have a basic understanding of conventional morality, and reason with an understanding that norms and conventions are necessary to uphold society. They tend to be self-identified with these rules, and uphold them consistently, viewing morality as acting in accordance with what society defines as right.

A. Xie & X. Huang (Eds.): Advances in Computer Science and Education, AISC 140, pp. 77–82.
springerlink.com

2 Background Theory and Research

Fortunately, systematic research and scholarship on moral development has been going on for most of this century, and educators wishing to attend to issues of moral development and education may make use of what has been learned through that work. Early 20th century American pragmatist philosopher John Dewey, one of the most influential figures in the philosophy of education, also emphasized the moral element in education. Piaget viewed moral development as the result of interpersonal interactions through which individuals work out resolutions which all deem fair. Piaget concluded from this work that schools should emphasize cooperative decision-making and problem solving, nurturing moral development by requiring students to work out common rules based on fairness. Durkheim, similar to Piaget, believed that morality resulted from social interaction or immersion in a group. However, Durkheim believed moral development was a natural result of attachment to the group, an attachment which manifests itself in a respect for the symbols, rules, and authority of that group.

However, Durkheim believed moral development was a natural result of attachment to the group, an attachment which manifests itself in a respect for the symbols, rules, and authority of that group. Piaget rejected this belief that children simply learn and internalize the norms for a group; he believed individuals define morality individually through their struggles to arrive at fair solutions. Given this view, Piaget suggested that a classroom teacher perform a difficult task: the educator must provide students with opportunities for personal discovery through problem solving, rather than indoctrinating students with norms.

Kohlberg's point is that moral reason does not emerge spontaneously as a result of environmentally evoked hard wired modules or Platonic forms, nor is the capacity for moral reasoning the result of the gradual building up of habits, but rather the construction and reconstruction of forms of understanding that emerge through processes of cognitive equilibration as outlined within Piaget's (1932) genetic epistemology.

Morality and convention, then, are distinct, parallel developmental frameworks, rather than a single system as thought of by Kohlberg. However, because all social events, including moral ones, take place within the context of the larger society, a person's reasoning about the right course of action in any given social situation may require the person to access and coordinate their understandings from more than one of these two social cognitive frameworks.

3 Occupation Moral Educations Is an Important Part of University Education

Today, as society became increasingly multicultural, universities were confronted with professors and students from widely varying cultural and religious backgrounds, and hence widely varying moral beliefs. The diversity of moral beliefs in the university community continues to present challenges to those who wish to see a return to moral education. I think that a return to moral education is just what a fractured university, and a fractured community, need.

Now, Let us see the following survey results:

Do you think it is necessary to set up occupation moral courses in Higher Vocational Colleges?

Very agree	Agree to	Uncertain	Do not agree	Very disagree
36.4%	35.1%	16.2%	8.4%	3.9%

Do you think you have known very much about your future occupation moral standard ?

Very agree	Agree to	Uncertain	Do not agree	Very disagree
6.1%	10.1%	37.9%	36.9%	9%

Do you think it is necessary to teacher occupation moral case with multimedia?

Very agree	Agree to	Uncertain	Do not agree	Very disagree
40.6%	30.1%	15.3%	10.4%	3.6%

The situation facing schools of education is strikingly similar to that which confronted business schools just a few years ago. After a number of high-profile business people found themselves in court and in jail, some business school leaders confessed they were accomplices in such criminal conduct thanks to their failure to instill in their students the appropriate moral values. Moral education therefore forms -- or should form -- the core of the professional school curriculum. And while the broader university community presents additional challenges, there is no reason why members can't agree that inculcating a sense of civic responsibility is a proper part of the university curriculum. Indeed, many universities do assess professors, not merely on their research and teaching performance, but on their service to the community.

If this is so, then moral education might well be capable of providing a unifying vision for the modern university. And in so doing, it will benefit not only the community, but the university itself.

4 How to Carry Out Occupation Moral Education

4.1 The Ethics Course

Somebody believe that an emphasis on applied ethics courses could help "make students more perceptive in detecting ethical problems when they arise, better acquainted with the best moral thought that has accumulated through the ages, and more equipped to reason about ethical issues they will face in their own personal and professional lives."

But it is exceptionally difficult to achieve in a secular university, given the disparate moral beliefs of community members. Universities could hire faculty members on the basis of shared moral commitments, but that would likely be viewed as a gross violation of academic freedom: Recall the controversy earlier this year over Trinity Western University's use of a statement of faith.

And this returns us to the professional schools. Most professions now follow, and are required to follow, codes of ethics. Consequently, most professional schools require students to take applied ethics courses. But more than that, students -- in fact, all members of the profession -- are expected to conduct themselves in a certain manner. And they can learn this behavior, not only through applied ethics courses, but by following the example of their professors and other members of the profession.

After all, even if applied ethics courses help students to develop moral sensitivity and skill at identifying and resolving moral dilemmas, there is little evidence they will instill in students a desire to act in an ethical manner.

4.2 The Role of Persuasion and Example

Some scholars argue that moral formation also requires exhortation and example -- that is, students must be encouraged to behave in a certain fashion, and professors must set an example through their own behavior.

But education begins with a message, followed by a role model, both of which, once combined, may amount to a kind of spiritual power. Choosing the right message and right model for the right group of targets is very important. The media and educators can begin by "telling" the stories of heroes basically on a par with the target audience, who should then feel that it is possible to emulate such examples. Organized actions should then be taken in "doing" good deeds in the spirit of the heroes. Finally, and most decisively, the spirit of the heroes needs to be internalized by the educated as a part of their "being". Only in such consummation of words into actions and understanding do we fulfill the highest goal of moral education - to let nobility become our nature. Certainly, this is easier said than done.

4.3 Moral Dilemma Discussion

The topics used for discussion may be fictitious, or real; they may originate from occupation moral case, the core task, the dilemma, must be very carefully prepared. Moral Dilemma Discussion has been found to be consistently effective for various age groups, As with any teaching, the basic principle of this method is to expose students to tasks which are optimally challenging, difficult enough to create cognitive activities, but not so much to overwhelm or frighten students. A really moral dilemma must involve some moral principles. These principles should be of about equal importance but imply mutually exclusive courses of action. The dilemma should always be demanding, but not too complex. Also, there should be no easy way out.

The difficulty of the moral dilemma should always be adapted to the past experience and the maturity of the students. For the teacher, it takes considerable competence and experience to design good dilemma discussion units.As the teacher gets more experienced he or she may vary this time schedule.

STEP	ACTIVITY	GOALS
1	Teacher presents the Occupation moral case, then makes sure that all are tuned in by asking students to Analysis the case and present their views from the case.	1. Learn to get the facts straight 2. To fully understand the nature of the moral dilemma case.
2	Students proposed to moral dilemma solution. The teacher collected the representative solutions. Teachers are encourage to share their experience with others by inviting colleagues to sit in class and by sending me written reports about working dilemma discussions.	1. Express your right or wrong opinion on a controversial issue 2. know the diversity of opinion.
3	Big group discussion: The two big groups challenge each other's opinion. The teachers explain the principles and rules. In the discussion the teacher confines him/herself to the role of a moderator. Everyone has the right to speak freely about anything	1. Learn to present the reasons for your opinion succinctly. 2. Learn to distinguish between the quality of an argument and the quality of a person .
4	Divided into small groups to discuss:" How do you feel about the arguments of the other side? How do you judge them? Which were the best? Did some make you re-assess your opinion?"	1. Learn to appreciate good reasons even when given by your opponents.
5	Teacher wrote down the main points of view on the blackboard. Final vote:" We have now considered many sides of the problem. Some of us may now even think differently about the problem that was presented. Who believes now that it was [more or less] right to ... and who believes it was [more or less] wrong to ...?"	1. Learn to estimate the opportunity to deliberate on your own opinions and to have exchange with opponents. 2. Learn that discussions on occupation case develop the quality of occupation moral
6	Teacher briefly comments on the class: "I am surprised how sophisticated and convincing many of you made your point……"	1. Get the occupation moral development

5 Reflect

In the following ,let see the perfect occupation moral self cultivation. As a way of beginning to think about how this might apply to the area of morality, let's think for a moment, in the higher vocational college students how to apply their sense of self to moral development. How is it that we move from knowing right from wrong to acting in relation to that moral understanding? It turns out that a students' view of themselves in terms of occupation capabilities is necessarily tied to students' general sense of self-worth. The business of leading a good life is to move toward the human goals and to achieve eudemonia or a flourishing. With respect to ethics, the process of flourishing entails the gradual development of virtues, or personal characteristics that will support ethical conduct. In youth, this process involves the building up of habits that in time translate into ways of being that constitute virtuous conduct.

Because moral character has been shown to be a complex and difficult problem, moral psychology is necessary. Such a morality person will be "concerned about friendship, justice, courage, moderation and generosity; his desires will be formed in accordance with these concerns; and he will derive from this internalized conception of value many ongoing guidelines for action, pointers as to what to look for in a particular situation."

So, The development of a morality person involves the cultivation of the right set of habits, ethical values, and a conception of the good human life as the harmonious pursuit of these.

Research Projects Supported by Education Department in Hainan Province: "The Study of Efficiency of Habituation-oriented Education for Higher Vocational Students 'Professional Ethics" (Serial Number:Hjsk2011-88)

References

1. Nussbaum, M.: The fragility of goodness. Cambridge University Press, Cambridge (1986)
2. Wainryb, C., Turiel, E.: Dominance, subordination, and concepts of personal entitlements in cultural contexts. Child Development 65, 1701–1722 (1994)
3. Turiel, E.: Equality and hierarchy: Conflict in values. In: Reed, E., Turiel, E., Brown, T. (eds.) Values and knowledge, pp. 41–60. Lawrence Erlbaum, Hillsdale (1996)
4. Blasi, G., Glodis, K.: The development of identity: A critical analysis from the perspective of the self as subject. Developmental Review 15, 404–433 (1995)
5. Wertsch, J.V., Toma, C.: Discourse and learning in the classroom: A Sociocultural approach. In: Steffe, L., Gale, J. (eds.) Constructivism in Education, pp. 159–174. Lawrence Erlbaum, Hillsdale (1995)

Dynamic Analysis and System Design based on Diluted Asymmetric Discrete-Time Hopfield Network

Li Tu and Chi Zhang

Department of Computer Science, Hunan City University,
Yiyang, Hunan, 413000, China
tulip1907@163.com, carol_zc@126.com

Abstract. In this paper the dynamics of the asymmetric discrete-time Hopfield networks are studied, and the sufficient conditions for endowing the network with retrieval properties are proposed. In addition, a method for designing efficient diluted networks is proposed based on the matrix decomposition and connection elimination strategy. Numerical simulations show that the designed diluted network can act as efficient neural associative memories.

Keywords: diluted asymmetric discrete-time Hopfield network, matrix decomposition, connection elimination strategy.

1 Introduction

The asymmetry of synaptic connection is an unavoidable problem in neural network studying. In this paper the dynamics of the asymmetric discrete-time Hopfield networks are studied, and the design of fully connected and s diluted asymmetric discrete-time Hopfield network are given.

2 Network Model

The following is a n-neuron Hopfield network xi(t) = ±1 is the i-th neuron state. Set $X(t)=[x_1(t),x_2(t),\ldots x_n(t)]^T$, T is the network state at time t, then the network state in time t+1 is determined by the following formula:

$$X(t+1)=SGN(WX(t)) \tag{1}$$

in this formula $\mathrm{SGN}(WX(t))=[\mathrm{sgn}(f_1(t),\ldots,\mathrm{sgn}(f_n(t))]^T$, and $\mathrm{f_i}(t)=\sum_{j=1}^{n} w_{ij}x_j(t)$, function $SGN(\cdot)$ is a sign function, $W=[w_{ij}]\in R^{n\times n}$, it is the matrix of synaptic connection,here W is an asymmetric matrix,and $w_{ij}\neq w_{ji}$.

A. Xie & X. Huang (Eds.): Advances in Computer Science and Education, AISC 140, pp. 83–87.
springerlink.com

3 Dynamics of the System and System Design

Energy function of network(1) is:

$$EX = -\frac{1}{2} X^T WX \tag{2}$$

$H = \frac{1}{2}(W + W^T)$, because $x \in R^n$, $X^T WX + (X^T WX)^T = 2X^T WX$, $X^T WX + X^T W^T X = 2X^T HX$, so $X^T WX = X^T HX$, set the degree of asymmetry as $A_s = \max_{ij} |w_{ij} - w_{ji}|$. When the network status updates, the energy change is :

$$\Delta E = E(X(t+1)) - E(X(t)) = -(\Delta X)^T HX(t) - \frac{1}{2}(\Delta X)^T H\Delta X \tag{3}$$

and $\Delta X = [\Delta x_1, \Delta x_2, ..., \Delta x_n]^T, \Delta x_i = x_i(t+1) - x_i(t), i = 1,2,...,n$.

$X^{(j)} = [x_1^{(j)}, x_2^{(j)}, ..., x_n^{(j)}]^T, (j = 1,2,...,m)$ is the vector mode need to be recognized by the Hopfield neural network. We need to design the appropriate parameter W to meet the following conditions.

$\Delta E \le 0$, it is $W \ge 0$, (i)

$X^{(j)}$ is the balance point , (ii)

$X^{(j)}$ is the local minimum point of formula $E(\cdot)$, j = 1,2,...,m, (iii).

$$F(t) = WX(t) = [f_1(t), f_2(t), ..., f_n(t)]^T, G(t) = HX(t) = [\breve{g}_1(t), \breve{g}_2(t), ..., \breve{g}_n(t)]^T$$

if A_s is small enough, and H > 0,then

$$WX^{(j)} = \lambda_1 WX^{(j)}, \lambda_1(W) > 0, j = 1,2,...m \tag{4}$$

the condition (i) and condition (ii) can be met.

In the formula $\Delta x_i = \begin{cases} 0, & x_i(t) = \text{sgn}(f_i(t)) \\ 2, & x_i(t) = 1, \text{sgn}(f_i(t)) = -1 \\ -2, & x_i(t) = -1, \text{sgn}(f_i(t)) = 1 \end{cases}$

the above equation shows , $-(\Delta X)^T F(t) \le 0$,because H > 0, if X = 0, There is a positive real number δ_1 ,it makes the following inequality holds:

$$\frac{1}{2}(\Delta X)^T H\Delta X > \delta_1 \tag{5}$$

Calculate $(\Delta X)^T F(t)-(\Delta X)^T G(t)$, because $|x_j|=1,|x_i|\le 2$, so $-(\Delta X)^T G(t)\le n^2 A_s - (\Delta X)^T F(t)\le n^2 A_s$, take another positive real number δ_2, which makes $0<\delta_2<\delta_1$, if A_s is small enough,then:

$$-(\Delta X)^T G(t) < \delta_2 \tag{6}$$

So E ≤ 0, it satisfies condition(ⅰ).

In addition, j = 1,2,...,m,so

$$X^{(j)}(t+1) = SGN(WX^{(j)}(t)) = X^{(j)}(t) \tag{7}$$

it satisfies condition (ⅱ),so $x^{(j)}$ is the balance poin.

So we propose the following system design:

Step 1) Set $\overline{Y}=[X^{(1)},X^{(2)},...,X^{(m)}]$;

Step 2) Carried out $\overline{Y}$ by singular value decomposition, $\overline{Y}=\hat{U}\Lambda\hat{V}^T$, $\hat{U}$ and $\hat{V}$ are the orthogonal matrix,Λis a diagonal matrix;

Step 3) Calcalute W $=\hat{U}L\hat{U}^T$, $L=[l_{ij}]\in R^{n\times n}$ is a lower triangular matrix,

$$l_{ij}=\begin{cases}\alpha, & if \quad i=j, i\le m\\ \varsigma_i, & if \quad i=j, i>m\\ rand(\beta_{ij},\gamma_{ij}), if & i>j, i>m, j>m\\ 0 & other\end{cases}$$

$\alpha>\varsigma_i>0.\beta_{ij},\gamma_{ij}\in R, rand(\beta_{ij},\gamma_{ij})$ is a random function in (β_{ij},γ_{ij}).

$W=\hat{U}L\hat{U}^T$ is the Schur decomposition of W,so α and ζ_i are the eigenvalues of Matrix W, and $\lambda_i(W)=\alpha$. The algorithm shows $L\Lambda=\alpha\Lambda$,.if $\beta_{ij}=\gamma_{ij}=0$, then L is a diagonal matrix,and W $=\hat{U}L\hat{U}^T$ is a symmetric matrix,so the asymmetry can be controlled by adjusting the parameter β_{ij},γ_{ij}. An Appropriate diluted asymmetric discrete-time Hopfield network can be controlled by adjusting the parameters of matrix L easily.

There are some unimportant connections in symmetric discrete-time Hopfield networks,and these unimportant connections can be removed.then we define the strength of the connections as $\hat{\varphi}(w_{ij}) = \frac{|w_{ij}|}{\max\limits_{1 \le j \le n} |w_{ij}|}$,if $\hat{\varphi}(w_{ij}) \le \hat{c}$, $\hat{c} \in [0,1)$ then the connection is unimportant.

4 System Simulation

Three characters need to be recognized are shown in Figure 1(A), each model composed by small grids of 7×8, and it corresponds to a 56-dimensional binary vector.System parameter is set, $\varsigma_i = 0.1$, parameter $l_{ij} (i > j, i > 3, j > 3)$ generated randomly in the interval (0.1, 0.1) .then an asymmetric discrete-time Hopfield network was designed. And $x^{(j)} (j = 1,2,3)$ can be considered as the local minimum point of E(X).

A. 3 Original characters

B. initial state of the network

Fig. 1. 3 original characters and initial state of the network

Then diluted asymmetric discrete-time Hopfield networks can be designed by setting different parameters $\hat{c}$. Figure 1(B) shows, the initial states are three damaged models. The recovery models are shown in Figure 2.

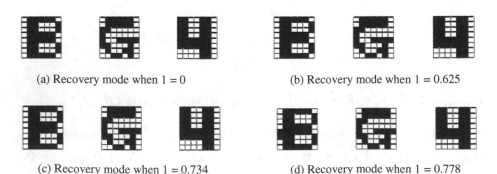

(a) Recovery mode when 1 = 0

(b) Recovery mode when 1 = 0.625

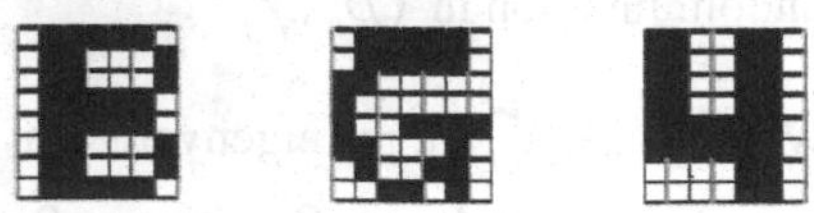

(c) Recovery mode when 1 = 0.734

(d) Recovery mode when 1 = 0.778

Fig. 2. Output states in different sparsities

The sparsity of the symmetrical network is :

$$\Gamma = \frac{\text{The number of zero weights dendritic connections}}{n \times n} \tag{8}$$

When c=0.30, Γ= 0.625, this means 62.5% of connections are removed. when c=0.40, Γ=0.734, about 73.4% of connections are removed. Figure 2(c) shows, although only 28% of connections were used, But the system has the same performance as an asymmetric Hopfield network.Figure 2(d) shows, when c=0.45, 77.8% of connections were removed, but thc nctwork can work effectively.

5 Conclusion

In this work, asymmetriy of Hopfield networks is an important factor affecting dynamics,and a sufficient condition of pattern recognition in these networks was proposed. Finally, through system simulation parameters and sparsity of networks on the performance was analysised. The simulation results show that these diluted asymmetric discrete-time Hopfield networks can be used as associative memories.

Acknowledgment. This work was supported by Scientific Research Fund of Hunan Provincial Science Department under Grant (No: 2011FJ3016).

References

1. Valle, M.E.: A class of sparsely connected autoassociative morphological mem-ories for large color images. IEEE Trans. Neural Networks 20(6), 1045–1050 (2009)
2. Lu, H.: Global exponential stability analysis of Cohen-Grossberg neural net-works. IEEE Trans. Circuits Syst. 52(8), 476–479 (2005)
3. Huang, W.Z., Huang, Y.: Chaos of a new class of Hopfield neural networks. Appl. Math. C. 206, 1–11 (2008)
4. Huang, Y., Yang, X.S.: Hyperchaos and bifurcation in a new class of four-dimensional Hopfield neural networks. Neurocomputing 69, 1787–1795 (2006)

Evaluation Method on the Scientific Research of Local University

Xiaomin Huang

Department of Science and Technology, Huanggang Normal University,
Huanggang, Hubei, 438000, P.R. China
hxm1998@sohu.com

Abstract. The problem of evaluating the scientific research level of local university is analyzed and researched. Firstly, the index system of evaluation is established, i.e., research team, research platform, research achievement, research funding and academic reputation. Then, an evaluation method is presented based on fuzzy multi-attribute decision making. Therefore, a novel and effective way is obtained to solve the problem of evaluating the scientific research level of local university under the uncertain information.

Keywords: scientific research level, local university, fuzzy multi-attribute decision making, fuzzy approach degree.

1 Introduction

Scientific research in university is a research activity for all the college teachers and students, including research and development, application of research achievement, the R & D activities and related technical services and technology promotion activities. Assessment of university research is an assessment activity by fund managers or management for the external performance evaluation is both an academic evaluation and a management evaluation [1]. A fair, scientific and rational assessment of scientific research activities can provide the government with necessary information and decision-making basis, while this competition assessment activity has the practical significance to improve the quality of scientific research, to promote the researchers' research level, to allocation the resource of resources reasonably and to develop the characteristics research fields [2].

In practice, many aspects to evaluate the scientific research level of a local university must be considered, for example, research team, research platform, research achievement, research funding and academic reputation, and so on. In practical decision making, because of the complexity of decision systems and the fuzziness of human's thinking, decision makers usually give the evaluation results with the type of linguistic fuzzy variables, such as "worst, worse, bad, common, good, better, best" or "lowest, lower, low, common, high, higher, highest" [3-5] when they make evaluation on the research team, academic reputation, and so on. Considering this uncertain information, the problem of evaluating the scientific research level of local university can be converted to a problem of fuzzy multi-attribute decision making.

A. Xie & X. Huang (Eds.): Advances in Computer Science and Education, AISC 140, pp. 89–95.
springerlink.com

Multi-attribute decision making is the process of making evaluation and ranking order for many alternatives by decision maker. It is widely applied in the fields of society, economy, and management and so on. Refs. [4-8] have studied these problems of multi-attribute decision making with the attribute values in the forms of linguistic fuzzy numbers, but they solved them by using the method of fetching values approximatively. This method is so easy to lose decision information that it can't obtain precise decision results. In order to avoid losing the decision information in the multi-attribute decision making, Refs. [8] established a model based on the linguistic two-tuple. However, there exist some disadvantages in this model. For example, the manoeuvrability and practicability of this model is not very good, the structure of this model is very complicated.

This paper studies the problem of evaluating the scientific research level of local university under the uncertain information. Firstly, we establish the index system of evaluation. Then the problem of valuating the scientific research level is converted to a problem of fuzzy multi-attribute decision making. Thirdly, an evaluation method on the scientific research of local university is proposed based on fuzzy math theory and decision making method. In this decision making model, a scale of the linguistic fuzzy numbers is defined and the corresponding representation of fuzzy interval number numbers is given, and a new algorithm is given for ranking candidates.

2 The Index System Evaluation

In the problem of evaluating the scientific research level of local university, we construct the index system of evaluation firstly.

According to the construction principles of evaluating the scientific research level of local university, and referring to existing research on evaluation of university research, we propose the following index system of evaluation, i.e., research team, research platform, research achievement, research funding and academic reputation. These indexes are called level-one index. They are composed of level-two indexes as follows.

(1) Research team
University research team is composed of a few research staff who have the common research objectives, and the complementary skills. The research team is the core and main force for university research and innovation capacity. Research team is reflected by four factors, i.e., title structure, degree structure, hierarchy and the number of graduate students. In practice, the decision maker usually gives the evaluation results with the type of linguistic fuzzy variables, such as "worst, worse, bad, common, good, better, best".

(2) Research platform
Scientific process requires certain experimental conditions, relying on the laboratory equipment and instruments. For example, the key laboratories, engineering training centers, and so are necessary conditions for the researchers and students in research training. Concretely, research platform includes doctor station, master station, and key national disciplines, national laboratories and national engineering centers and national humanities social sciences research base and so on. The evaluation value of research platform is usually precise information given by appraiser.

(3) Research achievement
Research achievement of local universities mainly includes four aspects, i.e., academic papers, monograph, patents and scientific research awards. Books and papers are basic research results, a patent is an important index which reflects the science and technology work, and scientific research awards show the research level of science and technology. The evaluation value of research achievement is usually precise real number given by appraiser.

(4) Research funding
Research funding is the foundation and prerequisite conditions for all kinds of scientific and technological activities, and reflects the scale of the task of science and technology research institutions and scientific prowess. In actual research, all universities are very focused on research funding and the research projects' important role in scientific research. The evaluation value of research funding is usually precise real number in the evaluation.

(5) Academic reputation
Academic reputation is the depth and breadth of academic influence of a university' comprehensive strength. Although it is a subjective indicator, almost is one of standard academic indicators for the most prestigious universities in the world. It is essential to accurately reflect the university's position and influence. Looking at today's university evaluation system, the academic reputation is playing an increasingly important role as a university evaluation system for construction of basis points [12].

3 Problem Description

Suppose that an educational management (hereinafter referred to as appraiser) needs to evaluate the scientific research level of m local universities (hereinafter referred to as candidate), the set of candidates is denoted as $X=\{X_1,X_2,\cdots,X_m\}$. The index system of evaluation is given by educational management, i.e., research team(p_1), research platform(p_2), research achievement(p_3), research funding(p_4), and academic reputation(p_5). These five indexes are called five attributes. The set of attributes is denoted as $A=\{A_1,A_2,\cdots,A_5\}$, and the vector of attribute weights is denoted as $W=(w_1,w_2,\ \cdots,w_5)$, where w_j satisfies $0\le w_j\le 1$ and $\sum_{j=1}^{5} w_j=1$.

Suppose that the evaluation value of candidate x_i $(i=1,2,\cdots,m)$ under attribute p_j $(j=1,2,\cdots,5)$ is a_{ij}. When $j=1,5$, a_{ij} is a linguistic fuzzy number, such as "worst, worse, bad, common, good, better, best" or "lowest, lower, low, common, high, higher, highest ". When $j=2,3,4$, a_{ij} is a precise real number. So there exists one evaluation matrix given by the appraiser, it is denoted as R, where $R=(a_{ij})_{m\times n}$, is composed of linguistic fuzzy numbers and precise real numbers.

Now our task is make evaluation and rank order for all candidates based on the information of evaluation matrix.

4 Evaluation Method

Considering the evaluation matrix includes uncertain information, thus the problem of evaluating the scientific research level of local university can be converted to a problem of fuzzy multi-attribute decision making. Next we give the detail decision process.

4.1 Definitions and Data Processing

Since the valuation value in matrix $R = (a_{ij})_{m \times n}$ are given not only in the form of precise real numbers, but also in the form of linguistic fuzzy numbers, such as "worst, worse, bad, common, good, better, best" or "lowest, lower, low, common, high, higher, highest ". In order to make comprehensive evaluation better, the attribute values are all transformed into fuzzy interval numbers. Firstly, the definition of fuzzy interval numbers and its operations are given.

Definition 1. Suppose that $a = [a^L, a^U] = \{x | 0 \le a^L \le x \le a^U; a^L, a^U \in R\}$, then a is called a fuzzy interval number. If $a^L = a^U$, then a just becomes a real number.

Definition 2. Suppose that $a = [a^L, a^U]$ and $b = [b^L, b^U]$ are two fuzzy interval numbers, $k > 0$, then the operations for a, b are defined as:

$$a + b = [a^L, a^U] + [b^L, b^U] = [a^L + b^L, a^U + b^U],\ a \times b = [a^L b^L, a^U b^U]$$

$$a \div b = [\frac{a^L}{b^U}, \frac{a^U}{b^L}],\ ka = k[a^L, a^U] = [ka^L, ka^U],\ \frac{1}{a} = \frac{1}{[a^L, a^U]} = [\frac{1}{a^U}, \frac{1}{a^L}]$$

Definition 3. Suppose that $a = [a^L, a^U]$ and $b = [b^L, b^U]$ are two fuzzy interval numbers, then the distance between a to b is defined as

$$D(a,b) = (|a^L - b^L| + |a^U - b^U|)/2$$

Secondly, the transformation method of how to transform the linguistic fuzzy numbers into fuzzy interval numbers are given in the following Definition 4.

Definition 4. Suppose that S_a ={worst, worse, bad, common, good, better, best}([3]) or S_b ={lowest, lower, low, common, high, higher, highest}([5,8]), and they are denoted as $S_a = \{s_1, s_2, \cdots, s_7\}$ or $S_b = \{s_1', s_2', \cdots, s_7'\}$, then S_a and S_b are called linguistic evaluation scales, and their corresponding fuzzy interval numbers are defined as follows:

$$s_1 = [0, 0.1],\ s_2 = [0.1, 0.25],\ s_3 = [0.25, 0.4],\ s_4 = [0.4, 0.6],\ s_5 = [0.6, 0.75],$$
$$s_6 = [0.75, 0.9],\ s_7 = [0.9, 1],$$

where $s_7 \succ s_6 \succ s_5 \succ s_4 \succ s_3 \succ s_2 \succ s_1$.

The corresponding grey interval numbers for S_b are the same as S_a.

By Definition 4, all the linguistic fuzzy number in matrix R can be transformed into fuzzy interval numbers. In addition, when the attribute value is precise real number a, it can be transformed into fuzzy interval number directly, i.e., a becomes $[a,a]$. When all attribute values are transformed into fuzzy interval numbers, they are denoted as B, where $B=(b_{ij})_{m\times n}$, $b_{ij}=[b_{ij}^L,b_{ij}^U]$, $i=1,2,\cdots,m$, $j=1,2,\cdots,5$. We need to normalize the matrix B. The following algorithms [7] are given to normalize the matrix B. Suppose that the matrix B is transformed into the decision matrix Y, where $Y=(y_{ij})_{m\times n}$, $y_{ij}=[y_{ij}^L,y_{ij}^U]$, $i=1,2,\cdots,m$, $j=1,2,\cdots,5$.

If the attribute is a type of cost attributes, then

$$y_{ij}^L=\frac{1/a_{ij}^U}{\sum_{i=1}^{m}(1/a_{ij}^L)} \qquad y_{ij}^U=\frac{1/a_{ij}^L}{\sum_{i=1}^{m}(1/a_{ij}^U)} \tag{1}$$

If the attribute is a type of revenue attributes, then

$$y_{ij}^L=\frac{a_{ij}^L}{\sum_{i=1}^{m}a_{ij}^U} \qquad y_{ij}^U=\frac{a_{ij}^U}{\sum_{i=1}^{m}a_{ij}^L} \tag{2}$$

where $i=1,2,\cdots,m$, $j=1,2,\cdots,5$.

4.2 Algorithm of Ranking Candidates

From the decision matrix $Y=(y_{ij})_{m\times n}$, the appraiser will make evaluation and rank order for all candidates. Here we propose an algorithm of ranking candidates based on fuzzy approach degree.

Definition 5. Suppose that $Y=(y_{ij})_{m\times n}$ is a decision matrix, where $y_{ij}=(y_{ij}^L,y_{ij}^U)$, $i=1,2,\cdots,m$, $j=1,2,\cdots,5$. Then $f=(f_1,f_2,\cdots,f_n)$ and $g=(g_1,g_2,\cdots,g_n)$ are called fuzzy positive ideal solution and fuzzy negative ideal solution, which satisfies

$$f_j=[f_j^L,f_j^U)=[\max_i(y_{ij}^L),\max_i y_{ij}^U] \tag{3}$$

$$g_j=[g_j^L,g_j^U)=[\min_i(y_{ij}^L),\min y_{ij}^U] \tag{4}$$

Based on the fuzzy positive ideal solution and fuzzy negative ideal solution given in Definition 5, the fuzzy approach degree is defined in Definition 6.

Definition 6. Let fuzzy positive ideal solution $f=(f_1,f_2,\cdots,f_5)$ and fuzzy negative ideal solution $g=(g_1,g_2,\cdots,g_5)$ be the reference sequence respectively, and alternative points $X_1,X_2,\cdots,X_m$ be the compared sequences, where $X_i=(y_{i1},y_{i2},\cdots,y_{i5})$, $i=1,2,\cdots,m$. If

$$r(f_j,y_{ij})=\frac{0.5\max_i\max_j\Delta(f_j,y_{ij})}{\Delta(f_j,y_{ij})+0.5\max_i\max_j\Delta(f_j,y_{ij})}, \tag{5}$$

$$r(g_j, z_{ij}) = \frac{0.5 \max_i \max_j \Delta(g_j, y_{ij})}{\Delta(g_j, y_{ij}) + 0.5 \max_i \max_j \Delta(g_j, y_{ij})}, \tag{6}$$

$$r(f, X_i) = \sum_{j=1}^{n} w_j r(f_j, y_{ij}), \tag{7}$$

$$r(g, X_i) = \sum_{j=1}^{n} w_j r(g_j, y_{ij}), \tag{8}$$

where $\Delta(f_j, y_{ij})$ is the difference information at the *j*-th point of X_i to f , and

$$\Delta(f_j, y_{ij}) = D(f_j, y_{ij}) = \frac{\left|\max_i (y_{ij}^L) - y_{ij}^L\right| + \left|\max_i y_{ij}^U - y_{ij}^U\right|}{2}.$$

$\Delta(g_j, y_{ij})$ is the difference information at the *j*-th point of X_i to g , and

$$\Delta(g_j, y_{ij}) = D(g_j, y_{ij}) = \frac{\left|\max_i (y_{ij}^L) - y_{ij}^L\right| + \left|\max_i y_{ij}^U - y_{ij}^U\right|}{2}$$

w_j is the weight of the *j*-th attribute p_j , $j = 1,2,\cdots,5$. Then $r(f, X_i)$ is called the fuzzy approach degree between fuzzy positive ideal solution f and X_i , and $r(g, X_i)$ is called the fuzzy approach degree between fuzzy negative ideal solution g and X_i .

For the fuzzy approach degree $r(f, X_i)$ and $r(g, X_i)$, the smaller the value of $r(f, X_i)$ is, the better the candidate X_i is, and the greater the value of $r(g, X_i)$ is, the better the candidate X_i is. Based on the conclusions, an algorithm of ranking candidates based on fuzzy approach degree is given as follows,

$$S_i = \frac{r(g, X_i)}{r(f, X_i) + r(g, X_i)}, \quad i = 1,2,\cdots,m \tag{9}$$

In practical evaluation, we can rank candidates according to the values of S_i. The greater the value of S_i is, the better the candidate X_i is.

Based on the analysis in above sections, the detail steps of evaluation method are concluded as follows. If an educational management (appraiser) needs to evaluate the scientific research level of m local universities, he can follow the following steps.

1) For m local universities (candidates), determining the original matrix $R = (a_{ij})_{m \times n}$.

2) Using the method given by Definition 4 to transform all attribute values into fuzzy interval numbers, and the matrix B is obtained.

3) Using algorithm (1) and (2) to normalize the matrix B , and matrix B is transformed into the decision matrix $Y = (y_{ij})_{m \times n}$.

4) Using (3) and (4) to determine the fuzzy positive ideal solution $f=(f_1, f_2, \cdots, f_n)$ and fuzzy negative ideal solution $g=(g_1, g_2, \cdots, g_n)$ respectively.

5) Using (5), (6), (7) and (8) to calculate the fuzzy approach degree $r(f, X_i)$ and $r(g, X_i)$ respectively, $i=1,2,\cdots,m$.

6) Using (9) to calculate the value of S_i, $i=1,2, \cdots,m$.

7) Ranking all candidates according to S_i, and giving the evaluation results for the scientific research level of m local universities.

5 Conclusions

In this paper, a new evaluation method based on multi-attribute decision making is proposed to evaluate the scientific research level of local university. This method has the advantages of scientific in decision principle, simple in computing and easy to carry out in computers. Therefore, it has a great theoretic value and applied value in practice, and provides an effective way to solve the problem of evaluating the scientific research level of local university under the uncertain information.

Acknowledgments. This work was supported by the Key Project of Hubei Provincial Department of Education (D20112908).

References

1. Liu, H.: Design of Evaluation Index System of Evaluating Scientific Research Level of Local University. Chinese University Technology 5, 51–53 (2010)
2. Gao, H., Li, Z., Wang, Q.: The Evaluation Index and Empirical Study of Scientific Research in Universities and Colleges. Education Science 27, 71–75 (2011)
3. Xu, Z.: Uncertain Multiple Attribute Decision Making: Methods and Applications. Tsinghua University Press, Beijing (2004)
4. Bordogna, G., Fedrizzi, M., Pasi, G.: A Linguistic Modelling of Consensus in Group Decision making Based on OWA Operators. IEEE Trans. on Systems, Man, and Cybernetics 27, 126–132 (1997)
5. Li, R.: The Foundation and Application of Fuzzy Multiple Attribute Decision Making. Science Press, Beijing (2002)
6. Herrera, F., Martínez, L.: A Fusion Method for Managing Multi-granularity Linguistic Terms Sets in Decision Making. Fuzzy Sets and Systems 114, 43–58 (2000)
7. Fan, Z., Xiao, S., Hu, G.: An Optimization Method for Integrating Two Kinds of Preference Information in Group Decision Making. Computer & Industrial Engineering 46, 329–335 (2004)
8. Cordón, O.: Linguistic Modeling by Hierarchical Systems of Linguistic Rules. IEEE Trans. on Fuzzy Systems 10, 1–19 (2002)

A Random Endogenous Economic Growth Model and Equilibrium Analysis

Zhongcheng Zhang

School of science, Wuhan Institute of Technology,
Wuhan 430073, Hubei, P.R. China
zzcheng63@126.com

Abstract. In this paper, a new random endogenous economic growth model is proposed by introducing an elastic labour supply to a simple AK model. The economic equilibrium of decentralized economy is analysed, and the influences of distorted taxes, government public spending and public consumption on fiscal policy are discussed. Based on these results, equilibrium growth rate and the corresponding economic rules are obtained, and the optimal fiscal policies are given.

Keywords: endogenous economic growth, decentralized economy, equilibrium analysis, fiscal policy.

1 Introduction

Recent endogenous growth models [1-6] believe that the fiscal policies play a key and decisive role to long-term economic growth. However, these models have an identical limitation, that is, most models regarded labor supply as non-elastic, and thus decided to allocate the work time and leisure time abstractively. This treatment regards the individual's consumption tax and labor income taxes as a non-distorting lump-sum tax, so that these models can not describe the real tax policy better, and thus can not reflect the impact of fiscal policy on economy better. However, the endogenous labor supply not only can reflect the impact of taxes on the real output and capital, but also can reflect the impact of government public spending and public consumption on the real output and capital.

This paper introduces an elastic labour supply to a simple AK model[3, 6], and establish a new random endogenous economic growth model. The rest of this paper is organized as follows. Section 2 gives some assumptions and definitions, and presents the random endogenous economic growth model. Section 3 analyses the economic equilibrium for decentralized economy, and discusses the influences of distorted taxes, government public spending and public consumption on fiscal policy, and gives the optimal fiscal policies. Section 4 concludes this paper.

A. Xie & X. Huang (Eds.): Advances in Computer Science and Education, AISC 140, pp. 97–101.
springerlink.com

2 Model Analysis of Decentralized Economy

2.1 Model Assumptions

The individual participants in the decentralized economy buy the consumer goods by using their own disposable labor incomes and individual capital, the utility function is as follows.

$$U = \int_0^\infty \frac{1}{r}(cl^\theta G_c^{\ \eta})^r e^{-\rho t} dt \tag{1}$$

The constraint condition of individual resources is

$$\dot{k} = (1-\tau_w)w(1-l) + (1-\tau_k)r_k k - (1+\tau_c)c - T/N \tag{2}$$

where r_k is return on invested capital(ROIC), w is wage rate, τ_w is wage income tax, τ_k is capital income tax, τ_c is individual consumption tax, T/N is lump sum taxes of individual participant, k is the capital of individual participant, c is individual consumption level.

2.2 Equilibrium Analysis

Individual designer will choose K, C, l and Y to maximize his utility function (1) under the condition (2), this optimal decision problem can be expressed by

$$\max U = \int_0^\infty \frac{1}{r}(cl^\theta G_c^{\ \eta})^r e^{-\rho t} dt$$

$$s.t.\quad \dot{k} = (1-\tau_w)w(1-l) + (1-\tau_k)r_k k - (1+\tau_c)c - T/N \tag{3}$$

the transversality condition is $\lim_{t\to\infty} \lambda k e^{-\rho t} = 0$.

The Hamilton of problem (3) is

$$\begin{aligned} H = \frac{1}{r}c^r l^{\theta r} G_c^{\ \eta r} + \lambda[(1-\tau_w)w(1-l) \\ +(1-\tau_k)r_k k - (1+\tau_c)c - T/N] \end{aligned} \tag{4}$$

where λ is shadow price of individual capital.

The first order conditions are as follows.

$$c^{r-1} l^{\theta r} G_c^{\ \eta r} = \lambda(1+\tau_c) \tag{5}$$

$$\theta c^r l^{\theta r-1} G_c^{\ \eta r} = \lambda(1-\tau_w)w \tag{6}$$

$$\lambda(1-\tau_k)r_k = \rho\lambda - \dot{\lambda} \tag{7}$$

Since $Y = Ny$, $C = Nc$, then we have

$$\frac{1-\beta}{\Omega(l)}(\frac{C}{Y})=\frac{1-\tau_w}{1+\tau_c} \tag{8}$$

By (7), we obtain

$$\gamma_\lambda=\rho-(1-\tau_k)(1-\beta)(\frac{Y}{K}) \tag{9}$$

Since the optimal decision exists in the equilibrium state, together with (5)-(9), we have

$$\gamma^*=\gamma_c^*=\frac{1}{1-r(1+\eta)}[(1-\tau_k)(1-\beta)(\frac{Y}{K})^*-\rho] \tag{10}$$

From (1) and (2), we have

$$\gamma^*=\gamma_k^*=[(1-g_p-g_c)-(\frac{C}{Y})^*](\frac{Y}{K})^* \tag{11}$$

Based on (10) and (11), we can determine these equilibrium values: the leisure time in distribution l^*, the ratio of total consumption and total output $(\frac{C}{Y})^*$, the ratio of total output and total capital $(\frac{Y}{K})^*$, and equilibrium growth rate γ^*.

Proposition 1. The influence of wage income tax τ_w, capital income tax τ_k and individual consumption tax τ_c on economic growth rate and recreational utility are the same. When their values are higher, the economic growth rate will lower, and the recreational utility will higher.

Proof. From (10) and (11) we have

$$\frac{\partial\gamma^*}{\partial\tau_i}<0,\ i=k,w,c \quad \frac{\partial l^*}{\partial\tau_i}>0,\ i=k,w,c$$

We can easily get the conclusion in Proposition 1.

Proposition 2. The influence of g_c and g_p on economic growth rate and recreational utility are the same. When their values are higher, the economic growth rate will higher, and the recreational utility will lower.

Proof. From (10) and.(11), we can get

$$\frac{\partial\gamma^*}{\partial g_p}>\frac{\partial\gamma^*}{\partial g_c}>0,\quad \frac{\partial l^*}{\partial g_p}<\frac{\partial l^*}{\partial g_c}<0,$$

We can easily know the conclusion in Proposition 2 is true.

2.3 The Optimal Fiscal Policy

Now we consider the optimal fiscal policy, that is to say, in the decentralized economy, the policy maker how to determine the optimal government expenditure and tax to realize the unity for the centrally planned economy and decentralized economy, so as to achieve the optimal macro economic equilibrium.

Proposition 3. If conditions $g_c^* = \eta(1-\beta)\Omega(l^*)$, $g_p^* = \beta$, $\tau_k^* = 0$, $\tau_w^* = -\tau_c^*$ are satisfied, then the decentralized economy has the optimal fiscal policy.

Proof. When the growth path of decentralized economy is in accordance with the growth path of centrally planned economy, the economy will achieve the optimal equilibrium, and realize the optimal fiscal policy. At this time the two economic growth rates must be the same, so (10), (11) and (4) are satisfied. In addition, decentralized economy must satisfy the conclusion of Proposition 2, so we have

$$g_c^* = \eta(C/Y)^* = \eta(1-\beta)\Omega(l^*),\ g_p^* = \beta \tag{12}$$

From (10), we can get

$$1-\tau_k = \frac{(\mu/\lambda)}{(1-\beta)} = \frac{1-g_p-g_c+\eta(C/Y)}{1-g_p^*} \tag{13}$$

Together with (8) we have

$$\frac{1-\tau_w}{1+\tau_c} = \frac{(\mu/\lambda)}{(1-\beta)} = \frac{1-g_p-g_c+\eta(C/Y)}{1-g_p^*} \tag{14}$$

Since $g_c = g_c^*$, $g_p = g_p^*$, the government expenditure reach the optimal value, together with (13) and (14), we can obtain the optimal solution of τ_k, τ_w and τ_c are as follows.

$$\tau_k^* = 0,\ \tau_w^* = -\tau_c^* \tag{15}$$

In the synthesis may know, the conclusion in Proposition 3 is true.

Proposition 3 shows that in the centrally planned economy, $g_c = g_c^*$, $g_p = g_p^*$ will realize. In the decentralized economy, if there is no tax to capital income, and tax to labor income and consumption, then the optimal fiscal policy will be realized. In the meantime, in the absence of any externalities, if tax to consumption and leisure, then the optimal tax structure will reach.

Proposition 4. Under the optimal fiscal policy, when the lump sum taxes $T = 0$, the condition $\phi < (1-\beta)\Omega(l^*)$ must be satisfied, and

$$\tau_w^* = -\tau_c^* = \frac{\eta(1-\beta)\Omega(l^*)+\beta}{\phi-(1-\beta)\Omega(l^*)} < 0.$$

Proof. Under the optimal state, the following conditions must be satisfied

$$Nw^* = \phi Y^*/(1-l^*),\ C^* = (1-\beta)\Omega(l^*)Y^* \tag{16}$$

When $\tau_k^* = 0$, $\tau_w^* = -\tau_c^*$, together with (16), we have

$$\tau_w^*[\phi-(1-\beta)\Omega(l^*)]Y^* + T = (g_c^* + g_p^*)Y^* \tag{17}$$

Obviously, any τ_w^* and T which satisfy (17) will reach the economic equilibrium, and realize the optimal fiscal policy. In addition, when the lump sum taxes $T=0$, this condition is also hold, and must satisfy $\phi<(1-\beta)\Omega(l^*)$, together with (17), we can obtain $\tau_w^* = -\tau_c^* = \dfrac{\eta(1-\beta)\Omega(l^*)+\beta}{\phi-(1-\beta)\Omega(l^*)} < 0$, which means the conclusion in Proposition 4 is true.

3 Conclusions

This paper introduces an elastic labour supply to a simple AK model based on investment, and analyses the economic equilibrium for decentralized economy, and obtains equilibrium growth rate and the corresponding economic rules. In decentralized economy, the influence of wage income tax, capital income tax and individual consumption tax on economic growth rate and recreational utility are the same. When their values are higher, the economic growth rate will lower, and the recreational utility will higher. Moreover, the influence of and on economic growth rate and recreational utility are the same. When their values are higher, the economic growth rate will higher, and the recreational utility will lower.

Acknowledgments. This research is supported by National Research Project (No. 2009IM010400-1-32) of The National Higher Education Research Center, and the Hubei Province Key Laboratory of Intelligent Robot in Wuhan Institute of Technology.

References

1. Turnovsky, S.J.: Optimal Tax, Debt, and Expenditure Policies in a Growing Economy. Journal of Public Economics 60, 21–44 (1996)
2. Ireland, P.: Supply-side Economics and Endogenous Growth. Journal of Monetary Economics 33, 559–571 (1994)
3. Zhang, H.L.: Fiscal Policy and Economic Development. China financial and economic publishing house, Beijing (2004)
4. Zuo, D.P., Yang, C.X.: Biochemical Course of Economic Growth Theory. China Morden Economics Publishing House, Beijing (2007)
5. Turnovsky, S.J.: Fiscal Policy, Elastic Labor Supply, and Endogenous Growth. Journal of Monetary Economics 45, 185–210 (2000)
6. Gomez, M.A.: Optimal Tax Structure in a Two-sector Model of Endogenous Growth. Journal of Macroeconomics 29, 305–325 (2007)

Try to Talk about the Background of the Information Network Moral Construction

LiYan Yang

Abstract. The widc application of electronic networks for human information communication convenient and share the information resources, and provide technical support also produced some new moral problems, it is urgent to strengthen moral construction through the network to solve.

Keywords: information, the network moral, construction.

21 century is the era of information, network information have already walked into thousands of families. According to data showed that in 2011, China's Internet penetration rate of 34.3%, the number 457 million. The network is like a double-edged sword, in bring convenience for us quickly and, at the same time, to traditional moral put forward the torture, make the moral field appeared some new problems.

1 The Network Moral New Problems

1.1 Addicted to Network

Network culture colorful a unique charm. Some in the life the unpleasant people often yearn for the virtual network world find you sense of achievement, and then to cause the part of all people addicted to online virtual world rather than face real life. As time passes, from the reality of the world people will develop excessive freedom of the lax bad style, to the lack of necessary social reality of moral responsibility, relax them to the moral restraint.

1.2 The Network Crime

The network crime information society is a kind of new type of crime, is generally refers to the use of information technology intentionally or unintentionally implementation in the serious harm society and harm of citizens' legitimate rights and interests and shall bear criminal responsibility behavior. Its basic types: information steals fraud and information network damage. Among them, steal information including piracy, spy and hackers; Internet fraud crime bill, including card crime and criminal procedure; Information including physical damage and destruction program .The set-up of the platform of network for some of the network technology has engaged in the network crime created opportunities. Some people in line with fun try to do the psychological network hacker, do STH without authorization changes of the account of others password, into someone else's web site, or modify the information published online by the false information that causes a bad influence.

A. Xie & X. Huang (Eds.): Advances in Computer Science and Education, AISC 140, pp. 103–106.
springerlink.com

1.3 Invading Their Privacy

Personal privacy is a basic right, is also a basic social moral requirements. In the network in the space, people's study, the life, entertainment and communication, etc, will leave traces of digital, about their name, sex, physical condition, family status, property status, personal privacy online information will be reflected. The development of the network culture, make information has become a commodity, can be reasonably to buy and sell, and it makes individual privacy being of unprecedented aggression and threats. Part of the people give in to temptation may use advanced network technology to spy on, steal or distort the privacy of others.

1.4 Deceit Sentiment

Chatting online dating has increasingly become an important people online activities. In the network communication, the basic information of the individual can be part of the lack of integrity and virtual, sense of responsibility person, often use the media to defraud the trust each other, even someone transform their gender in online porn or cheat money, cheating the color, cheating the feelings of the net love behavior, great harm to other people's body health, hurt the feelings of others.

1.5 Network of Plagiarism

The great advantage of the Internet is to make the information extends in all directions, convenient and quick, it intervene accelerated the information transfer and communication, greatly saves economic and social activities, improve the overall cost of the quality and efficiency, and made the social wealth get rise dramatically. And on the other hand, because of the existence of network, is that anyone can get information at any time by means of the network, and it makes some people are used to the network information the advantage of rich resources, excessive dependent on the network. Need material, meet the problem is to search for the network. The answer sometimes is waking, lack of individuality. Make our creativity and SiWeiLi don't get very good play. People called wrote before the text is "paste stories", and is now "the mouse and keyboard". In today's academic researchers' plagiarism papers, the infringement of intellectual property rights of the phenomenon has not surprising. This lack of independent thinking, independent the ability to create, for the development of innovative talents, the construction of the innovative country is very bad.

2 Strengthen the Network Moral Construction Way

2.1 Build Network Ethics System

"The social network" of relative autonomy and freedom, request the behavior of people with a more strong self-discipline and responsibility consciousness, that is, people need no society and others in the supervision and circumstance, still can keep awake, stick to comply with self-discipline of social standard moral belief, consciously act according to the moral standards, this is the important guarantee of building network moral. And because of the development of the network, people exchanges and activities in the way substantive change, ideas, moral emotion and value orientation system change inevitably occurred, and create some new moral requirements. In order to adapt to the "social network" of a new space, for the past some of the scattered no system network ethics become less or out of date, and so require in the traditional moral basis to develop more system maintenance electronic space order, the people's behavior to constraints, and to guide the new ethics system.

2.2 Strengthen the Core Values of Propaganda Education

The network moral education must take on traditional moral education as the foundation. Must carry forward the finc culture of the Chinese nation and the fine tradition; Must guide users to set up the correct world outlook, the outlook on life and values; Strengthening patriotism, collectivism and socialist sense of honor and disgrace education; We must strengthen the social responsibility of education, make the Internet users a correct understanding of individual and collective, individual and social relations; Internet users must pay attention to improving the aesthetic interest and moral sentiment, the determination resists the rule of culture erosion.

2.3 Improve Network Moral of the Main Body of the Self-discipline Ability

In the influence of social environment and conscious education condition, main body need to keep learning network, according to the network moral principles and the requirement of the specifications for moral cultivation consciously, constantly improve the network of the main body of the moral self-discipline. Chinese ancient thinkers have paid great attention moral standards. Xun zi in the psalms, cultivate one's morality "first words were:" see good, repair but will not in itself also; see, QiaoRan will also; good body of self-examination, interface with the good also ran will not, will the body; the evil also." Also is to through the inner reflection, to develop and cultivate their moral quality. The main body to strengthen moral self network, to develop healthy, noble moral accomplishment, still has a reference. The process of network moral need training, which in essence is a main body self constraint network, and self process of ascension. Only by constantly teach network subject, practice, thinking, can be a real network moral man, to have higher self-discipline ability.

2.4 Increase Network Supervision

(1) Strengthen the network supervision of public opinion. Government departments and other relevant departments shall intensify the online opinion examination and supervision of strength, to false information on the Internet, spreading rumors in their products or services and for false propaganda and deliberately to others insult, libel, against men, and to strictly investigate its responsibility, and increase the strength of the punishment.

(2) strengthen the network technology supervision. On the one hand, to strengthen the government's capital, innovation network supervision technology, or introducing new technology, new products to the hackers and viruses and other Internet crime with effective prevention and control. On the other hand, to

innovation network supervision method, installation information filter, the system automatically the supervision, and the identity card "Internet" Internet system, strengthen supervision level.

2.5 Improve Network from Personnel Quality

Network builder is to strengthen self-discipline, and practitioners are to consciously abide by the relevant laws, regulations, and consciously construction site civilization. Internet cafe and other network management place must strictly abide by the provisions of the network, for the majority of Internet users to provide healthy and civilized Internet sites and network environment, make its be beneficial to health of people body and mind network garden.

Reference

[1] Chen, B.: Some of the thinking about the network moral. Fudan university (5) (2008)

Construction of Library Information Resources Depending on Internet Environment

LiYan Yang

Abstract. The paper propose appropriate strategies through discussion the effect produced by internet environment on the library's information resources, reestablish the new criteria for the collection of library ; Speed up the construction of collection; Reinforce the characteristic construction of the virtual collection; Promoting exchange and cooperation in libraries.

Keywords: Internet environment, Library, Information, Construction.

So-called internet environment refers to build broadband, high speed, integrated, wide-area digital electronic internet through combination of computer with modern communications technology. Rapid development of the network for the construction of library information resources has laid a strong foundation, which brings an impact and challenge to the traditional collection of library. How to strengthen the construction of library information resources depending on internet environment are important issues to be solved.

1 The Effect Produced by Internet Environment on the Library's Information Resources Construction

Information resources construction is the basic conditions of the library's development. As the society to information and internet, development of the library into the digital age, the library's literature resources in terms of structure, type, vector also has undergone great change. Influence internet environment on the library information resources construction is mainly reflected in the following areas:

1.1 The Effect Produced on the Contents of Library Collection

Traditional library collections mainly refer to the result of the collection, collation, storage, and the owning document resources, main features are physical entities. Such as printing books, microform, audio-visual materials, CD-ROM, etc. As network and storage of information, and other modern information technology increasingly to the penetration of the library, the library's collection has gone beyond the scope of this, extend to networked information resources. That is, the collection of information resources of library depending on the internet environment includes not only the substance collection, you can use the library's service for readers of all collections, also includes virtual collection,

A. Xie & X. Huang (Eds.): Advances in Computer Science and Education, AISC 140, pp. 107–110.
springerlink.com

through the library of computer systems and communications equipment can be shared outside information resources of the library, outside the collections database of the library, electronic publications, network information. Realistic and virtual collection that as information carrier and communication style in different ways, bring differences in use, determines the meaning of realistic and virtual collection of reality co-exist. Virtual collection greatly expands the range of information resources, Provide increased opportunities for library's information service, have a positive significance for collection development.

1.2 The Effect Produced on the Way to Get Resources

The library's traditional way to get information resources is subscribing, exchanging, accepting donation and so on. After literature into the reservoir, the library will have the permanent right to use, and ownership of the document. Information resources get from internet often are not entity but right to use. Right and access capabilities will become more important than the document entity library owns. Obtaining methods of Library information resources in addition to ordering, it also includes internet access, online use, rental, free access to a variety of ways.

1.3 Transformation of Objects Under the Information Resources Processing

In internet environment, information resources processing objects have changed by the information processing of the object transfer. From the past simple printed books, audio-visual materials, electronic literature and other vector processing change for the network information collection and processing. New information sources on the internet become the main way of access to information and knowledge. But library staff on should arrange and organize scattered and disorderly information, and filter out the complicated information, make it become an integral part of library information resources system. Particularly on the base of investigation and study, according to the specific user needs, deeply process information, make the information on your order, concentration and refinement. Expand and supplement their collection through using of internet information resources.

1.4 The Effect Produced on the Document Selection Criteria

Developed countries are now presented in three traditional document criteria for selection, document they are the fame of authors and publishers, documentation, content and breadth of features, other forms. On the internet environment, authors and publishers will be extended for maker of diagramming, developer of software, designer of screen and home page. About features in the form of extensions to the screen display, the system reflects and helper functions of information content structure. Format compatible and easy to use and retrieval method is also very important selection criteria, retrieve steps, screens and friendly interface, document formats, software updates are all factors to consider when you select.

2 Strategy of Library Information Resources Construction on the Internet Environment

Face of the impact and challenges of the new information environment, the traditional library must face the reality with an objective attitude, rethinking their advantages and disadvantages, ordain correct strategies of information resource construction.

2.1 Reestablish the New Collection Development Criteria

In the process of collection information resources construction, above all, establish the new collection development criteria. Library basic resources beyond traditional library collection and extended to the entire internet. The task, status and role of the traditional library has being changed greatly. So establish the collection development criteria suited to the characteristics of the library, it is necessary to construct information resources on internet environment.

2.2 Accelerate the Reality of Digital Library Construction

Real digital collection is that put library printed literature and other types into digital form, made into a variety of databases. Each library put their collection into digital form and upload to internet, will greatly enrich the internet information resources, most adequately improve easy use and sharing of resources. This is an important work of the library information resource construction on internet environment. Due to the limit of technical, financial, human and other conditions, It is unnecessary to put all collection into digital form, determine the scope and extent of collection resources digitization and improve quality of database, that is the key issues of the library digitization construction. Library should firstly put characteristic and lasting value collection especially rare, orphan and other valuable documents into digital form, make it the most intelligence value on internet information resources.

2.3 Strengthen the Characteristic Construction of Virtual Collection

The virtual library is an effective and affordable form of the organization information resources on the internet. It can be disordered cyber source into an orderly, to save time and improve the utilization rate of internet information. Specifically, the virtual library characteristic construction set a goal based on the library professional characteristics, selected information data category for closing false , keeping true, extracting the essence and organizing reasonably, make it became information collection that can provide searching, browsing and linking. For example, according to specific target of library and demand that user proposed, with “Yahoo”, “Excite”, from searching engine such as “Vista”, look for, and collect information source required, find various corresponding information through resources guide. Then arrange the retrieved information, filter, breakdown, comb, summarize, generalize in accordance with the certain theme, compile secondary literature matched, become serialized and effective information resources that meet the needs of a particular user.

2.4 Communicate and Collaborate between Libraries

Information internet provides the advantage to break self close, communicate and collaborate between libraries. In construction and utilization of information resources and collection information resources, as the important information resource base, library should strengthen lateral Exchange and cooperation, establish a multi-level, interwoven information network, take advantage of library community as a whole, walk the joint construction road of "distributed database building, centrally-attached, decentralized service, the library resource sharing" . This can save manpower, reduce costs, avoid duplication and realization joint construction and sharing of internet information resources.

In short, for many new changes that library information construction have emerged on internet environment, in order to do better on information construction, not only update idea of construction and clear target of construction, but also optimize the method of construction and grasp the important factor of information construction. Only in this way, library information construction can adapt to development of the internet. Lay a good foundation for library digital construction.

References

1. Zhang, L.: Utilization of information resources of library on the internet environment. Library theory and practice (1) (2005)

A General Fuzzified CMAC and Its Function Approximation

Zhipeng Shen and Chen Guo

College of Information Science and Technology,
Dalian Maritime University, 116026, Dalian, China
s_z_p@263.net

Abstract. Combined CMAC addressing schemes with fuzzy logic idea, a general fuzzified CMAC (GFAC) is proposed, in which the fuzzy membership functions are utilized as the receptive field functions. The mapping of receptive field functions, the selection law of membership with its parameters and the learning algorithm are presented. By using GFAC, the approximation of complex functions can be obtained which is more continuous than using conventional CMAC. The simulation results show that GFAC has good generalization, proper approximate accuracy and capacity to calculate function differential output.

Keywords: General fuzzified CMAC (GFAC), receptive field function, learning algorithm, function approximation.

1 Introduction

As known to all, within brain many creature control constructions to apperceive motion are composed of nerve cell which adjusts partly and mutually covers the receptive field [1]. According to this kind of thought, Albus brought up the Cerebellar Model Articulation the Controller (CMAC) [2].It is a local learning neural network, and has some merits such as fast learning speed, good generalization and easiness to implement.In recent years, combining CMAC with fuzzy logic is an important orientation [3-6], but till now there is no an integrated framework. Currently the study about Fuzzy CMAC is being concentrated on the following three aspects. First, based on fuzzy logic and neural network ratiocination, applying BP learning method, a common fuzzy neural network was constructed [3], which does not use the CMAC peculiar addressing mode and has slow training speed. Second, a self-organizing fuzzy CMAC was introduced based on similarity measure and competitive learning [5], whose process is complicated for the competitive learning scheme. Third, integrating conventional CMAC structure and fuzzy logic, a fuzzy CMAC was presented [4], which utilize the CMAC peculiar addressing scheme, but its fuzzy membership value is set random, and doesn't own universality. So it is necessary to make systematic researches on the FCMAC structure and learning algorithm.

This paper preserving CMAC local learning and addressing schemes, integrating fuzzy logic idea, proposed a general fuzzified CMAC (GFAC), in which the fuzzy

A. Xie & X. Huang (Eds.): Advances in Computer Science and Education, AISC 140, pp. 111–116.
springerlink.com

membership functions are utilized as the receptive field functions. Compared with conventional CMAC and FCMAC, applying GFAC some simulations on non-linear function approximation are made.

2 GFAC Structure and Its Learning Algorithm

The proposed network uses fuzzified language to define input variables, integrates fuzzy membership function μ(.) into association unit, so it has the fuzzy logic property. At the same time, it uses CMAC addressing method as mapping, so the input space can be demarcated more fine, which is different from conventional fuzzy CMAC [4]. Therefore, this network is called general fuzzified CMAC, abbreviated as GFAC. In fact its implementation is similar with traditional CMAC, converting the mapping $X \Rightarrow A$ into tow sub-mapping, $R{:}X \Rightarrow M$ and $E{:}M \Rightarrow A$. Mapping $R(x)$ decides the excited unit position of middle variable M for input vector x, and calculates the value of the corresponding membership function. E is a synthesized function, which mapping the excited unit of input to association location unit A. Fig. 1 shows the GFAC structure with double inputs, single output and generalization size 3. Where A 、A_p are, respectively, conceptional memorizer and practical memorizer, $\hat{y}$ 、 y are, respectively, expected output and practical output, and $m_{i,j}$ is the mapping address in middle variable M of input vector x_i.

2.1 Network Structure and Mapping

The input state space can be demarcated by fuzzy language. Suppose input vector is $x = [x_1 \quad x_2 \quad \dots \quad x_N]^T$, where $x_i \in I_i$, and I_i is a finite space and defined as:

$$I_i = \{x_i : x_{i,\min} \le x_i \le x_{i,\max}\} \tag{1}$$

The section is demarcated as

$$x_{i,\min} \le \lambda_{i,1} \le \lambda_{i,2} \le \cdots \le \lambda_{i,N_i} \le x_{i,\max} \tag{2}$$

Where $\lambda_{i,j}$ is the jth inner node of x_i, corresponding the fuzzy language positive big(PB), positive small(PS), negative big(NB), negative small(NS) etc. N_i is the demarcated node number. Similarly $\lambda_{i,j}$ can also be defined as outer node:

$$\cdots \le \lambda_{i,-1} \le \lambda_{i,0} \le x_{i,\min}, x_{i,\max} \le \lambda_{i,N_i+1} \le \lambda_{i,N_i+2} \cdots \tag{3}$$

Every outer node corresponds the fuzzy language as same as inner node, which can be considered as general fuzzy language. While different from fuzzy structure, how the input variable corresponds the right mapping section is decided by the quantification function and the CMAC peculiar mapping. The quantification function may be linear or non-linear. Here choose the linear function as

$$q_i(x_j) = \left\lfloor \frac{x_{i,j} - x_{i,\min}}{x_{i,\max} - x_{i,\min}} N_i \right\rfloor \tag{4}$$

Where $q_i(x_j)$ is quantified value of ith input x_i, and $\lfloor \cdot \rfloor$ is a floor function.

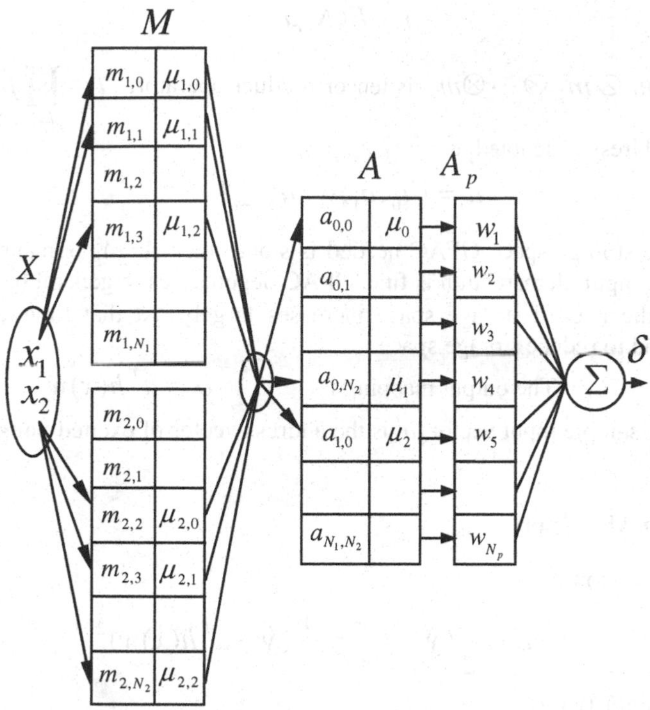

Fig. 1. The structure of GFAC

Define $D_i = \dfrac{x_{i,\max} - x_{i,\min}}{N_i}$ as distinguish rate of ith input variable. Then formula (4) can be transformed as

$$q_i(x_j) = \left\lfloor \frac{x_{i,j} - x_{i,\min}}{D_i} \right\rfloor \tag{5}$$

Mapping $X \Rightarrow M$ can be different. Through inducing, here adopt the following mapping rule.

$$m_{i,k} = \left\lfloor \frac{q_i + N_g - k}{N_g} \right\rfloor \cdot N_g + k \tag{6}$$

Where $m_{i,k}$ is the address of mapping from input vector $q_i(x_j)$ to middle variable m, N_g is the number of excited units, k is the ordinal number of excited unit and k=0-(N_g-1).

Different from simple fuzzy CMAC, only fewer membership function of excited units have to be decided in GFAC. Then the function's calculating consumption is reduced greatly. In order to describe the decision process of membership functions clearly, the membership functions are divided into two categories: single-variable and multi- variables.

When the mapping $X \Rightarrow M$ is determinate, the position a of input vector in A is given by searching the table, and shown as

$$a = E(\Lambda, \mu) \tag{7}$$

Where $\Lambda = m_1 \otimes m_2 \otimes \cdots \otimes m_n$ is tensor product operator, $\mu = \prod_{i=1}^{n} \mu_i$. And the excited unit address is denoted as

$$a_i = \{a_0, a_1, \cdots, a_{N_g - 1}\} \tag{8}$$

In general, the storage space GFAC needed is small, then $A \Rightarrow A_p$ is mapping one by one. When the input demarcation is fine, GFAC degenerates to general basis function CMAC, and the needed storage space increases largely. At that time, $A \Rightarrow A_p$ uses hashing method to reduce storage space.

The output mapping is $$\delta = a^T h(x) w \tag{9}$$

Where x is the sample input vector, a^T is the address vector of excited units. $h(x)$ is the weight vector.

2.2 Learning Algorithm

Define error function as

$$E = \frac{1}{2}(\hat{y} - y)^2 = \frac{1}{2}(\hat{y} - a^T h(x) w)^2 \tag{10}$$

The learning algorithm is

$$\Delta w_k = \frac{\beta}{a^T h(x) h^T(x) a}(\hat{y} - a^T h(x) w) a_k \mu_k \tag{11}$$

Where β is learning rate, and k is the ordinal number of excited unit and k=0-(N_g-1).Therefore, the GFAC differs from general basis function CMAC on weight value assignment. It determines the weight adjustment degree according to the membership function value of excited unit and the sum of all excited membership function square value, so it has adaptability, and better accuracy than fuzzy CMAC described in [4].

In order to improve the approximate accuracy of GFAC , the center c_j and width σ_j of membership function can also be adjusted during learning. Algorithm equations are

$$\Delta c_{k,i} = -\frac{\beta}{\rho}\frac{\partial E}{\partial c_{k,i}} = \frac{\beta}{\rho}(\hat{y} - a^T h(x) w) a_k w_k \mu_k \frac{2(x_i - c_{k,i})}{\sigma_{k,i}^{\;2}} \tag{12}$$

$$\Delta \sigma_{k,i} = -\frac{\beta}{\rho}\frac{\partial E}{\partial \sigma_{k,i}} = \frac{\beta}{\rho}(\hat{y} - a^T h(x) w) a_k w_k \mu_k \left[\frac{2(x_i - c_{k,i})^2}{\sigma_{k,i}^{\;3}}\right] \tag{13}$$

3 Simulation Study

In order to demonstrate the validity of GFAC, some simulations on non-linear function approximation applying GFAC are made, compared with conventional CMAC[2] and FCMAC[4]. Select a classical function below.

$$\hat{g}(\theta_1,\theta_2)=(\theta_1^{\ 2}-\theta_2^{\ 2})\sin(5\theta_1) \tag{14}$$

Where $\theta_1 \in [-1,1]$ and $\theta_2 \in [-1,1]$.

For clear compare, define the critical function as

$$e=\sum_{\text{all sample S}} |y-\hat{y}| \tag{15}$$

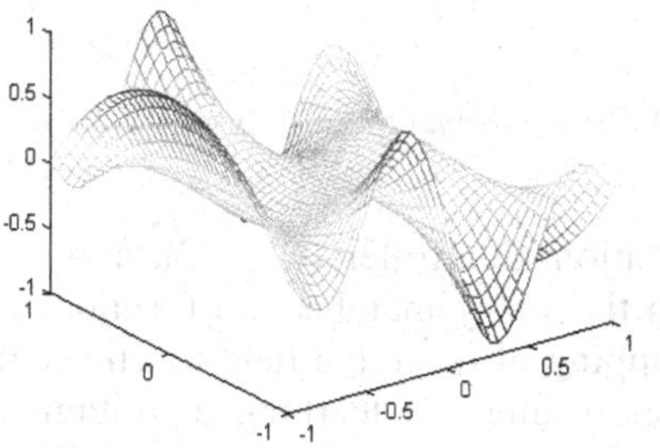

a. Function output graph

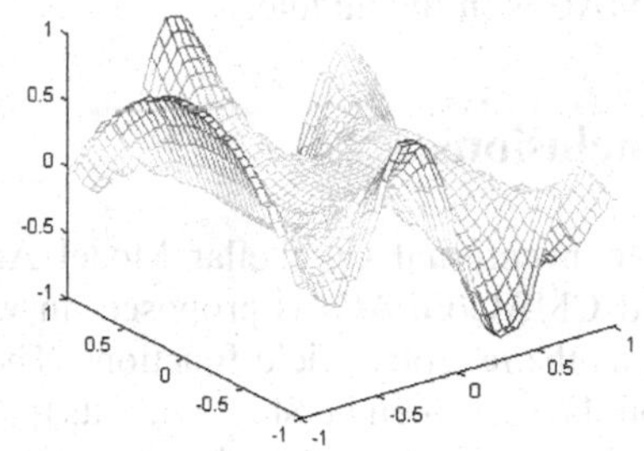

b. CMAC output graph

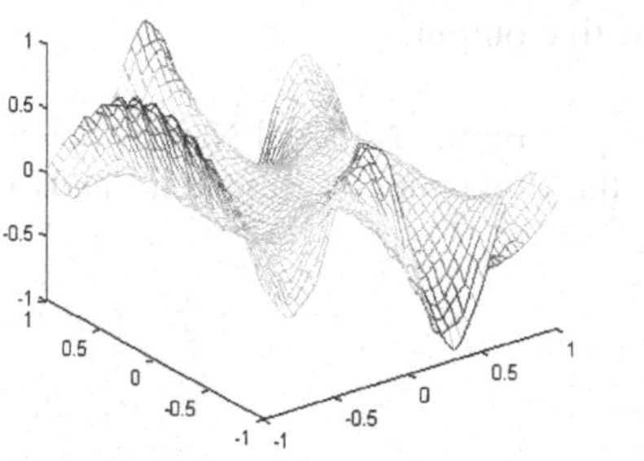

c. FCMAC output graph

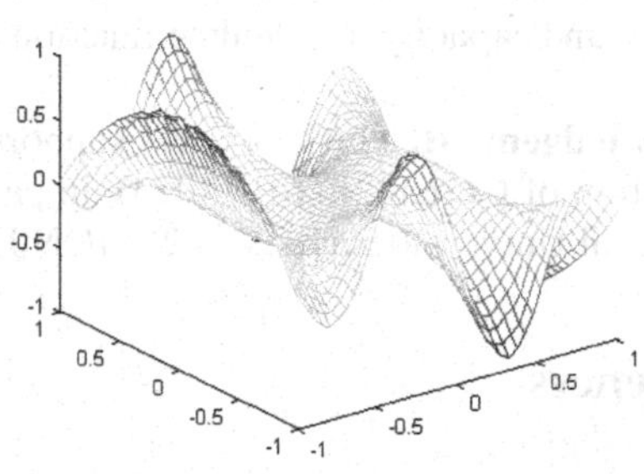

d. GFAC output graph

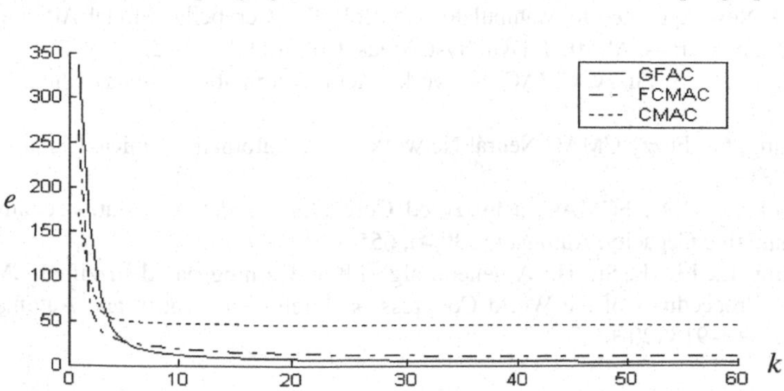

e. Compared curves of convergence speed

Fig. 2. Simulation function $\hat{g}(\theta_1,\theta_2)=(\theta_1^{\ 2}-\theta_2^{\ 2})\sin(5\theta_1)$

For function $\hat{g}(\theta_1, \theta_2)$, the parameters of GFAC are selected N_a=8, μ_c=0.986, and when training D_i=0.1, when testing D_i=0.05. While in terms of CMAC approximation for this function, N_a is large, and the learning speed is influenced greatly, so select. Fig. 2 shows the simulation results. Where $\hat{g}$ and g are respectively, target function output and practical output of neural network.

Compare the above curves, we can draw the conclusion: GFAC has good generalization, better approximate accuracy and capacity to calculate function derivative output; CMAC has faster learning speed, but worse approximate accuracy; and FCMAC is in the middle.

4 Conclusions

Aim at conventional Cerebellar Model Articulation Controller (CMAC), a general fuzzified CMAC(GFAC) is proposed, in which the fuzzy membership functions are utilized as the receptive field functions. The mapping of receptive field functions, the selection law of membership with its parameters and the learning algorithm are presented also. Compared with conventional CMAC and FCMAC, applying GFAC some simulations on non-linear function approximation are made. The results show that GFAC is not only effective, but also has good generalization, proper approximate accuracy and capacity to calculate function derivative output.

Acknowledgement. This work was supported in part by the National Natural Science Foundation of China (No.61074053), in part by the Fundamental Research Funds for the Central Universities (DLMU-2011QN029).

References

1. Hubel, D., Wiesel, T.N.: Receptive Fields, Binacular Interaction and Functional Architecture in Cat's Visual Cortex. J. Physiology 160(106), 5–10 (1962)
2. Albus, J.S.: A New Approach to Manipulator Control: The Cerebellar Model Articulation Controller(CMAC). Trans. ASME-J. Dyn. Syst. Meas. Control 97, 220–227 (1975)
3. Deng, Z., Sun, Z.: A Fuzzy CMAC Network. Acta Automatica Sinica, China 21(3), 288–293 (1995)
4. Zhou, X., Wang, G.: Fuzzy CMAC Neural Network. Acta Automatica Sinica, China 24(2), 173–177 (1998)
5. Nie, J.H., Linkens, D.A.: FCMAC: a Fuzzified Cerebellar Model Articulation controller with Self-organizing Capacity. Automatica 30(4), 655–664 (1994)
6. Guo, C., Wang, L., Li, H., Su, H.: A genetic algorithms learning based fuzzified CMAC controller. In: Proceedings of the World Congress on Intelligent Control and Automation (WCICA), pp. 915–918 (2000)

Discussion on Young Teachers' Teaching Method

Peng Qin, Yinghua Wang, and Zhihai Qin

Zhejiang Water Conservancy and Hydropower College, Hangzhou, China
qp021625@163.com

Abstract. Young teachers are lack of teaching experience, but their main work is teaching in class. So how to teach well and achieve the teaching purpose, which are every young teachers experiencing. Lesson preparation, teaching, mentoring are three very important teaching links. When prepares, teachers should emphasize on teaching material; when lectures, after introducing the subject, teachers should strive to achieve the target, take the teacher as the leadership, take the student as the main body, make full use of teaching resources, taking actual project as the carrier, linking theory with practice, also the previous course review and summary of the course cannot be ignored; timely tutoring, solve the students' problems.

Keywords: preparation and teaching, guidance, teaching practical engineering, vector.

1 Introduction

As the saying goes: There's no fixed way in teaching, but in practice, different ways of teaching can lead to quite different results. Teaching is an art that we need to pursue. Only when we are good at learning, diligent in thinking and practice, ready to practice, we will get the best reward. Young teachers are lack of teaching experience, but their main work is teaching in class. So how to teach well and achieve the teaching purpose, which are every young teachers experiencing. Making good lectures is the purpose for every teachers. Lecture preparation, teaching process, after-school tutoring are three very important teaching links, When prepares, teachers should emphasize on teaching material; when teach, teachers should strive to build a harmonious relationship between teachers and students [1].

Mobilize the enthusiasm of the students, make the teaching goal as the center, take the teacher as the leadership, take the student as the main body, reasonably use of the teaching resources, take actual project as the carrier, link theory with practice, not to ignore the link between the previous course review and summary of course; timely tutoring, solve the students' problems in learning.

2 Lecture Preparation

Lecture preparation is a re-creation process. First, we should understand the teaching material. The first step for lecture preparation is to study the teaching material.

A. Xie & X. Huang (Eds.): Advances in Computer Science and Education, AISC 140, pp. 117–121.
springerlink.com

The purpose for studying teaching material is to be familiar with the teaching material, fully understand the teaching material. Therefore, studying teaching material is not simply read the textbook, but in reading, completely understand the teaching content, grasp the material substance, and conform the teaching materials, recombine the knowledge system required in class, which is called thinking conformation. When finished this course, it means completely understanding the teaching material. After it, we can assure the important and the difficult part. Teaching stress is the important knowledge points according to teaching content and teaching purpose. Teaching difficulty is the knowledge hurdle in teaching process according to the cognitive theory and the teaching content. It is also very important to make certain about stress and difficulty. When teaching, teachers can achieve the teaching goal by standing out stress and breaking through difficulty. Lecture preparation stage, teacher should not only prepare knowledge, but also start from the reality of the students, completely understand their repertory, try hard to search the point which can stimulate their learning interest [2].

In the grasp of the teaching materials, first teachers should give the right disciplines of the subject, always understand the theory front dynamics and development, can bring new knowledge to the students, and be good at taking actual project as the carrier, combine the knowledge and the practical engineering with the students real life. Teaching is an interactive relationship between teachers and students, in order to achieve a good teaching effect, one of the important principles of teaching that teachers need is student-centered. Ignore students can not achieve good results. Therefore, teachers in the teaching preparation must embarks actually from the student, teach students in accordance with their aptitude, treat different students by designing different teaching requirements. In the higher education from elite education to mass education today, part of the students learning methods and learning habits, knowledge ability is not just as one wishes, therefore the design of teaching to achieve the goal also cannot make rigidly uniform. Good students to fill, slightly worse students can learn, and some students can understand. That is to teach students in accordance with their aptitude.

3 Teaching Process

3.1 The Establishment of Harmonious Relationship between Teachers and Students

Harmonious relationship between teachers and students should be the teachers and the students in the personality of equality, in the interactive activities in a democratic, the atmosphere in cooperating is harmonious. In class teacher should love his students, understand their personality, when teaching, teachers should really find, understand and recognize the students in all aspects. Care the students' personality and create a pleasant learning atmosphere, more communicate with students, make the boring class into an interactive learning space full of freedom and wisdom. Communicate with the students as a friend and student of equal exchange, rather than superior, face the teacher's dignity. Especially young teachers should keep this in mind.

3.2 To Mobilize Students Learning Enthusiasm

Classroom is a platform for interaction between students and teachers, teachers should fully mobilize the student 's subjective initiative, let the students to actively play an active role in teaching, teachers should play leading role, students are the real main body, only the student's subjective initiative be fully mobilized, the task of teaching can be done well. Give the students the method of fish, reason thoroughly, especially the stress and difficulty. Teachers should not confuse the fish and fish. Students should grasp the fish way, and then he can achieve the life goal. If only a fish, there's no easy way. So the soul of higher education should be to teach him to fish [3].

3.3 With the Teaching Goal as the Center

By multiple teaching steps constituting the teaching process is an integral whole, which goes through the whole line is the teaching goal, teaching process in all the teaching activities should be based on the central goal---teaching objectives, and to achieve this goal. To learn in a lesson seems like a lot, but there are only a few difficulties. Therefore, in order to effectively achieve teaching goals, in the teaching process we must highlight the teaching focus. And the difficulties of teaching are the hurdles in teaching process, we should take appropriate choice or adopt corresponding measures to break through according to the actual situation and the different requirements of the students.

3.4 Take the Actual Project as the Carrie

Although teachers can teach well, but if only remain in the textbook, to college students is also empty talk. Based on fully understanding of the teaching material, teachers should take the actual project as the carrier, combine what to teach with practical engineering, to make substance in speech. Learning the knowledge link with the specific parts of the project, what are their effects in practical enginnering should be made clear, so students can initiate the interest of learning.

3.5 Take Teacher as the Leading Factor

In the teaching process, teachers play a leading role. The leading role is giving a correct direction to teaching goal through multiple teaching steps. To make students' thinking activities smoothly follow teachers teaching steps designed from shallow into deep, step by step to get close to the teaching goal. Teaching activities designed by teachers must be equal to the students' acceptability.

3.6 Take the Student as the Main Body

The classroom is not the teacher's stage, it also should not only for teachers teaching, students listening. All the teaching activities in class must take the student as the center, the development of teaching activities must make the students understand the teaching material, digestive knowledge, acquire ability. The nature of teaching process is the teacher's teaching, student learning is not only learn knowledge, more important is to learn knowledge acquisition method. Therefore, teaching in class, the teacher 's mind must have students, always think what students thinking. Classroom teaching should not pay more attention on the teacher's teaching, but also on the students' learning. To make the students understand is the principle to show the classroom teaching that makes the student as the centre.

3.7 Rational Utilization of Education Resources

Good writing will leave a good impression to the students and it will stimulate the students interest in learning. On the contrary it will have bad effect on the students if the blackboard is very chaotic, and it gives poor handwriting. With the rapid development of science and technology, multimedia teaching has been applied in many schools, and it plays a very important role in the teaching activities. Rational use of multimedia teaching does good to the improving the young teachers' teaching effect. It can be extremely rich, vivid, so it must be preferred by young teachers. We must emphasize that we should make rational use of multimedia teaching means. Although multimedia teaching is lively and intuitive, its shortcoming is very obvious. We should be cautious to deduction of the theoretical formulas. The process of theoretical formulas deduction is a logical process. We should think more about even a sign and a step, if we improve the speed of teaching by multimedia teaching, lose the humanization and the students' mind is lag behind, image the teaching effect.

3.8 The Review before Class and the Summary after Class

The review before class and the summary after class is also a teaching link which should not be neglected. Concisely review the previous teaching content, it can not only deepen the knowledge, but also play an important role for the classroom teaching, to pave the way for it. Summary after class can use very simple language but it should not be neglected. Summary after new class is very important. Review the newly learned knowledge again in a short period can intensify the memory and it will decrease the forgetting rate. Therefore, summary after class should be an unforgettable link in class teaching. Teachers should summarized the teaching content and knowledge points again to enable students to consolidate knowledge.

4 After-School Tutoring

Learning is a process of receiving, digestion and absorption process. Teaching in class the students must accept knowledge, after-school review is a digestion process, this process is very important, if not digest, then not absorb, teaching process will fail. So after-school tutoring is very important. If the students have problems in digestion, they should ask the teachers to help solving or they can't master the knowledge. After-school tutor in an important teaching link which can't be neglected.

5 Conclusion

Every teacher should be careful to give class. Only by this, one can continue to learn, continue toreflect, sum up, utilize and it can rapidly improve, grow, mature. So it is worth learning for every teacher.

To grasp the class language to encourage, guide and inspire. Fully prepared for teaching, meticulously prepared for teaching, give the learning right to students, let the students to study knowledge, establish a good relationship with students, let the students believe their teachers, be familiar with the subject, understand the students, encourage students to participate in, put the teaching quality into the first place, create a good learning environment. Being a qualified teacher.

Acknowledgment. The study was in part supported by the Research Found of Zhejiang province (No.22169-1),Zhejiang Water Resources Department (No.RC1028)and Zhejiang Water Conservancy and Hydropower College (ZJWCHC) (No. xky-201005), P.R.China.

References

1. Cai, K.: In the 21st Century the trend of China's higher education. The Reference for Publicity and Education 2, 28–32 (2004)
2. Yang, D.: Innovation also needed in the classes of Colleges and universities. Journal of Henan Business College 6, 82–83 (2003)
3. Zhou, X.: Implementation of the creative education in class teaching. The Present Education Forum 5, 97 (2003)

Reform of Basic Computer Teaching in Universities and Its Measures under New Circumstances

Jianguo Yu and Shuanghong Liu

Zhengzhou Institute of Aeronautical Industry Management, 450005
{yjg,cloudyni}@zzia.edu.cn

Abstract. Computer basic course is a public compulsory course in computer application for the non computer major students. According to the current situation of basic computer teaching in universities and problems in teaching reform under new circumstances, the necessary of basic computer courses reform is emphasized. Then some concrete measures on curriculum system and contents, teaching and examination methods, teaching management are proposed in order to solve the problems and improve teaching qualities and teaching effects of the curriculum.

Keywords: basic computer teaching, teaching reform, universities.

1 Introduction

With the rapid development of scientific technology, the society has entered into information era. The information socialization has become a key index to evaluate the comprehensive strength of a nation. As a necessary tool in information society, computer has penetrated into every field; it propels the informationization of the whole society forward and becomes a culture progressively. The application of computer is one of the basic capabilities for people in current society. Therefore, basic computer courses are offered in universities successively; students can grasp necessary computer knowledge and obtain the ability of computer application through the study of this course, so that they can serve the social economy better.

2 Current Situation of Basic Computer Teaching in Universities

Valuable explorations on how to develop basic computer teaching, such as its curriculum system and contents, teaching material, teaching methodology, teaching method, etc. have been carried out in many universities throughout the country. As early as in 1997, the Ministry of Education proposed Several Suggestions on Strengthening Basic Computer Teaching in Non-computer Specialties of Engineering Faculty (hereinafter referred to as Suggestions). In the Suggestions, it pointed out that basic computer courses for non-computer specialties had strong instrumental and practical properties; attention should be paid to foster students' abilities of having confidence to and being able to apply computer, as well as their consciousness of utilizing computer to analyze and solve

A. Xie & X. Huang (Eds.): Advances in Computer Science and Education, AISC 140, pp. 123–128.
springerlink.com

problems on their own initiative; foster students' abilities of self-learning and receiving new techniques and methods that emerged in endlessly. In practice, teaching may be organized in three levels (computer culture, technique and application bases)

The first level is computer culture basis, including basic computer knowledge and operation, as well as the development and the characteristics of computer culture. It aims at helping students grasp necessary basic computer knowledge and operational skills in order to work, study and live better in information society; fostering their consciousness of computer culture.

The second level is computer technique basis, including advanced language programming, fundamentals of software design, and fundamentals of hardware technology. It aims at helping students lay a foundation of necessary technical knowledge and abilities for application and development of software and hardware in this specialty.

The third level is computer application basis. In accordance with the demands of different specialties, relevant courses on computer application are offered. It aims at enhancing students' abilities of building up a computer application system in this specialty and relevant fields.

In accordance with the spirit of Suggestions, most of universities keep identical with the Suggestions on curriculum system and structure, teaching organization and contents, etc. In the first level, the course of Computer Culture Basis is generally offered. In the second level, difference is obvious: some universities offer one or two courses of programming language, such as C Language, Visual Basic Programming, etc. and some universities offer one basic course of software technology and one basic course of hardware technology, such as Fundamentals of Computer Software Technology and Fundamentals of Computer Hardware Technology. In the third level, one or several computer courses relevant to this specialty are offered. On teaching method, most universities adopt multi-media and projection. Some universities develop CAI teaching courseware independently and some purchase from other universities and presses. On the checking method, most universities still adopt the way of traditional written examination. On specific teaching organization, universities try different approaches in consideration of their own actual conditions. Taking Computer Culture Basis as an example, the teaching hours vary from 10 to 50, computer operation hours from 20 to 30 and credit is generally between 1 and 3.

3 Problems and Measures under New Circumstances

Objectively, the current model of basic computer teaching in universities plays an important role for the universal education of computer in a certain period. However, computer science and technology is a discipline that changes with each passing day and its knowledge updates fast; if universities follow the old model without any innovation, various problems will arise. Therefore, it is an urgent demand to satisfy the requirement for computer development under new circumstances in the reform of basic computer teaching in universities.

3.1 Reform of System and Contents of Basic Computer Courses and Its Measures

In recent years, residents' economic and cultural levels have been further improved with further development of society; computer is more and more popular and has entered into school and family. In Guiding Suggestions on Accelerating the Construction of Information Technology Courses in Elementary and Middle Schools (Draft) issued by Department of Basic Education subordinated to Ministry of Education on November 26, 1999, it clearly pointed out that information technology, as a compulsory course, would be offered in regular senior middle schools throughout the country from the new academic year in 2001. In terms of the document, the teaching arrangement for this course almost covers all contents of basic computer courses currently offered in universities, or even richer.

Currently, it seems that most universities do not pay enough attention to the above situation. If corresponding measures were not adopted, the basic computer teaching in universities may get into a very passive position in one or two years later, which has taken on among fresh students in recent years. Quite a few of them have grasped basic computer knowledge and application skills when they studied in senior middle schools, which have achieved or even exceeded the requirements for the curriculum of Computer Culture Basis in the first level of basic computer teaching in current universities; some of them even have passed the national computer rank examinations with the results of grade 2 or 3. But most of universities still adopt the old curriculum, and take Computer Culture Basis as the first basic public course without modifying any of the contents. Consequently, even the students who have grasped the knowledge and skills required by the curriculum have to study them once again, which have exerted much negative impact on the growth of students.

To solve the above problem, universities may consider reforming the curriculum system and contents on one hand, e.g. compile a series of textbooks that are connected to the contents of information technology course in elementary and middle schools as soon as possible, or even alter the three-level teaching model mentioned in Suggestions in consideration of actual situations; on the other hand, universities may consider testing fresh students' abilities of computer culture basis when they are enrolled. For the students that pass the examination, they are exempted from Computer Culture Basis, the first-level course and acquire corresponding credits, and they may take another elective computer course which keeps continuity with Computer Culture Basis on its contents and depth; universities may make independent choices on curriculum name and content (e.g. we call it Computer Information and Introduction for the present) to improve the students' level of computer culture. For the students that do not pass the examination, teaching will be organized in accordance with normal teaching order. In this way, for the students who have a better foundation of computer culture, they can avoid many unfavorable factors brought about by repeated study of the same knowledge as well as study more abstruse knowledge on computer culture sooner, so they can improve their zeal and efficiency on the study of computer.

3.2 Teaching and Examining Methods Should Satisfy the Requirements for the Reform

It has been recognized by most educational experts, educators and teachers that computer aided instruction (CAI) should be developed vigorously, CAI courseware should be developed and applied actively, and the audio-visual teaching should be carried out in basic computer teaching, i.e. the teaching method based on CAI instead of that based on blackboard and chalk. However, the teaching results are affected adversely in practical teaching since some teachers do not know or use CAI properly. For example, quite a few of teachers think that teaching in multi-media classroom only means playing CAI courseware. Then the teacher turns to a courseware player merely from a comprehensive performer in the whole class. In addition, students are easy to be tired if watching CAI courseware for a long time.

In fact, CAI itself is merely an approach of "computer aided instruction". The effect of multi-media CAI is confined to several factors, such as previous knowledge, motivation, studying contents, course background and so on. Whether teaching media can help students promote their study is closely related to the way of using the media. Thus it can be seen that only teaching rules are adhered to and all conditions are satisfied in the process of teaching, can effective teaching be carried out and the potential advantages of CAI be brought into full play on information display, individual teaching, instant feedback, human-computer dialogue, and students' participation into the teaching, etc. Therefore, teachers should change the idea that CAI teaching may be instead of them and CAI courseware is used throughout the whole teaching process; they should fully realize that "human" is the best "multi-media". In the process of teaching, it is "human", i.e. "teachers" that are the leading roles, so they should bring "human's" subjective initiative into full play and improve teaching effect better with the help of CAI multi-media teaching system in the process of teaching.

Moreover, most universities actually do not have the ability to develop qualified CAI software for factors such as the development technique of CAI courseware, cognition of educational theories, etc. But it will be a force not to be ignored if several universities unite to develop by means of bringing their respective advantages into play. The modern teaching environment, such as electronic library, computer room, campus network, Internet, etc. should be brought into full play, for example, set up CAI web sites and CAI teaching CD ROM database for students to make full use of modern teaching environment.

Basic computer course covers a wide range and abundant knowledge, and there are usually thousands of students take part in the examination in most universities every academic year, so the traditional testing method undoubtedly brings great pressure for examining departments of universities. Particularly, basic computer course is a discipline with strong operational and practical properties, it is obvious that the traditional written examination cannot check the actual situation how students grasp the course. For example, the operational ability of software such as Windows, Word, Excel in the course of Computer Culture Basis and the ability of debugging in programming language courses cannot be checked in written examination. Therefore, we suggest developing a "paperless examination system for basic computer courses" based on Browser/Server network environment, including modules like question bank and its management, examinee information management, online examination, automatic scoring and inquiry of results, statistics and analysis, etc. so as to realize the checking of operational and debugging abilities; and all links such as examinee's login and registration, generation of examination paper, testing of examinee, collection of results, inquiry of scores, statistics, etc. are completed automatically in the system. Students are allowed to access to any host computer in the campus network to take part in the examination in the system through browser. Obviously, its efficiency is improved greatly comparing to the traditional examination method. Universities, like Tianjin University, have realized online paperless examination for the course of Computer Culture Basis in non-computer specialties.

It is worthy to be mentioned that the examination system for the course of Computer Culture Basis in non-computer specialties of universities has been spread in some provinces; students cannot acquire the credit for the course unless they pass the uniform examination. It is particularly worth noting that we should avoid striving for one-sided pass rate and entering into mistakable area of exam-oriented teaching model in the application process of the system.

3.3 The Management of Basic Computer Teaching Should Be Reformed Simultaneously

In recent years, the enrollment quantity soars at the speed of more than double every year with the acceleration of higher education development in China. The quantity of students in quite a few of universities has exceeded the maximum limit for the resources such as teachers, classrooms and teaching equipment, so problems emerge like the deficiency of teachers and classrooms, the shortage of computer resources for students' operational exercise. Many universities have to adopt the methods that only seek temporary relief regardless of the consequences, for example, reduce class or computer practice hours, or even reduce basic computer courses to relieve current pressure, whose premise is the cost of sacrificing teaching quality.

To solve these problems, on one hand, universities need to increase the input of teaching facilities; on the other hand, the current shortage may be preliminarily relieved by means of reforming the teaching system or managing limited teaching resources effectively; the example is the one we mentioned previously that separating fresh students when they are enrolled through the examination on their abilities of computer culture basis. For the shortage of computer resources, it is particularly worth noting that the reduction of actual operational hours should be avoided in basic computer courses since they emphasize the training of operational and practical abilities; but it is an advisable method to reduce operational hours in class and increase the opportunities after class, whose premise is to realize 24-hour open computer laboratories controlled by electronic cards automatically. The reduced operational hours may be converted into students' electronic cards, who are free to use the computers after class in accordance with their own actual conditions. In details, the management model of open computer laboratories in Tsinghua University is regarded as a reference.

4 Conclusion

The reform of basic computer teaching in universities is a complicated task, which requires the active participation of computer teachers, laboratory personnel and teaching management personnel. The problems in the process of reform should be paid attention to and solved in time. Only in this way, can it satisfy the demand of the rapid development of computer science and technology and serve the social economy better.

References

1. Liu, Y.: Exploration of Efficient Mechanism for Improvement of Teaching Quality under New Circumstances (EB/OL) (June 5, 2001), `http://www.edu.cn/20010827/208980.shtml`
2. Ministry of Education, Circular of the Implementation of "Higher Education Reform Project in New Century" (EB/OL). January 26, 2000), `http://www.fosu.edu.cn/laws/edu2.html`
3. Xu, A.: Establishment of Monitoring System for Improvement of Teaching Quality Adapting to the Credit Reform. China Advanced Mathematics (1), 40–41 (2011)
4. Su, B., Hua, X.: Initial Exploration on Practical Teaching of Computer Basis. Journal of Tarim University (3), 77–79 (2011)
5. Wan, W.: An Exploration Based on Practical Reform of Basic Computer Teaching Contents. Computer & Communication (2), 198–199 (2011)
6. Yu, H., Feng, K.: Study on Evaluation of Teaching Quality in Higher Universities. Heilongjiang Foreign Economic Relations & Trade (9), 138–140 (2010)
7. Liu, X.: New Orientation for Construction of Teaching Quality in Higher Education- A Microscopic Study on Construction of Teaching Quality in Higher Education, 89–91

Exploration of the Building Model about Theme_Based Learning and Shared Community Based Python Add Django

Xiang Li, Mei Zhu, and Xuan Zhan

Department of Software Engineering, East China Institute of Technology, Nanchang, Jiangxi Province, China

Abstract. Theme-based learning refers to a way in which the students learn around one or more of the structured themes. As a new way, this kind of learning constantly has impacts on the traditional way. This paper combines the web2.0 technology development, focusing on the construction model of a new theme-based learning, with Django as bottom framework of Web, Python as the main developed language, and MySQL as the backend database. At the same time, the paper also combines theme-based online learning with open resources-sharing, in order to maximize initiative of user-learning, improve the access efficiency of resources and strengthen communication and collaboration between users. With these done, it will reflect the freedom and openness of Internet age.

Keywords: theme-based learning, shared community, Python, Django.

1 Introduction

Web 2.0 applications make people understand the current change of Internet -- from a series of static Web site to a sophisticated network applications for end users of the service platform. Web 2.0 is not a technical standard, it contains the technical architecture and application software, which is a user-centric, mainly user-created contents, and can decide how to show them. Two trends aroused in Web2.0 development, one is pairwise integration between Web2.0 sites, such as watercress network and services among kinds of book purchase sites, various micro-blog and instant messaging tools as well as the integration of mobile terminals, a variety of Web2.0 applications and e-mail, SMS and map integration. While the purpose of integration is still to promote each other, integration is an important step towards Mashup. Another trend is aggregation, google reader or shrimp itself is a polymer, which was more concerned about the aggregation of article information, the key feature is Digg. With wide applications of Web2.0, while putting aside the idea of pairwise polymerization, bus-style aggregator sites can be introduced, which combines all your Web2.0 activities and friends together.

Thematic learning and sharing communities is a Web2.0 system which is similar to this innovative social network. The main groups for this system is the IT industry enthusiasts and learners, focused on more IT Sector and fast renovation of technology,

A. Xie & X. Huang (Eds.): Advances in Computer Science and Education, AISC 140, pp. 129–134.
springerlink.com

this system combines theme-based online learning and open sharing of resources with the purpose of maximize User learning initiative and improving the efficiency of access to resources and strengthening communication and collaboration between users. All information provided by the system will be user-driven, which reflects the freedom and openness of the Internet age.

2 Technology Introduction

2.1 Python Introduction

Python is an open source, interpreted, interactive, and object-oriented programming language. As a dynamic scripting language, it emphasizes the development speed and code clarity. Python can be used to develop a variety of procedures, from simple script task to the application of object-oriented role. Rather than other object-oriented language, Python emphasizes space and coding structure, which allows developers to reuse the code, the other is to emphasize the dynamic nature, that is no need to compile Python code. Python is easy to extend, it can call the corresponding library functions, besides Python itself can also be embedded in other programming language or platform, such as using Python interpreter written in Java and based on. NET platform, IronPython, etc. . as a binder, developers can use C, C + + or Java, to write new modules for the Python, such as functions and so on. Python can be compiled directly with, or dynamic library is loaded to achieve and so on. Python currently supports the integration of almost all the key technologies, while documents, agreements, dynamic link library and COM objects can work together. In addition, Python provides a wide range of libraries to help developers get any kind of data. There are many other excellent Python features, no matter junior and experienced developers or experts, will find Python is a very flexible and powerful programming language, by using it can develop a wide variety of applications.

2.2 Django Introduction

Django is an open source Web application framework, written in Python. MTV design pattern is used, that is model M, the template T and the View Controller V. Django's primary goal is to make complex, database-driven sites development easier. Django focuses on components reuse and "pluggable", agile development and DRY principle (Don't Repeat Yourself). In Django, Python including configuration files and data model is widely used. The core framework including: an object-oriented mapper, as a data model (in the form of Python class definition) and the relationship between the media database; a regular expression based URL dispatcher; a view system to process the request; and a template system. The core framework includes: 1) a lightweight, standalone Web server for development and testing. 2) a form serialization and validation system for HTML forms and database to store the data for conversion. 3) a cache framework, and several ways to choose the cache. 4) middleware support, which allows the various stages of processing the request to intervene. 5) built-in distribution system allows application components to use pre-defined signal communicates among each other. 6) a sequence of system that can generate the Django model instance described by XML or JSON. 7) a template engine for the expansion of the capacity of the system.

Django includes a number of applications in its "Contrib" package including: 1) a scalable authentication system; 2) dynamic site management page; 3) a set of tools for generating RSS and Atom; 4) a flexible Comment System; 5) Google Site Map (Google Sitemaps) tools; 6) tools preventing cross-site request forgery (cross-site request forgery); 7)a group of supporting lightweight markup language (Textile and Markdown) template library ; 8) a package of creating a geographic information system (GIS) framework.

3 System Architecture Design

3.1 System Running Mode Design

The system application server uses the Django framework, all written by python language. System operation mode uses MVC (Model-View-Controller), shown in Figure 1, events caused by user lead to changes in Controller or Model View, or both changed. All View are automatically updated as long as Models' data or properties changed, Similarly, as long as the Controller changed the View, View has the potential to obtain data to refresh themselves. Controller accepts user's request, according to this URL to find a specific function to perform, the function is a dispatcher, which will determine the user's permissions. And then passed to View, View render the template to form a web page.

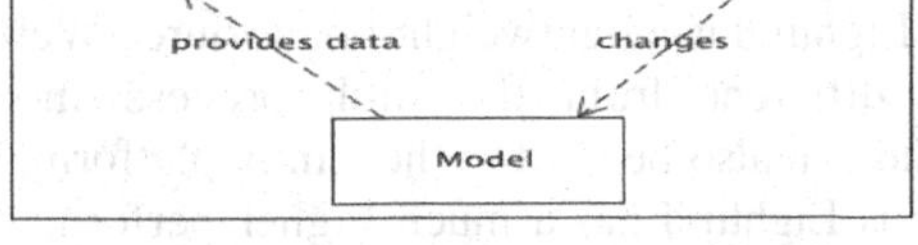

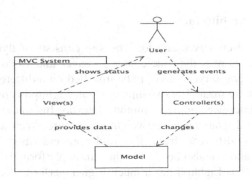

Fig. 1. The system MVC mode

3.2 System Function Modules Design

Divided from the functional modules, and maintain "functional independence"is the basic principle of modular design. Because "functional independence " can reduce the development, testing and maintenance phases cost. Meanwhile, a task can be achieved that requires cooperation between the various modules, then the exchange of information between modules is necessary.

Theme-based collaborative learning and resource sharing community can provide the theme-centered collaborative learning and open sharing of resources. Exchanges and cooperation in organization learning are theme-centered, learning content and activities around specific themes. Resource sharing was open, which can be used for all Internet users. System is divided into four modules: thematic learning modules, resource sharing modules, social networking module, the background management module.

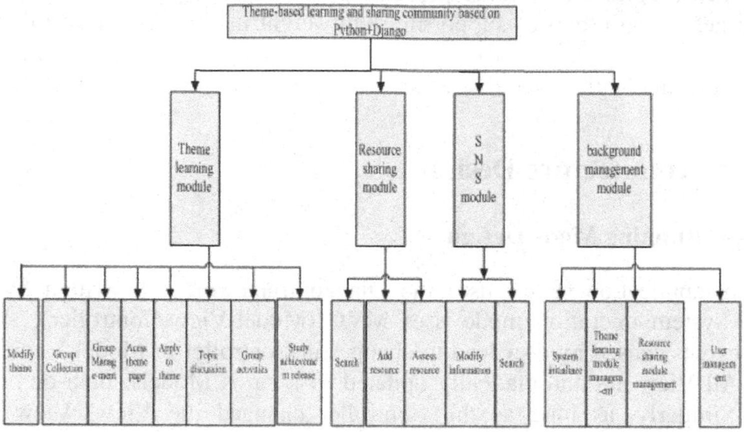

Fig. 2. System function structure

4 Technical Analysis and System Implementation Feature

4.1 Server Cluster Architecture

This system made a Web server division, the site consists of dynamic content and static content. All the html is dynamic content, images and video are static content, which is divided into two servers, and can be installed on a different Web server and different adjustments are made. For dynamic content, Lighttpd is used, SCGI (Simple Common Gateway Interface simple Common Gateway Interface) is connected with the application server. Lighttpd is a lightweight open source Web server. Compared to Apache, Lighttpd is different from the multi-process model of Apache, IO multiplexing is used and can also be used in the Linux platform, which is kernel-level event-driven model, thus Lighttpd has a much higher performance than Apache that can support more than tens of thousands of concurrent connection requests on a single physical server. For static content, Nginx is used. Nginx is a more lightweight Web server, and provides even higher performance than Lighttpd. According to a simple stress test, Nginx is roughly 10% -15% faster than Lighttpd, as well as CPU and memory consumption is even lower than Lighttpd.

After user's request is get, then analyse user's URL, and make use of external resources such as databases, finally html is combined and return. Because database access is slow, and has a large number of IO, so we use Memcached as a cache. Memcached is a distributed memory-based caching system, such as a lot of servers as a cache, each machine provides its own memory for system use. Due to the limited laboratory environment, where no separate opening for Memcached server, which can be extended later if needed. In addition, because the system has a lot of data and file resources, search engine users is relatively high, inefficient SQL LIKE statements are no longer used. A C + + based open source search engine is used, through the Web service interface to the user's request, for example, to access a file. For the database server, plan to use three database servers, one master, one slave, were classified according to the application, so you can ensure that the load is not too high. Another slave, on the one hand as a backup, one for data mining, because the data can not be done directly with the online.

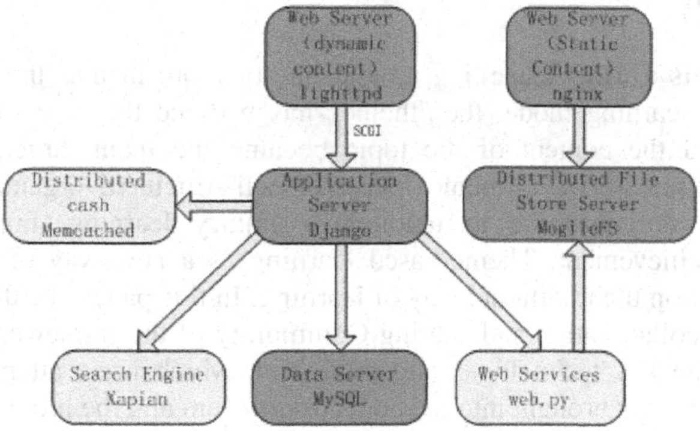

Fig. 3. Server architecture

4.2 Special User Authority Design

User rights system mainly consists of the user (Users) and permissions (Permissions). Authority is not binding on the user object, but as objects exist independently. This design is more compact, but also more convenient when control the user's authority. When the model is created, if "class Admin" is included, then will automatically create a corresponding object model 3 basic permissions – add, create and delete, which generate the three Permission instance, stored into a database auth_permission table. In addition to these three types of permissions, you can also use "model Meta attribute_" to specify custom permissions. Custom permissions are similarly stored into auth_permission table. User permissions and Permission has many relationships. Each user can have the authority to operate many types of objects, each object can also be operated by multiple users. When assign permissions, it will create an instance of User Permission instance and mapping, the mapping will stored into auth_user_user_permissions table.

4.3 DRY Principle Reflection

The system uses many decorator, including language Python provides and Django framework provides, as well as in developing custom. A large number of functions in the same operation needs to be done, decorator is a good solution, which avoids duplication and waste of time, reflecting the DRY principle.The main decorators are as follows: (1) @ staticmethod, Python provided, set the modified function as a static function, no need to define the function instance, and multiple instances to share the function. (2) @ login_required, Django provides, ensure that the user must log in to access the requested view. (3) @ models.permalink, Django provides, prevent the url in the hard-coded. (4) @ property, Django provides, make the instance methods can be modified in a similar way to read or assign the property. (5) @ build_form, custom decorators, used in the view before form creation. (6) @ render_to, custom decorators, specify template file used outside the view to achieve soft-coded url.

5 Conclusion

Student learning is a way of learning around one or more themes through structured learning. In this learning mode, the "theme" has become the core of learning, and structured around the content of the topic became the main target. Theme-based learning has the feature of a “theme focused, well-structured organization, easy to explore", which is conducive to autonomy, inquiry learning, improve learning efficiency and achievement. Theme-based learning as a new way of learning has a constantly impact on the traditional way of learning. In this paper, the theme of Python + Django-based collaboration and sharing Community of the framework are analyzed systematically, the key technologies are discussed, which is an attempt of the new information technology brought into the education system and the products.

References

1. Zhang, Y.-L., Gao, A.-H., Zhang, J.: The model of Graph Collaboration based on VRML and JAVA3D. Microcomputer Information (3) (March 2010)
2. Wang, Q.Y., Tian, L.: A systematic approach for 3D VRML model based assembly in Web-based product design. Int. J. Adv. Manuf. Technol. (33) (2007)
3. Lian, J.: A VRML Based Virtual PDS Lab. Shandong Science (2010)
4. Xiang, K.: The Key Tech of Java3D Applied in Modern Remote Education. Education Information (10) (2006)
5. Liu, S.H.: Application of Virtual Simulation Engine in 3D Fly Reappear. National Defense Technology University (12) (2004)

Principles and Methods in Teaching English with Multimedia

YiFei Zhu

School of Foreign Languages, Henan University of Technology, Lianhua Street, Zhen Zhou, Henan Province, China
hautzyf@163.com

Abstract. some principles and practical Methods about how to use multimedia in English class were elaborated in this paper. These principles and Methods were posed based on the author's teaching experience and students' feedback. The efficiency and quality of English teaching can be highly improved with the assistance of these principles and Methods.

Keywords: multimedia, English teaching, principles, methodology.

In the modern society, information has greatly influenced all aspects of people's life. As one means of information transmission, multimedia has been used in English teaching for more than a decade in China. Because of the function of transmission the text, graphics, audio, video and animation, multimedia can make the teaching process more direct, active, rapid and convenient so that it has been accepted by many university English teachers and become more and more popular. But besides the advantages, there are also some disadvantages in using. For example, some teachers use it too much but without any necessary and personal instructions; some teachers can't elaborate the key points very well with it; some teachers pore huge information into students but the students can't understand very well because of the fast speed. How to make a good use of this modern technology to improve English teaching and learning? Some basic principles and Methods should be noted in English teaching process.

1 Principles in Using Multimedia in English Teaching

1.1 Service Principle

The service principle is a guide in using multimedia in English teaching. English teachers should be very clear that multimedia is one means of assistance for English teaching but not the whole teaching. The purpose of using it is to make a good service and optimize the teaching, which request teachers should firstly classify and refine the information to insure that demonstrated in class can be the most necessary information for students. Second, when making class wares, instead of copying the contents and structures of teaching materials, teachers should make the organization according to students' cognitive regulations and features. Besides, some other important details such as the opportunities and ways of demonstration, teaching key

A. Xie & X. Huang (Eds.): Advances in Computer Science and Education, AISC 140, pp. 135–139.
springerlink.com

points and difficult points should also be taken into great consideration. English teaching is an active process in which the students are the centers, targets and teachers are organizers, instructors and supervisors. With the assistance of multimedia, students can make a selective study and teachers can give different guidance to different students. In this way students can get better teaching service from teachers' individual instructions. What's more, the use of multimedia should go around students and be able to provoke their learning interest, which requires English teachers should not only improve their traditional teaching methods but also make good use of multimedia in the teaching process.

1.2 Diversity Principle

Multimedia is superior to the traditional teaching in sound, image, color and shape. These modern factors make multimedia teaching more vivid, active and infective. It can transform those rather abstract and dull tasks such as leaning new words, phrases and grammar into vivid words, pictures and video which can stimulate students' hearing and visual senses at the same time. It is a good way to promote students' learning interest and efficiency. Besides course wares provided by the publishing house, teachers can make their own ones based on the students' practical conditions. For example, when teaching the text Smart Car (Unit 2, An Integrated English Course, Book 3), the teacher can edit some video clips about the advanced intelligent cars such as atomic car, solar energy car and etc. Students are very glad and feel curious when watching these video clips. At the same time the standard pronunciations of the new words are provided. These direct and specific video clips make an effective complement to the traditional teachers' oral and blackboard instructions. Consequently, the passive teaching and learning are transformed into active teaching and learning.

1.3 Mutuality Principle

The modern teaching should train learners' abilities of gathering, analyzing, disposing and applying to the information they need. Using multimedia is an effective way to train these capabilities. It can change the information into diversities and multi-angles but not only from the teacher to students or students to the teacher. So an optimal teaching and learning result can be realized by integrating the information from teachers, students, teaching materials and environments. English teaching and learning should be a repeated and mutual communicative process. Through the human-computer interaction instead of the only spoon-feeding education students will think, discover and quest more actively. In using multimedia teachers should take the leading role to inspire and at the same time they should leave enough freedom for students to think and study independently. For students this teaching process is a providing-and-receiving process in which both teachers and students are dynamically changing. The multimedia plays a role both as a tool for teachers' inspiration and students' independent thinking and study. The theoretical basis of this mode combines the "teacher-centered" teaching and "student-centered" teaching skillfully together.

2 Methods to Use Multimedia in English Class

As we all know, there are many advantages of multimedia. We should take a good use of it to make English teaching more effective and efficient. Based on the author's own teaching experience, a practical "five-step" method is posed. In the following, the author takes one intensive reading class (non-English major, the second year students) as an example to illustrate.

Step 1: To stimulate students' interest with vivid pictures. Interest is the most powerful motivation in students' learning. Therefore, at the beginning some interesting pictures and audio links should be presented to students for them to perceive the relative information about the text such as the background, the writing purpose, the artistic conception and so on. Compared with the information written on the blackboard or told by teachers, these vivid and colorful pictures are very attractive and students' attention can be strongly attracted.

Step 2: To set up a clear and detailed teaching purpose for each class. No matter it is the traditional background teaching or teaching with multimedia, teachers should set up a clear and detailed teaching purpose for each class and they should be very clear about the abilities that students should achieve in each class in listening, speaking, reading, writing and translating. Taught according to the clear purpose, students can make a long-standing and steady progress.

Step 3: To use multimedia properly in English teaching to focus on the key points and overcome the difficult points. It is essentially important for teachers to understand and teach teaching materials properly. The teachers' skills of solving the difficult and key points are an important factor to promote the teaching and learning effect. In most cases, multimedia will assist teachers to teach better. Properly-made multimedia teaching software will use video and audio links to imitate the similar context or more interesting situations in which students can be inspired and learn better. In this aspect, multimedia is an effective complement to the traditional teaching. Taught by multimedia students will feel very relaxed so that they can fulfill the learning tasks easily.

Step 4: Students should be centered in using multimedia. When making teaching multimedia software, teachers should pay much attention to their students' present levels and their different learning needs. Only based on these facts, the multimedia software can be suitable to the students. For example, when teaching skills in listening and speaking, after the practical investigation, teachers should make the software step by step. Besides, they can use audio and video links according to the practical context and needs.

Step 5: To use suitable teaching methods. The use of multimedia enriches the class teaching. It is able to make the instruction and demonstration in class more attractive and colorful so that students will be glad to study independently and increase their study enthusiasm and efficiency. Let's take Unit 8 in Book 3 for example to illustrate these 5 steps.

1. Revision: Using pictures in the computer to review Unit 7 which is about how a sturdy disabled salesman to sold products in a rainstorm day. When looking at those real pictures, students are greatly shocked and are very active to express their

feelings. At the same time, students are required by the teacher to use the new words like self-confidence, optimistic, strong, persistence and etc. so that they can have a good command of them.

2. Presentation: To provide some related information about the cloning technology with multimedia. Such as the information about cloning Dolly, the spiral column structure of DNA and its relationship with gene. As the information and knowledge in this unit is rather abstract, the concrete and direct pictures and video links in multimedia can help students learn and understand easily
3. Drill: Teachers can demonstrate sentence structures and the change of tense with different colors. It is very clear and understandable. What's more, demonstrating with multimedia is not that time-demanding than basketball demonstration.
4. Discuss: To gather some topics about the advanced biological technology through the internet. Let students discuss the advantages and disadvantages about cloning. Teachers should give a proper guidance in their discussion.

3 Some Points Should Be Noted in Teaching with Multimedia

The use of multimedia in English teaching greatly increases the information in class, broadens students' views, promote their learning enthusiasm and finally enhance the teaching effect. But in the use of multimedia, a couple of points should be noted. First, teachers should provide proper information to students in one class. Avoid too much at a time because students are unable to understand quite well. Second, teachers should pay attention to students' different personalities, especially to those dependent ones. More encouragements and individual instructions should be given to them to study more independently. Third, students should have a good use of computer to save time and increase efficiency. Besides, they should also be able to distinguish the information from the internet and not be obsessed with chatting or playing computer games.

4 Conclusion

The use of multimedia in English teaching is a great progress of English education. It is an effect of the renewing and development of the ideas in English education. By using the multimedia not only the teachers increase their qualities and abilities but also the students broaden their views, interest and knowledge. It is valuable to put multimedia into English education.

References

1. Bu, Y.: Using Teaching Multimedia to Promote Teaching of English Pronunciations. Media in Foreign Language Instruction (2003)
2. He, K.: Constructivism, the Basis of the Revolution of Traditional Teaching. Research of Educational Technology (1997)
3. Qin, X.: The Revolution of English Teaching Measures. Foreign Language World (1998)

4. Zhang, H.: The New Mode of English Teaching under Multimedia and Internet. Media in Foreign Language Instruction (2003)
5. Ding, Y., Chen, D.: Instructions and Evolutions of the Reach Study. WuHan Publishing House, WuHan (2005)

Research on the Difference of Attitudinal Changes between Ambivalent Consumers

Minxue Huang, Xiaoliang Feng, and Chao Wang

School of Economics and Management,
Wuhan University
Wuhan, China
ebusiness@whu.edu.cn

Abstract. This study focuses on difference of attitudinal changes between high and low ambivalent consumers. The result shows that high ambivalent consumers will change their affective attitude easily as influenced by the information; however, their cognitive and conative attitudes are difficult to be changed. Low ambivalent consumers have the opposite trends. This study explains the controversies on ambivalence research and contributes to theory as well as to management.

Keywords: Ambivalence, Affect, Cognition, Conation, Changes.

1 Introduction

There are a lot of ambivalent choices for consumers to make decision. Take examples, consumers will be hesitating about whether to buy a discount product, because it has both price appealing and quality deficit. It is also difficult for consumers to make a choice between high brand product and common goods, because of different price and prestige. Ambivalence is common among consumers and has influence on consumer behaviors. Though ambivalence is existing as far as our minds, it still is a newly research area in psychology and marketing (Otnes, Lowery, & Shrum, 1997; Conner & Sparks, 2002; Priester, Petty, & Park, 2007; Olsen, Prebensen, & Larsen, 2009).

Ambivalence is the positive and negative attitude existing independently on the attitudinal target in simultaneously, which is different from bipolar attitude. Ambivalence has moderating effect on the relationship between attitude and intention. Consumer ambivalence is the simultaneous or sequential experience of conflicting emotional states to an attitudinal object (Huang, Feng, & Xie, 2010). When consumer has attitudinal ambivalence, he or she will both like and hate the product at the same time. As different from social ambivalence, consumer ambivalence can be changed easily by the marketing environment. Seller can employ specific promotion tactics on ambivalent consumers according to different levels of ambivalence. However, there are some controversies in ambivalent area research, some scholars hold that high ambivalent person are much easier to change their attitude than low ambivalent, others find that it is easier to switch low ambivalent people' attitude. This article wants to explore the reason of controversies and study the difference of attitudinal changes between ambivalent consumers.

A. Xie & X. Huang (Eds.): Advances in Computer Science and Education, AISC 140, pp. 141–146.
springerlink.com

Based on the tripartite model of attitudes (Rosenberg & Hovaland, 1960), we conduct a 2 (high vs low ambivalence) X 2 (positive information vs negative information) between group experiment to study the process of attitude change. We find that high ambivalent consumers will change their affective attitude easily, but their cognitive and conative attitudes are difficult to change. The low ambivalent consumers have the opposite trends.

2 Literature Review

2.1 Consumer Ambivalence

Marketing research on consumer ambivalence started from Otens, Lowery and Shrum (1997), they adopted case study method to observe ambivalent attitude on wedding-related items and revealed antecedents of consumer ambivalence. They offered a definition of consumer ambivalence that is the simultaneous or sequential experience of multiple emotional states, as a result of the interaction between internal factors and external objects, people, institutions, and/or cultural phenomena in market oriented contexts, which can have direct and/or indirect ramifications on prepurchase, purchase or postpurchase attitudes and behavior (Otens, Lowery, & Shrum, 1997). Ambivalence has a negative effect on strength and stability of attitude, high ambivalent people has a low strength and stability in attitude compared to low ones (Thompson, 1996). Prior ambivalence research focused on moderation effect (Conners et al., 2002), information process (Zemborain & Johar, 2007) and so on. The difference of attitude change between ambivalent consumers is seldom studied.

2.2 Tripartite Model of Attitudes

Attitude is the evaluation about the object and hot topic noticed by the scholars and practitioners. However, the insistency between attitude and behavioral intentions is a trouble for academic research on attitude. To recognize the whole elements on attitude is the baseline to improve accuracy of research. Tripartite model of attitude (Rosenberg & Hovaland, 1960) connects the object with personal perspectives, which can offer a holistic view on attitude. It considers that attitude has three dimensions, including cognition, affect and connation. Cognition is the mental process or faculty of knowing, including aspects such as perception and judgment, which is astricted by personal ability and knowledge. Affect is a feeling or emotion of attitudinal object and can be easily influenced by environment stimuli. Connation is the behavioral intention to object, as formed by past experience and observation of others' behavior. It is difficult to change one's connation, because connation needs a lot of inputs. To understand the consumer's real attitude, we should analyze his or her cognition, affect and connation together, and then we can predicate his or her behavioral intention accurately.

Prior research on ambivalence cannot get a clear answer, whose attitude could be easily changed between ambivalent people. Because they take attitude as whole part and measure it just by like or dislike, ignoring that there are some distinguishes among attitudinal components between high and low ambivalent people. So we adopt tripartite model of attitudes to explore the difference of attitudinal changes between ambivalent persons.

3 Study

3.1 Framework

This article wants to study the differences of attitudinal components' change under influence of outside information, and to offer a holistic view on attitude change process. The research framework is as follow figure1.

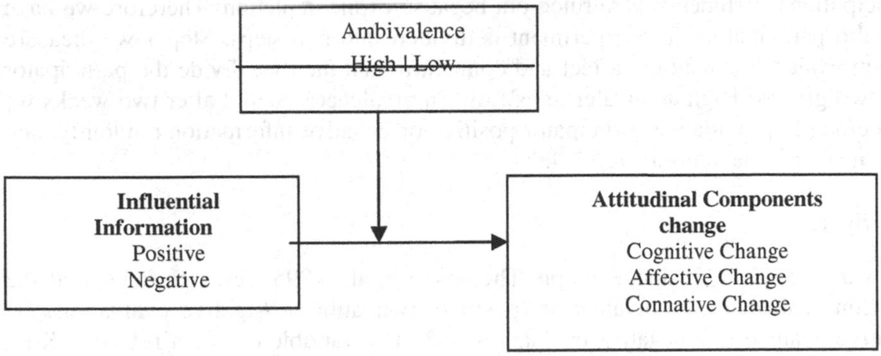

Fig. 1. Framework

3.2 Hypotheses

As lack of information or knowledge, consumer would be hesitate to make a decision and in ambivalent. Ambivalence can reduce the strength and stability of attitude (Thompson et. al., 1995). When consumer has ambivalent attitude, his or her evaluation on the object will be low. It is not difficult to understand that high ambivalent consumer's attitude will be lower than low ambivalent consumer.

H1: Compared to low ambivalent consumer, high ambivalent consumer will have a lower (a) cognitive, (b) affective and (c) conative evaluation.

The characteristics of attitudinal components are different, it is would be easier to change affect than cognition and connation. Affect can be changed just by personal feeling; however cognition and connation need a lot of inputs to change them, because they are formed on the knowledge of object elements. Cavazz and Butera (2002) find that high ambivalent people are easier to change their direct attitude and more difficult to change their indirect attitude, low ambivalent people are opposite. Direct attitude looks like affect, both are performing opinion to outside; Indirect attitude looks like cognition and connation, they are inside knowledge and opinion. As for ambivalent consumer, the motivation of high ambivalent consumer is to reduce cognitive dissonance and notice valence of information, because ambivalence brings up conflicting feeling; low ambivalent consumer cares about postpurchase risk and credibility of information. So high ambivalent consumers can easily change his or her affect, low ambivalent consumer is rational to focus on cognition and connation.

H2: Compared to low ambivalent consumers, it is much (a) easier to change high ambivalent consumer's affective attitude, and more difficult to change high ambivalent consumer's (b) cognitive and (c) conative attitude.

3.3 Methodology

We conduct an experiment to test the attitudinal change difference between ambivalent consumers. 91undergraduates were invited to take part in the investigation of stimulated purchasing 3G mobile phones. 10 yuans were given to encourage their participations. 3 students were ruled out because of incompletion. Therefore we have 88 valid participators. The experiment is divided into two steps. Step 1 we measure the ambivalence, cognition, affect and connation, and then we divide the participator into two groups (High ambivalence vs Low ambivalence). Step 2 after two weeks we used email to provide the participator positive or negative information randomly, and then measured the items as step 1 did.

3.4 Scale

The variable of ambivalence adopts Thompson et.al. (1995) research, and used the "Griffen" calculator, ambivalence= (positive evaluation + negative evaluation)/2 - |positive evaluation – negative evaluation| + 2. The variable of affect refers to Kim, Allen and Kardes (1996) research's scale. The variable of cognition used the Crites, Fabrigar and Petty (1994) work's scale, and connation adopts Peutrevu, Sanjay and Lord (1994) measurement scale.

3.5 Reliability and Validity

We check the reliability and validity in step1 and step2, both display good qualification. The KMO are 0.832 and 0.843 respectively, which is higher than 0.6. So the validity is acceptable. We check the reliability by the Cronbach's α. The information of each variable can be seen in table 1. The Cronbach'sαis higher than 0.80, which display good reliabilities.

Table 1. Information of each item

Item	Ambiv-alence	Affect	Cogniti on	Cona-tion	Mean	SD	Reliabi-lity
Ambiva-lence	—	-.39** /-.32*	-.25* /-.21*	-.49** /-.27*	4.34 /4.29	1.31 /1.25	.83/.82
Affect		—	.45** /.38*	.55** /.62**	4.69 /4.63	.87/.86	.87/.81
Cogni-tion			—	.47** /.54**	4.73 /5.06	1.12 /1.03	.85/.83
Conation				—	4.25 /4.29	1.09 /1.21	.86/.89

3.6 Hypotheses Test

Hypotheses 1 predict that high ambivalent (HA) consumers have a lower evaluation on cognition, affect and connation than low ambivalent (LA) consumers. We used F variance test to compare the difference, and find the difference of ambivalence (M_{HA}=5.40, M_{LA}=2.73, $F(1, 86)=476.09$, $p<.001$), affect (M_{HA}=4.15, M_{LA}=4.89, $F(1, 86)=22.78$, $p<.001$), cognition (M_{HA}=4.90, M_{LA}=5.21, $F(1, 86)=4.03$, $p<.04$) and connation (M_{HA}=3.92, M_{LA}=4.75, $F(1, 86)=15.42$, $p<.001$) are significantly. Hypothesis1 are supported.

In hypotheses 2 we want to prove that high ambivalent consumers will be easier to change their affect, but more difficult to change their cognition and connation than low ambivalent consumers. Positive information will enhance evaluation of attitudinal components (cognitive changes: M_{HA}=0.31, $t(21)=1.61, p>0.10$; M_{LA}=0.45, $t(20)=2.37$, $p<0.05$; affective changes: M_{HA}=0.87, $t(21)=5.76, p<0.000$; M_{LA}=0.47, $t(20)=1.55$, $p>0.14$; conative changes: M_{HA}=0.17, $t(21)=0.96, p>0.40$, M_{LA}=0.28, $t(20)=1.77$, $p<0.09$). Negative information will reduce the degree of attitudinal elements (cognitive changes: M_{HA}=-0.07, $t(22)=-0.35, p>0.71$; M_{LA}=-0.17, $t(21)=-2.34$, $p<0.05$; affective changes: M_{HA}=-0.33, $t(22)=-2.17, p<0.05$; M_{LA}=-0.10, $t(21)=-0.86$, $p>0.40$; conative changes: M_{HA}=-0.04, $t(22)=-0.28, p>0.80$, M_{LA}=-0.73, $t(21)=-2.83$, $p<0.01$). We can see from above statics that high ambivalent consumers' affective changes are significant; however their cognitive and conative changes are not significant. Low ambivalent consumers' trends are just opposite. So that hypotheses 2 are also supported.

4 Discussion

About the question that whose attitude is easier to be changed, prior research on ambivalence cannot give a clear answer. Because that they took the attitude as a whole part, and did not consider the components of attitude. So the results are conflicting. Each component has its own attribute and needed changing condition is different. Ambivalent consumers have various motivations with different kinds of ambivalence. For high ambivalent consumers, their first task is to reduce the uncomfortable cognition dissonance; low ambivalent consumers pay more attention to purchasing risk. The variety of ambivalence will lead consumers to change different attitudinal element. Our research find that high ambivalent consumers' affect are easier to be changed than low ambivalent consumers, but their cognition and connation are more difficult to change. This study offers a clear answer to the controversy and enriches the theory of ambivalence.

Acknowledgment. This research was supported by the Hubei Provincial Department of Education under Grant 060164 and the National Natural Science Foundation of China under Grant 70972091. Xiaoliang Feng is the correspondent.

References

1. Ajzen, I., Brown, T.C., Carvajal, F.: Explaining the Discrepancy Between Intentions and Actions: The Case of Hypothetical Bias in Contingent Valuation. Personality and Social Psychology Bulletin 30(9), 1108–1121 (2004)
2. Conner, M., Sparks, P., Povey, R., James, R., Shepherd, R., Armitage, C.J.: Moderator Effects of Attitudinal Ambivalence on Attitude-Behavior Relationships. European Journal of Social Psychology 32, 705–718 (2002)
3. Cavazza, N., Butebra, F.: Bending without Breaking: Examining the Role of Attitudinal Ambivalence in Resisting Persuasive Communication. European Journal of Social Psychology 38, 1–15 (2008)
4. Crites Jr., S.L., Fabrigar, L.R., Petty, R.E.: Measuring the Effective and Cognitive Properties of Attitudes: Conceptual and Methodological Issues. Personality and Social Psychology Bulletin 20, 619–634 (1994)
5. Huang, M., Feng, X., Xie, T.: New Understanding of Consumer Attitude: Dualistic Attitudinal Ambivalence. Advances in Psychological Science 18(6), 987–996 (2010)
6. Kim, J., Allen, C.T., Kardes, F.R.: An Investigation of the Mediational Mechanisms Underlying Attitudinal Conditioning. Journal of Marketing Research 33, 318–328 (1996)
7. Otnes, C., Lowrey, T.M., Shrum, L.J.: Toward an Understanding of Consumer Ambivalence. Journal of Consumer Research 24(3), 80–93 (1997)
8. Priester, J.R., Petty, R.E., Park, K.: When Univalent Ambivalence? From the Anticipation of Conflicting Reactions. Journal of Consumer Research 34(2), 11–21 (2007)
9. Putreu, S., Lord, K.R.: Comparative and Noncomparative Advertising: Attitudinal Effects Under Cognitive and Affective Involvement Conditions. Journal of Advertising 23, 77–90 (1994)
10. Rosenberg, M., Hoveland, C.: Cognitive, Affective and Behavioral Components of Attitudes. In: Rosenberg, M., et al. (eds.) Attitude Organization and Change, Yale University Press, New Haven (1960)
11. Thompson, M.M., Zanna, M.P., Griffin, D.W.: Let's Not Be Indifferent about (Attitudinal) Ambivalence. In: Petty, R.E., Krosnick, J.A. (eds.) Attitude Strength: Antecedents and Consequences, pp. 361–386. Lawrence Erlbaum, Mahwah (1995)
12. Zemborain, M.R., Johar, G.V.: Attitudinal Ambivalence and Openness to Persuasion: A Framework for Interpersonal Influence. Journal of Consumer Research 33, 506–514 (2007)

Study on Chinese Error Checking

Chongwen Wang and Bo Yuan

School of Software
Beijing Institute of Technology
Beijing, China
wcwzzw@bit.edu.cn, niumzh@163.com

Abstract. The word-level error checking in Chinese has been discussed. During words Segmentation, the algorithm is divided into two steps. Firstly, the longest match algorithm of forward heuristic, reverse backtracking and the recursive word segmentation algorithm of left and right sub-segment have been used to divide the text into more small loose strings. Secondly, the forward longest matching algorithm has been used to merge casual strings backward as far as possible, and the casual strings being segmented are the basis of error checking operation later. In the system of error detecting, an algorithm based on similar pronunciation strategy has been introduced. This strategy uses large-scale lexicon (340 millions) as the basis of data analysis. Then, error checking algorithm that based on similar shape which includes similar character table, Wubi repeat-code table, and Zhengma repeat-code table has been introduced to check character error. Experiments show satisfactory results.

Keywords: Chinese error, lexicon, similar pronunciation, similar shape.

1 Introduction

Chinese error checking, which is also known as Chinese auto-correction, is a process that computer automatically analyses electronic text, finds, identifies and corrects errors according to information contained in the language itself. Compared with traditional manual proofreading, automatic error checking using computer is only modifying but not necessarily right as there is no original. Text error detection is an important application of natural language processing area. Early in 1960s, people in foreign countries have done some research on error detecting of English text, and both IBM Watson Research Center and Stanford University did some pioneering research. For English, there are some common errors such as non-word error, real word error and semantic error in syntax. Different with English, Chinese are coded characters which need input method such as Wubi-shaped or Pinyin to input to computer ,so they exists in Chinese character library(GB2312, GBK) and there is no non-word error. The common type of Chinese error is context-related error caused by loss, replacement, insertion or translocation of Chinese character. For example, as "花前月下" and "花钱月下" have the same pronunciation, mistaking "花前月下" to "花钱月下" when using Pinyin input method. Now, there are many methods detecting word

A. Xie & X. Huang (Eds.): Advances in Computer Science and Education, AISC 140, pp. 147–154.
springerlink.com

errors in Chinese text, such as dictionary method, word shape distance method, misspelling dictionary method, the minimum edit distance method, neural network technology, similar key method, skeleton key method, rule-based method and etc, but recall rate and accuracy are low. The main content of this paper is to design a Chinese error checking algorithm based on large-scale lexicon, including two core parts: word segmentation over Chinese text and error checking over segmented string.

2 Construction of Lexicon

Lexicon in this paper, whose entry capacity and scope of knowledge determine accuracy of word segmentation and error checking, is data source table providing data to word segmentation and error detecting modules. We create GB2312 table, GBK table, main dictionary table, similar character table, Wubi repeat-code table and Zhengma repeat-code table. GB2312 and GBK table are mainly used to detect same pronunciation of different words in Pinyin input method; Main dictionary this paper uses contains about 3.4 million entries, including people name, organization name, entries of daily life and professional fields, mainly from thesaurus of many input methods. It supplies segmentation algorithm with string matching and provides error detecting algorithm standard entry for proofreading. Similar character table calculates acquaintance degree between characters after changing character to picture using image recognition. Similar table used this paper consists of the top 50 most similar characters of each character among 6763 Chinese characters in GB2323 library. Wubi repeat-code table and Zhengma repeat-code table are used to detect errors generated by inputting with shape input method.

3 Chinese Word Segmentation

Word segmentation algorithm used in this paper is based on string matching. It uses heuristic backtracking algorithm to match Chinese character string to be analyzed with main dictionary containing 3.4 million entries. Matching succeeds if finding a string in the dictionary; else continue to split up string to single characters. Basic flow of the algorithm is shown in Figure 1.

As shown in Figure 1,

a. If matching number is greater than 0 during forward heuristic search, continue; else turn to reverse backtracking search;
b. If matching number is greater than 0 during reverse backtracking search, continue; else turn to recursive of left sub-segment search;
c. If matching number is equal to 0 during recursive of left sub-segment search, continue; else turn to recursive of right sub-segment search;
d. If matching number is equal to 0 during recursive of right sub-segment search, continue; else stop text segmenting;
e. Get remain word in the remain table after word segmentation and combine it with next character of heuristic search then turn to a

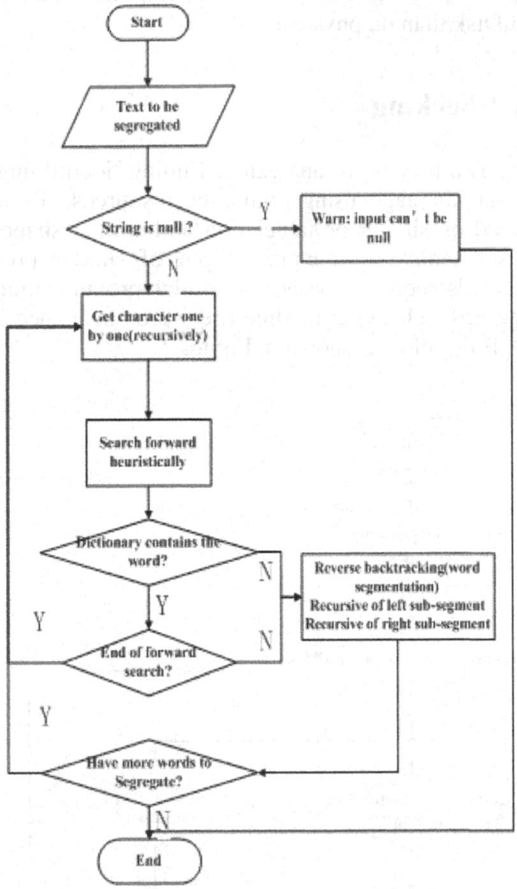

Fig. 1. Heuristic backtracking word segmentation flow chart

When Chinese text analysis ends, recursive algorithm will end. Algorithm above will try to keep the longest word string while dividing words that have ambiguities or errors into loose string or even single character. For example, when ABC is text and it can be segmented as AB, BC and ABC, it will be segmented to ABC based on the longest string match rule; If a word can match with both front and rear word, when ABC can be segregated as AB and BC, BC will be reserved and A will be segregated alone, so BC can attend recursive segmentation later.

It should be noticed that there will be merge operations to combine words segregated by mistake after word segmentation. It uses the longest backward match algorithm merging words as long as possible. It can re-combine words segregated by mistake during word segmentation to a single word.

During the implement of this algorithm recursive scan of each word are done 4 times, so word segmentation of large Chinese document will take a long time. To enhance its practical value, we use memory database to speed up progress of word segmentation solving system bottlenecks caused by database I/O. After actual test, segmentation efficiency can be improved 3 to 10 times putting data table on memory disk simulated by RamDisk than on physical disk.

4 Chinese Error Checking

Chinese error detecting is a process of analyzing, finding, identifying, and correcting errors in text of natural language using computer resources. There are two error detecting strategies based on similar pronunciation and similar shape. With character sequence after word segmentation as input, output of similar pronunciation error checking module are words sequence based on similar pronunciation proofread, and output of similar shape error checking module are words sequence based on similar shape proofread. Work flow below is shown in Figure 2.

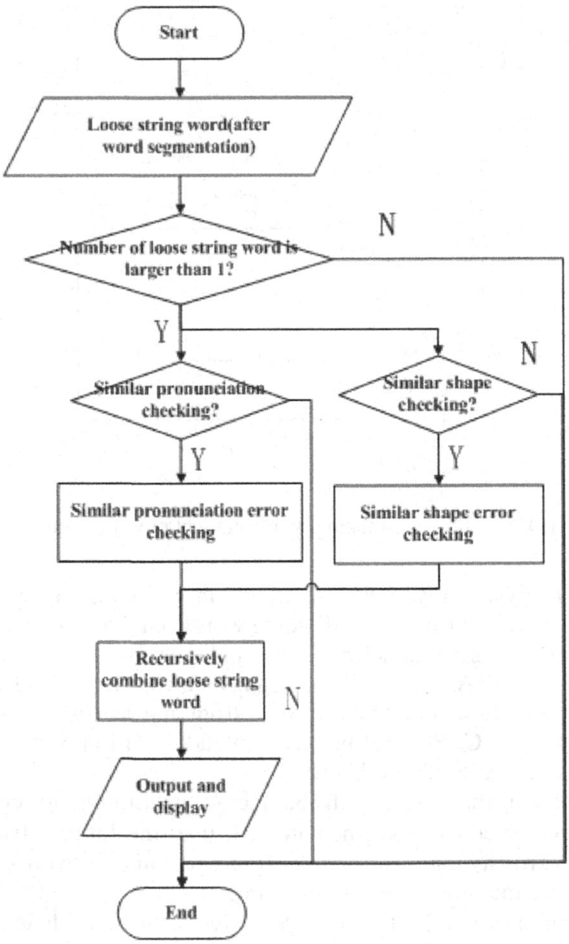

Fig. 2. Work flow of Chinese error checking

4.1 Strategy Based on Similar Pronunciation

It is common for us inputting a wrong character with the same pronunciation using Pinyin input method as there are a lot of same pronunciation characters with same initial, final, and intonation. Error checking based on similar pronunciation is to solve such problems, and similar pronunciation here means similar pronunciation characters with same initial, final but different intonations.

Error checking based on similar pronunciation is one of the most important error checking methods of Chinese text. Process is shown in Figure 3.

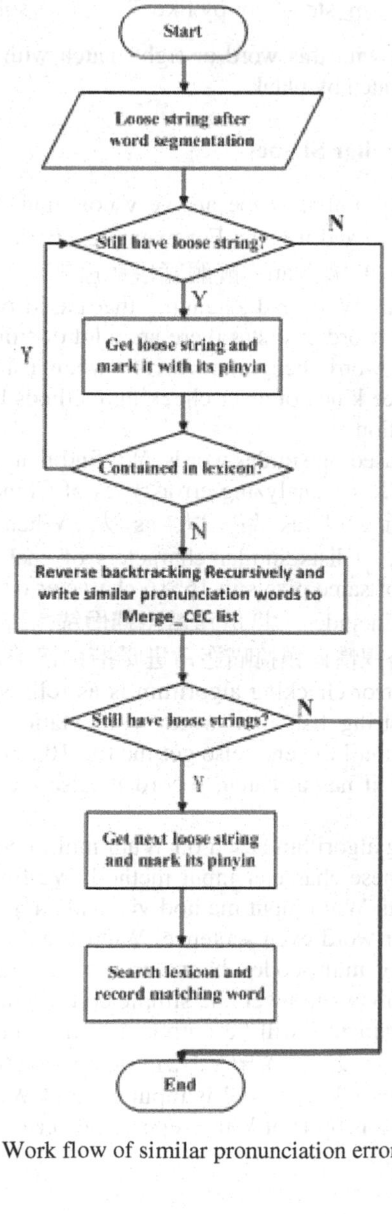

Fig. 3. Work flow of similar pronunciation error checking

Similar pronunciation checking program fist gets loose string from list and mark it with its pinyin. If it can be right include (same as right match, for example, wo meng is right include by womeng de zu guo), repeat get string until the end; else search similar pronunciations recursively and record all words have the same pronunciation. Then recursive to the beginning of program if there is more loose strings.

In Mysql right include is achieved by using % wildcard:

sqlstr = "select term from "+ dict_file + " where py = '"+ word_sql_str + "' or py like '"+ (word_sql_str + " ") + "%' limit 0, 1" ;

There is a where condition in sql statement of right include :

" where py = '"+ word_sql_str + "' or py like '"+ (word_sql_str + " ")

It means completely match with this word or right match with this word added with blank as all pinyin are separated by blank.

4.2 Strategy Based on Similar Shape

In Chinese, characters with similar shape are very common. Similar characters are words that have similar shape and written. For example, similar words of '哀' are '衷衰袁衮柬'; similar words of '舨' are '舰舸艋艉艄廒舻叩船舭艏胀畈舳'. When using input methods such as Wubi and Zhengma that based on shape of characters, there will always be similar word errors as there are a lot of similar words in Chinese; also there will be similar word recognition errors when using OCR to recognize characters. We propose three kinds of error checking methods based on similar shape according to different situations.

First is error checking based on similar words. We find that part of errors is caused during recognizition of OCR via analyzing error types of Chinese text. For example, OCR recognizes '巾' as '币' , '晶' as '品' , '大' as '太'. When building lexicon there is a similar character table, it lists similar characters of each characters measuring similarity with proportion of same pixels in 16*16 character dot-matrix. For example, similar characters of '错' include '惜借镨谱糌锈愶锚褙皆诌谐镥崔锌锫褶诺冉造钮锘馇锴偌偕僧增铭悟镍惰谵镇掎喵迮锖迸喏揩锋掊诰镤绢逞镒锆锵'. Work flow of similar character error checking algorithm is as follows: Program get loose string one by one from string list after word segmentation. If it is not a single character, repeat get string until the end; else get the top 10 most similar characters to match next loose string. If it has a match, record it ;else recursive to beginning of program.

Second is error checking algorithm based on Wubi multi-code. Wubi input method is a widely used rapid Chinese character input method. We find that part of errors is caused when user input with Wubi input method via analyzing error types of Chinese text. For every character or word even sentence, Wubi uses four codes to input, and some code will have many multi-code characters. For example, 25 common used characters correspond to 25 keys in level one simple code. When we input such code, many other repeat-code characters will be output. When input 'q' using Wubi input method, '我 、 金 、 多 、 爻 、 希望 ...' will output. Wubi repeat-code table has three fields, 'id' is the primary key, 'code' is input code of Wubi, 'term' is character corresponding to the code. It is basis of Wubi repeat-code error checking. The process is as follows: Wubi repeat-code error checking program gets loose string characters from loose character list after word segmentation one by one. If string is a single character, repeat getting loose string characters until the end; else get Wubi repeat-code characters of the single word to match next loose string. If it is matched then record the character, else recursive to beginning of the program.

The third is error checking algorithm based on Zhengma repeat-code. Zhengma code is 26 yards per pure image code, its basic principle of coding is a search sequence such as "stroke - root - character - word". Repeat-code characters of Zhengma input method are mainly similar shape Chinese characters. For example, repeat-code characters of 'pvgx' are '爱' , '爰'; repeat-code characters of 'rsuo' are '灸' , '炙'. Zhengma repeat-code table also has three fields, 'id' is the primary key, 'code' is Zhengma input code , 'term' is Chinese character responding to the code. Characters with same code will lead to repeat-code error. For example, repeat-code characters of 'xmrr' are '尼', '屁'. These repeat-code characters are disorderly arranged, so we have to get all to check. Process of Zhengma repeat-code error checking algorithm is as follows: Zhengma repeat-code error checking program get loose strings one by one. If it is not a single character, repeat getting loose strings until the end; else record the Zhengma code of it and match it with next loose string. If it is matched then record it, else recursive to the beginning of program.

5 Experiments and Results

As a limit of time, we do not experiment on large samples. After a number of small sample experiments, accuracy of word segmentation is around 95%, recall rate of error checking is larger than 90%, precision rate is about 60-70%.It has some advantages compared with other Chinese error checking method. It will have a broad market prospect as it can proofread text before post of new content saving the cost of manual proofread in industries based on electronic document such as press, publishing, electronic books, electronic newspapers. In future we will do research on storage of main dictionary, error checking of flat cocky, before and after nasal, polyphone error to improve the efficiency of this method.

References

[1] Song, Y., Cai, D., Zhang, G., Zhao, H.: Apply Normalized Accessor Variety in Chinese Word Segmentation. Journal of Chinese Information Processing 20(9) (2009)

[2] He, S., Wang, X., Dong, Y., ZhangT., B.X.: Apply Normalized Accessor Variety in Chinese Word Segmentation. Journal of Chinese Information Processing 24(1) (2010)

[3] Liu, C.-L., Lai, M.-H., Tien, K.-W., Chuang, Y.-H.: Visually and phonologically similar characters in incorrect chinese words: Analyses, identification, and applications. ACM Transactions on Asian Language Information Processing 10(2) (June 2011)

[4] Chen, Y.-Z., Wu, S.-H., Yang, P.-C.: Improve the detection of improperly used Chinese characters in students' essays with error model. International Journal of Continuing Engineering Education and Life-Long Learning 21(1), 103–116 (2011)

[5] Zhu, J., Zhang, Y.: Research and implementation on a hybrid algorithm for Chinese automatic error-detecting. In: Proceedings - International Conference on Artificial Intelligence and Computational Intelligence, AICI 2010, vol. 1, pp. 413–417 (2010)
[6] Chiu, D.-Y., Lee, C.-C., Pan, Y.-C.: An automated error detection for news webpages of chinese portal. Journal of Software 5(12), 1334–1341 (2010)

Double-Sources Dijkstra Algorithm within Typical Urban Road Networks

Shiming Wang[1], Jianping Xing[1,*], Yong Wu[2], Yubing Wu[2], Wei Xu[1], Xiangzhan Meng[1], and Liang Gao[1]

[1] School of Information Science and Engineering, Shandong University, 250100 Jinan, China
[2] Center of Bus Transmit Information, Public Traffic General Company, 250100 Jinan, China
wangshiming1191@yahoo.cn, xingjp@sdu.edu.cn

Abstract. A Double-Sources shortest path algorithm for urban road networks is proposed in this paper. In typical urban road networks, the probability that the ratio of the shortest path length to the Euclidean distance denoted by $|SD|$ between source station and destination station is smaller than 1.414, is larger than 95%. Based on Dijkstra algorithm and the characteristics of the typical urban road networks, this algorithm starts at searching for the shortest path from the source station and destination station respectively and simultaneously and ends at having found all stations which are less than $0.702|SD|$ far from the source station or destination station. Compared to the single-source Dijkstra algorithm, theory analysis and experimental results both show that the algorithm can great reduce the time-complexity, especially on condition that stations in urban road networks uniformly distribute.

Keywords: Dijkstra algorithm, Typical urban road networks, Double-sources shortest path algorithm, Restricted searching area.

1 Introduction

The Dijkstra algorithm [1]-[2], is a classical shortest path algorithm in city road networks and solves the single-source shortest path problem, the time-complexity of which is in direct proportion to the square of the stations' number[3]. Assuming that the stations' number is in direct proportion to the size of the searching area, one method of solving the problem is reducing the size of the searching area. Since the shortest path algorithm within ellipse restricted searching area (SPAERA) was first put forward in 1969 [4], there have appeared all kinds of optimization algorithms within restricted searching area.

Since when judging a station whether or not in the ellipse restricted area, lots of involutions and evolutions that cost much time are needed. A shortest path algorithm within rectangle searching area (SPARRA) is introduced in Ref [5]. However, the

* Corresponding author.

A. Xie & X. Huang (Eds.): Advances in Computer Science and Education, AISC 140, pp. 155–161.
springerlink.com

realization of SPAERA and SPARRA are both dependent on the statistical information of the road networks.

In Ref [6], a universal algorithm for typical urban road networks is proposed. This algorithm searches for the shortest path in an ellipse whose major axis length is 1.414|*SD*| while |*SD*|>8km, and 1.273|*SD*| while |*SD*|≤8km. However, the boundary, i.e.8km, of using the two kinds of ellipses, is not explained clearly. Besides, in urban road networks, especially for those relatively small cities, the Euclidean distance from the source to destination is usually smaller than 8km.

There are also other shortest path algorithms within restricted searching area. Some directly limits the searching area in a rectangle which contains the source station and destination station [7] and some other algorithm limits the searching area in a polygon [8].There is also a shortest route-planning algorithm within dynamic restricted searching area [9].

As the algorithms above are all based on Dijkstra algorithm and search for the shortest path from the single source station, here we denote this kind of algorithms as single-source Dijkstra algorithm (SSDA). In this paper we give a double-sources algorithm for typical urban road networks (DSDA). Typical urban road networks are first explained and then the double-sources algorithm will be described in detail. After that, we will analyze the performance of this algorithm in theory. At last, experimental results show that this algorithm can greatly reduce the time-complexity of the shortest path algorithm.

2 Typical Urban Road Networks

In this paper, a road network with the following characteristics is called a typical urban road network: The roads are crisscrossed and the longitude-latitude are relatively clear; There are no natural factors, such as large mountains or long rivers, which can greatly influence the distribution of a city road network; Stations uniformly distribute in the road networks and the stations' number is in direct proportion to the size of the searching area [10].

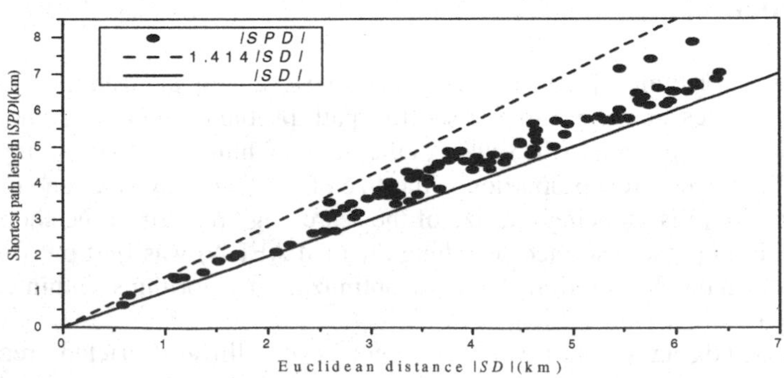

Fig. 1. The shortest path ratio within typical urban road networks

According to Ref [7], in typical urban road networks, given a source station *S* and a destination station *D*, let $|SD|$ and $|SPD|$ denote the Euclidean distance and the shortest path length from *S* to *D* respectively. Then the probability of $1.414|SD|>|SPD|$ is larger than 95%.

As one of ten initial national intelligent transportation pilot cities, Jinan has all characteristics of urban road networks. Here we define the ratio of shortest path length to the Euclidean distance as the shortest path ratio and denote r_0. Fig. 1 gives the value range of r_0 for 100 different Euclidean distances within urban road networks and it can be easily seen that the value of r_0 is between 1 and 1.414. That's to see, given source station *S* and destination station *D*, the shortest path length $|SPD|$ should be smaller than $1.414|SD|$.

3 Realization of DSDA

Considering the characteristics of typical urban road networks analyzed above, the flow chart of DSDA realization is displayed in Fig.2 and the realization steps of DSDA are shown below:

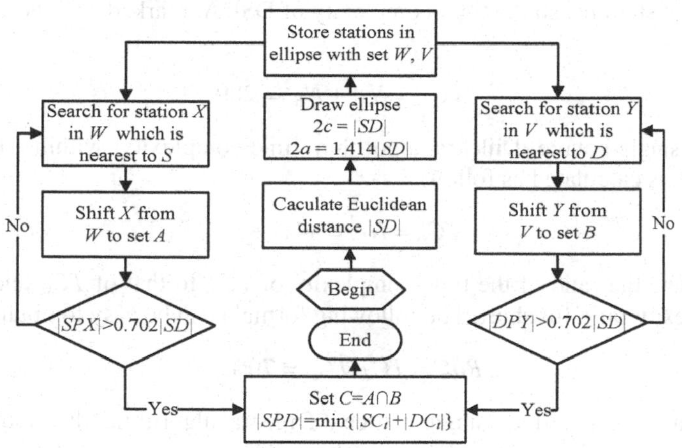

Fig. 2. The realization flow chart of DSDA

(1) Given a source station *S* and a destination station *D*. First, calculate the Euclidean distance $|SD|$ between *S* and *D*.

(2) Draw an ellipse with $|SD|$ as its focal length and $1.414|SD|$ as its major axis length.

(3) Search for the shortest, 2th shortest, … , *i*th shortest station(marked $A_1, A_2, \ldots, A_i$) from station *S* until $|SA_i|>0.702|SD|$; Simultaneously , search for the shortest, second shortest, third shortest, … , *j*th shortest station(marked $B_1, B_2, \ldots, B_j$) from station *D* until $|DB_j|>0.702|SD|$; Mark set $A = \{ A_1, A_2, \ldots, A_i \}$ and set $B = \{ B_1, B_2, \ldots, B_j \}$.

(4) Find the common stations of set A and set B and mark set $C = A \cap B = \{C_1, C_2, \ldots, C_k\}$, find the minimum of $(|SC_t|+|DC_t|)$ as $|SPD|$ and the corresponding path, which begins at S, goes throuth C_t and ends at D, as the final shortest path from S to D.

4 Analysis of DSDA

Take the road network shown in Fig. 3 as the example, we will analyze the effectiveness of the algorithm in theory.

In Fig. 3, the source station S is numbered 480 and the destination D 506;

Stations within the ellipse area are put in set E. E = {480, 471, 481, 482, 470, 485, 479, 486, 477, 494, 487, 495, 630, 490, 488, 493, 498, 499, 489, 497, 506} and the size of set E, marked N_E is 21.

Set A defined in part 3 is {480, 471, 481, 482, 470, 485, 479, 486, 477, 494, 487, 495, 630, 490, 488} and the size of set A, marked N_A is 15..

Set B defined in part 3 is { 506, 493, 498, 499, 489, 488, 497, 487} and the size of set B, marked N_B is 8.

Set C defined in part 3 is { 488, 487} and the size of set C is 2.

As the time-complexity of Dijkstra algorithm is in direct proportion to the square of the number of stations, so the time-complexity of DSDA, marked TC_{ds} is calculated as follows.

$$TC_{ds} = N_A^2 + N_B^2 = 289 \tag{1}$$

While the single-source Dijkstra algorithm time-complexity within ellipse area, marked TC_{ss}, is calculated as follows.

$$TC_{ss} = N_E^2 = 441 \tag{2}$$

let RoC denote the ratio of the time-complexity of TC_{ds} to that of TC_{ss} and call it the time-complexity ratio for short. The following formula can be easy to obtain.

$$RoC = TC_{ds}/TC_{ss} = 70\% \tag{3}$$

That's to say, compared to single-source Dijkstra algorithm, the double-sources algorithm reduces the time-complexity by about 30%.

When the source station is far from the destination station, which means that there are more stations in the ellipse area, assuming that the number of stations is in direct proportion to the size of the restricted searching area, the single–source Dijkstra algorithm's searching area is nearly the whole ellipse area whose area value is $\pi*a*b$, and the double–sources Dijkstra algorithm's searching areas are about two semi-ellipse areas, each area value of which is $\pi*a*b/2$, so the time-complexity ratio can be caculated in theory.

$$RoC = (\pi*a*b/2+\pi*a*b/2)^2/(\pi*a*b)^2 = 50\% \tag{4}$$

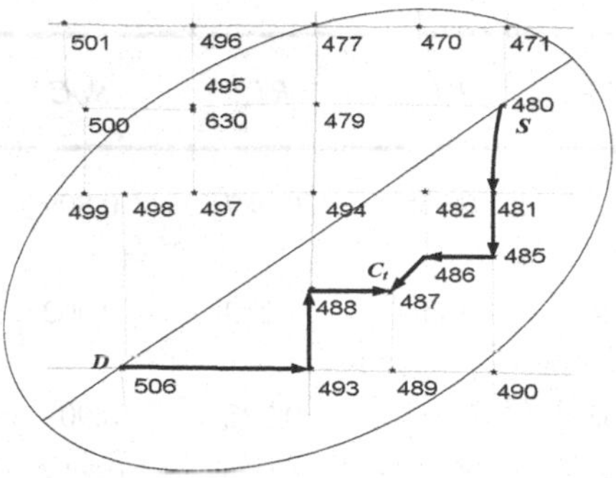

Fig. 3. Sketch map of DSDA

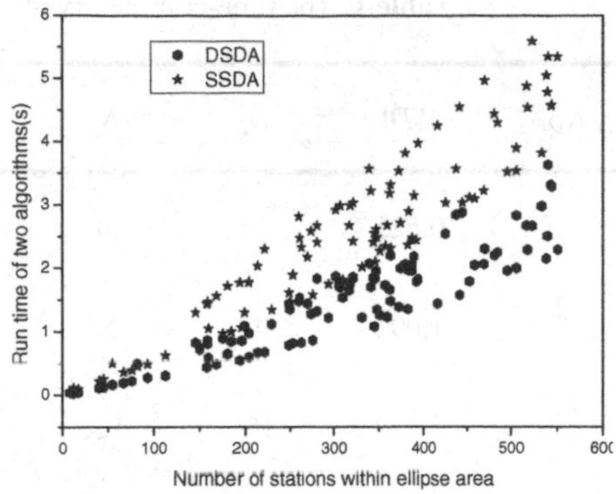

Fig. 4. Run-time of two algorithms

5 Experiment Validation

In this section, we will take Jinan as the example and compare the efficiency of SSDA and DSDA within restricted area through measuring run-time (*RT*) of different Euclidean distances. The computer program for implementing these two algorithms was coded in NetBeans IDE 6.5.1 and the program was run on Lenovo notebook PC y450a-TSI. Run-time data of two algorithms for different N_E is shown in Fig. 4 and part experimental data is displayed in Table 1.

From Fig. 4, It's can be seen that the query time of DSDA is overall shorter than that of SSDA and with the stations' number in the ellipse area increasing, the run-time reduction is more obvious and clear.

Table 1 gives 10 groups of data extracted from Fig. 4. All *RoC* values are less than 1, which means that the time-complexity of DSDA is lower than that of SSDA. Statistical result shows that the average *RoC* value of all 100 groups of data shown in Fig. 4 is 60.4%. The reasons for experimental result (60.4%) surpasses theory value (50%) are as follows. First, actually the restricted searching area of SSDA is smaller than $\pi*a*b$ and the restricted searching area of DSDA is larger than $\pi*a*b/2$, so the theory value of *RoC* should larger than 50%; Second, the statistical data amount is also not enough; Besides, stations in urban networks don't strictly uniformly distribute. For example, the N_A value and N_B Value of the first group data have big differences shows that stations don't uniformly distribute and as a result the *RoC* value is much larger; So is the third group data.

Table 1. The time-complexity ratio for different Euclidean distances

No.	$\|SD\|$	N_E	N_A	N_B	RT_{ds}	RT_{ss}	*RoC*
1	0.525	11	4	10	0.038	0.047	0.809
2	1.093	47	22	20	0.110	0.219	0.502
3	2.432	159	122	33	0.711	0.782	0.909
4	2.642	175	107	88	0.820	1.297	0.632
5	3.073	232	135	121	0.976	1.782	0.548
6	3.297	349	173	194	1.648	2.672	0.617
7	4.478	436	245	213	1.983	3.531	0.562
8	5.449	499	295	255	2.296	4.069	0.564
9	5.962	540	278	323	2.664	4.875	0.546
10	6.392	576	347	305	3.625	5.344	0.678

6 Conclusion

Based on the analysis of the single-source Dijkstra algorithm and the shortest path ratio in typical road networks, this paper presents a double-sources shortest path algorithm. Because this algorithm searches for the shortest path both from the source station and destination station at the same time, while the time-complexity of Dijkstra algorithm is in direct proportion to the square of the stations'number, this algorithm can greatly reduce the algorithm time-complexity. Experiment results also show that this algorithm performs well, especially when stations in road networks distribute uniformly.

Acknowledgments. This work is funded by Program for New Century Excellent Talents in University (Grant No.NCET-08-0333) and Independent Innovation Foundation of Shandong University (Grant No.2010JC011).

References

1. Dijkstra, E.W.: A note on two problems in connection with graphs. Numerische Math. 1, 269–271 (1959)
2. Ma, D.L.: Research of the shortest path algorithm of urbanized public traffic networks. Science & Technology Information 26, 15 (2008)
3. Fu, M.Y., Li, J., Deng, Z.H.: A Practical Route planning Algorithm for Vehicle Navigation System. In: Proceedings of the 5th World Congress on Intelligent Control and Automation, p. 5327. WCICA, Hangzhou (2004)
4. Nordbeck, S., Rystedt, B.: Computer Cartography: Shortest route Programs. The Royal University of Lund, Sweden (1969)
5. Lu, F., Lu, D.M., Cui, W.H.: Time Shortest Path Algorithm for Restricted Searching Area in Transportation Networks. Journal of Image and Graphics 4(10), 849–853 (1999)
6. Wang, S.M., Xing, J.P., Zhang, Y.T., Bai, B.H.: Ellipse-based shortest path algorithm for typical urban road networks. Systems Engineering Theory & Practice 31(6), 1158–1164 (2011)
7. Wang, Y.W., Wang, X.L., Cao, H.: Shortest Route-planning Algorithm within Dynamic Restricted Searching Area. Application Research of Computers 24(7), 89–91 (2007)
8. Fu, M.Y., Li, J., Deng, Z.H.: A Route Planning Algorithm for the Shortest Distance within a Restricted Searching Area. Transactions of Beijing Institute of Technology 24(10), 882 (2004)
9. Zhou, Y., Cao, J., Li, J.X.: A Shortest Route-Planning Algorithm within a Restricted Area. Microelectronics & Computer 24(8), 110–112 (2007)
10. Divitin, M., Simone, C.: Supporting Different Dimensions of Adaptability in Workflow Modeling. Computer Supported Cooperative Work 9(3/4), 365–397 (2000)

Development and Practice of Knowledge Service Platform Based on DSpace

Lu Han[1,*] and Yi Ding[2]

[1] Beijing Institute of Technology's Library, Beijing, 100081
hanlu311@bit.edu.cn
[2] China University of Geosciences, Beijing, 100083

Abstract. In this paper, a knowledge service platform according with the features of universities is established after second development and revision of DSpace. The main development process is described in detail, which includes the specification of metadata, the encapsulation of core service based on SRU, the OAI-based metadata harvesting and the user-centered service optimization.

Keywords: DSpace knowledge service, SRU, OAI, metadata harvest.

1 Introduction

More manners for science communication are required in the institution for scientific research with the development of knowledge and the increase of new scientific results. In China, the institution repositories are popular which has already become the primary carrier for science communication in university. However, there are few institutions to implement and the efficiency is very low although the institution repositories are established. Most institution repositories aim to save resources for long time and to access for public. But, with the improvement of information service level, the information service is not a simple resource search, but knowledge organization, classification and custom of users. The previous institution repository could not satisfy the needs of scientific researchers, and it is just a sole platform in most applications without combination with the entire environment of numerical library. We firstly have analyzed the difference between the function and service of institution repository and actual requirement of university with the view of technology, and secondly carried out the practice in our school.

2 Implement Scheme and Entire Design of System

DSpace[1] is applied widely in the field of numerical repository construction due to its relative complete functions, easy setup and customization. DSpace was developed

* Corresponding author at: Beijing Institute of Technology's Library
Address: 5 South Zhongguancun Street, Haidian District, Beijing 100081, P.R. China
Tel.: +86 01068918418
hanlu311@bit.edu.cn

according to the Apache basic protocol by MIT and HP Co. in 2002, which is a kind of complete open-source software. DSpace also calls for a great number of other open-source works, for example, its system framework adopted the OAIS reference model proposed by American CCSDS (Consultation Committee of Space Dada System). According to the statistics of OpenDOAR, there is almost one third open numerical repositories constructed referring to DSpace. Therefore, we have designed and improved a platform for knowledge service in our study.

The system designed by us also adopts the original three shells of DSpace which uses layered encapsulation and call-for mechanism ensuring the logic integrality and the system expansibility. We need to carry out expansion in the corresponding layer during the second development. The three-layered framework of DSpace respectively includes the storage layer, the operation logic layer and the application layer. We have improved this system according to this framework (Fig. 1).

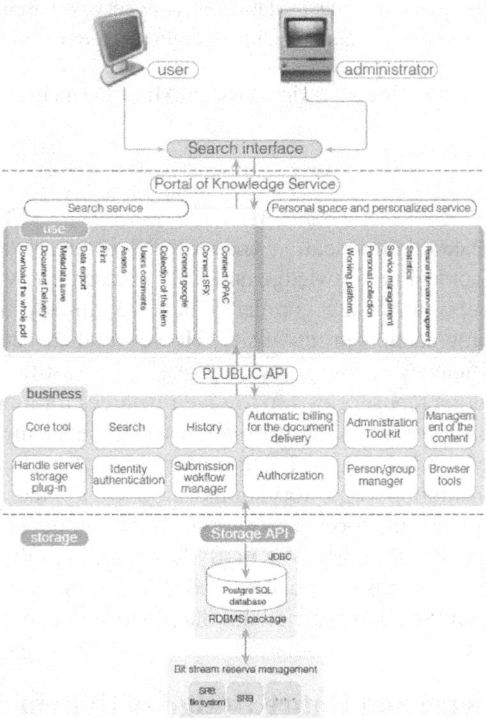

Fig. 1. The structure diagram of knowledge service platform after improvement

The storage layer is mainly in charge of physical storage. DSpace adopts the SRB (Storage Resource Broker) manner, which is available for operating background database and storing data flow. The storage of data flow is processed by the API at the bottom layer and every kind of data has different driver and storage mechanism.

The operation logic layer mainly contains a few system operation modules which supply public API for the application layer (Figure 1). We have added the data statistic module and the raw document transfer module. The application layer is an interaction interface with users, for example browse, search, batch input of data and output, and so on. The application expansion is realized by resetting or by calling for the API of the operation logic layer. The service style of application layer is very abundant. The user with different level possesses of different authorization to literature. We combined all services of application layer and established the scene model. The information system is driven by the scene sensitivity and guides users to use the service within their authorization. Figure 1 shows that there are two parts in the application layer, one is combined with resource search, and the other is the personal space of user. The institution repository constructed by DSpace is a service platform with the B/S structure and adopts the JSP+Sevlet to realize the interaction between the user and the system at the client terminal. The Sevlet focuses on attempering kinds of claim and calls for the corresponding public API of the operation logic layer. On the whole, the second development is accomplished at the operation logic layer and the application layer.

3 Key Problems and Their Solutions

There exists some common problems with the service platform of institution repository in the information environment of university. We have proposed their corresponding solutions as following:

1) Most institution repositories pay attention to the transfer of resources but ignore the classification of resources. Therefore, the classification according to the style of literature is more important than the organization of resources according to institutions as establishing the platform. For example, at present, the existing classifications mainly include the conference paper, the academic paper, the electric book and the study report of project. DSpace considers the DC (Dublin Core) metadata as the uniform metadata description template to describe the resource[2], however it has a huge limitation to separately describe different literature style especially the conference paper style. We have expanded the metadata of some literature styles mentioned above based on the DC metadata referring to the study result of CDLS (Chinese Data Library Standard) and established a metadata criterion to form different metadata description template for different literature styles. After that, the user can submit the data by choosing the right metadata template to describe the resource according to the literature styles.

2) The institution repository in high schools is not a sole platform and usually embedded into the entire digital library sestem to play its role efficiently. In addition, the information systems in some schools need to integrate the resource service. We have separately encapsulated the core search functions with the Web Service based on the SRU search protocol and embedded the search into other information system to carry out the search service by using the DSpace and SRU interfaces.

3) Most institution repositories belong to open-access resource which content and forms are correspondingly concentrated and they always support the protocol of OAI-PMH with lower level. Therefore it is available to harvest the metadata because it could magnify the present resource and realize the search in different database. Except

that, it is also useful to resolve the metadata to search and index the whole paper. We have developed a harvest tool based on the OAI by using the OAIharvester which is a kind of open-source software.

3.1 Constituting the Metadata Criterion

It is important to set the metadata criterion as constructing the institution repository. The uniform metadata can increase the operation ability between different databases. For this reason, a lot of countries established the uniform metadata when they practiced the institution repository, for instance, the search engine of the CARL Harvester in Canada[3] and the JAIRO in Japan[4]. Consequently, it is important to unify the metadata for founding an institution alliance or carry out the transverse harvest between different institutions in the future. The metadata of DSpace adopt the DC core metadata which is a standard of information resource description in different fields. Although it contains 15 core elements in the Version 1.1, all these core elements however could not satisfy all requirements for describing metadata. Hence, we have expanded the core elements. The principle of expansion is that the ornament word could be directly used if it has already supplied the expanded content in the Version 1.1 or some other elements should be added if not. We referred to Special Data Object Description Metadata Criterion of Numerical Library Standard Criterion in China published by CDLS [5]to constitute our metadata criterion. We have constituted templates for every kind of criterion and registered them in the system. The user just chooses the metadata template according to the literature style when he or she submits resource. At present, this service platform using this criterion is only used in our university and the correlative groups.

3.2 Web Service Package Based on SRU

In libraries of university there may be a lot of database platforms. From the need of integrating the database into a digital library management system for data interaction, the main integrating methods include two types: one is based on metadata; the other is based on system.

The data are usually stored in a distributed heterogeneous system and have not unified data structure, which however causes that the data interaction based on metadata is just partial .Therefore we could not adopt the first method, and only use the second method to embed the core service of the institution repository into the digital library management system. In this paper we packed the core search function into a web service based on SRU, and used the web service to combine the core function of institution repository into the digital library management system. SRU[6] is a search agreement to the internet environment, and is a simplified version of the Z39.50. SRU realized the standardization of web query and the structured query result. For this reason it is used widely in the retrieval service areas. We have presented the main development steps as following:

1) Packaging the core search function into a web service
   a. Packaging the public search API of Dspace into a Web service and providing a uniform standard search interface.

b. Converting the search condition into parameters and returning a string with SRU response schema.

2) Constructing the application based on SRU: conversion of query sentences, search, formatting of XML .

c. Converting the URL request containing query expression to SRU query sentences.

d. Obtaining response result by using web service.

e. Converting the result into XML format according to specific metadata schema.

After packaged, the core function of Dspace could be embedded into the web service with such kinds of application patterns as Gadget/Widget or JSHTML.

3.3 Development of Metadata Harvesting Tool Based on OAI

In the present paper the development of metadata harvesting tool is based on OAIHarvester 2.0[9], which is a Java application that provides an OAI-PMH harvester framework. Its main function is to get open metadata from digital repository which supports the OAI-PMH protocol. OAIHarvester 2.0 sends the http request to data providers and the data providers will return a XML document according to request content. The tool currently is applied in the related institutions which adopt the same metadata criterion. We have presented the main development steps as following:

1) Harvesting metadata: assigning URL of data providers, then marking the harvesting process by using the start timestamp and the end timestamp.

2) Parsing XML document: metadata harvested from different providers are various, so we have to parse the XML document and map parsed metadata with local metadata.

When harvesting metadata, the repeated elements will be mapped to corresponding local elements and qualifiers. During the process of realization of the data parse, the important thing is to analyze the URL for the full text, and then the full text could be correctly obtained.

4 Implement of User Context-Sensitive Information Service

Constructing information system according to the habit of user is the main trend of information service. In this paper, the institutional repository stores not only open access resource but also a lot of other types of literature, such as original reports, conference papers, and academic journals. So some literatures can be accessed freely, and others need document delivery, or borrowed in the library. In order to simplify user operations in a complex information environment, we construct a user context-sensitive model (Figure2) to correctly guide users to choose all best resource service according to comprehensive factors such as user permissions and resources type etc. The user context-sensitive model includes the following several parts:

1) User information: registration information, login information, and identity;

2) Resources information: authorization of resources classified and open access by managing its metadata;

3) Strategy of service: all services including PDF download, document delivery, OPAC link in library of university, metadata saving, data export, resources evaluation, etc.

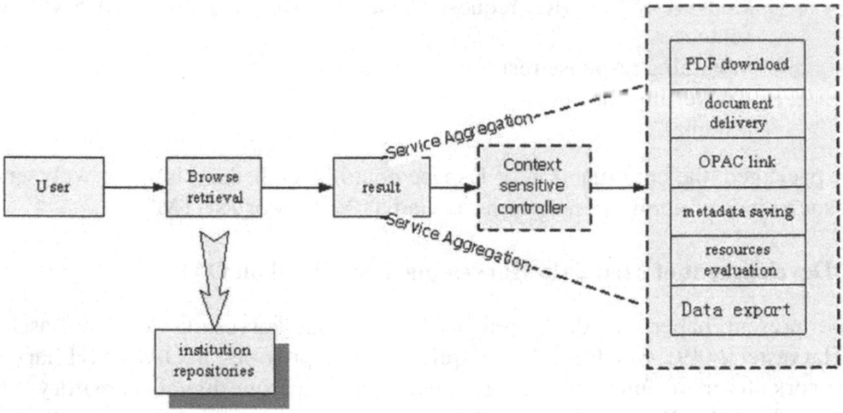

Fig. 2. User context-sensitive model

5 Conclusions

We have ultimately constructed a new knowledge service platform by the second development and improvement of DSpace. The new platform is able to organize the resources according to the literature type and to store the metadata harvested from other institutions into the database according to their classes. The uniform standard of metadata establishes a base for harvesting data and founding a unit institution. The package of core service based on SRU enables the system conveniently to be embedded into other information environments in the universities which availably generalize the application of core service provided in our new platform. During the development of application layer, we have paid more attentions to adopt the service pattern of focusing on the user and to realize the available match between the resource, the service and the user by using the context-sensitive model and the service combination.

Acknowledgment. We thank the financial support by the Ministry of Education P. R. China, under Grant 09YJC870005.

References

1. Tansley, R., Bass, M., Stuve, D., et al.: The DSpace institutional digital repository system:current functionality. In: Proceedings 2003 Joint Conference on Digital Libraries, pp. 87–97 (2003)
2. `http://dublincore.org`

3. Shearer, K.: The CARL institutional repositories project: A collaborative approach to addressing the challenges of IRs in Canada. Library HiTech 24(2), 165–172 (2006), doi:10.1108/07378830610669547
4. Zhu, L., Liu, C.: Status uo of the construction of the Japanese institutional repository. Information Studies:Theory & Application 32(8), 125–128 (2009)
5. http://cdls.nstl.gov.cn
6. searchRetrieve Operation:Binding for SRU 1.2, http://www.loc.gov/standards/sru/oasis/

Shearer, [illegible] C.A.R.: Institutional constraints [illegible] adaptive approaches addressing the challenges of [illegible] Change [illegible] Hazards 24(4) [illegible]
doi: 10.[illegible]
Zhang, [illegible] Status [illegible] information [illegible] Studies: Theory & Application [illegible]
[illegible]
[illegible]

Application of Hopfield Network in Grayscale Image Recognition

Li Tu, Zhiping Liao, and Chi Zhang

Department of Computer Science, Hunan City University, Yiyang, Hunan, 413000, China
tulip1907@163.com, 729627398@QQ.com, carol_zc@126.com

Abstract. The applications of neural associative memories in image retrieval are studied based on a class of reduced Cohen-Grossberg neural networks and continuous-time Hopfield network. Numerical simulations show that the designed networks can perform as efficient noise-reducing systems.

Keywords: reduced Cohen-Grossberg neural network, neural associative memory, image retrieval.

1 Introduction

Now the studying on neural memory mostly concentrates in binary model of memory and storage,however,the binary model of memory storage is just a small part of function in the human brain,most of the human brain memory modes are color image modes and non-binary modes.Simulating the processing of human brain on grayscale and color images has important biological significance.

2 Network Model

A class of reduced Cohen-Grossberg neural networks were uesd:

$$\dot{x}_i = a_i(x_i)\left[b_i(x_i) + \sum_{j=1}^{n} t_{ij} g_j(x_j) + \theta_i \right], i = 1,2,..n \tag{1}$$

And $x_i \in R$ is the state of i-th neuron, $a_i(\cdot)$ is a gain function, $b_i(\cdot)$ is the inhibition of neurons, $g_j(\cdot)$ is activation functions of neurons, θ_i is a real constant, t_{ij} is the connection weights from i-th neuron to j-th neuron, Here we assume that $t_{ij} = t_{ji}$.

In order to make the network (1) has ability of memory, these parameters $a_i(\cdot)$, $b_i(\cdot)$, t_{ij}, $g_j(\cdot)$ and θ_i should be selected, this makes the models need to be retrieved can be designed as the balance of network (1).

A. Xie & X. Huang (Eds.): Advances in Computer Science and Education, AISC 140, pp. 171–175.
springerlink.com

When the network's initial state closes to an equilibrium point enough, the final state of the network can converge to the equilibrium point.. Suppose these functions $a_i(x)=1, x\in R$, $b_i(\cdot)=-g_i(\cdot)(i=1,2,...n)$, $g_i(\cdot)(i=1,2,...n)$ are continuous and differentiable,and they satisfy the following conditions:

$$0<\frac{g_i(x)-g_i(y)}{x-y}\le k, \forall x, y\in R, i=1,2,...n \tag{2}$$

At this time network (1) degradated to network (3):

$$\dot{x}_i=-g_i(x_i)+\sum_{j=1}^{n} t_{ij}g_j(x_j)+\theta_i, i=1,2,...n \tag{3}$$

The simplified form of network 3 is:

$$\dot{X}=WG(X)+\Theta \tag{4}$$

and $X=[x_1,x_2...x_n]^T\in R^n$, $\Theta=[\theta_1,...,\theta_n]^T\in R^n$, $G(X)=[g_1(x_1),...,g_n(x_n)]^T$, $W=[w_{ij}]\in R^{n\times n}$ is a symmetric weight matrix, and $w_{ij}=\begin{cases} t_{ij}, & i\ne j \\ t_{ij}-1, & i=j \end{cases}$.

3 System Designing

$x^j\in R^n, j=1,2,...m$ is m-model need to be remembered, parameters W and Θ need to be designed rationally, these make $W\le 0$,$WG(X^j)+\Theta=0, j=1,2,...,m$,and in this work the following designing method has been developed:

Step 1) Caculate $Y^j=G(x^j), j=1,2,...,m$

Step 2) Set $\overline{Y}=[Y^1-Y^m,...,Y^{m-1}-Y^m]$,carried out $\overline{Y}$ by singular value decompsit -ion, $\overline{Y}=\hat{\Gamma}\Lambda\hat{V}^T$,and $\hat{\Gamma}\in R^{n\times n}$, $\hat{V}\in R^{(m-1)\times(m-1)}$ are orthogonal matrix, is a diagonal matrix, its elements are the singular values of matrix $\overline{Y}$;.

Step 3) Caculate $W=\hat{\Gamma}L\hat{\Gamma}^T$, and $L=\begin{bmatrix} 0 & 0 \\ 0 & l_1 I_{n-m+1} \end{bmatrix}$, $l_1\in R^-$, $I_{n-m+1}\in R^{(n-m+1)\times(n-m+1)}$ is a unit matrix;

Step 4) Caculate $\Theta=-WY^m$.

In order to achieve the retrieving and recoverying of images, first, the images need to be preprocessed,and the images are converted to vector network can be identified in Hopfield network.

Data preprocessing : Matrix can be used to represent the image, the data type of matrix is unit 8, and each element of the matrix corresponds to a pixel image, these pixel values are in the interval of [0,255].

In order to make function $g_i(\cdot)$ match the range of the pixel value, set $g_i(\cdot)$ as the following formula:

$$g_i(x) = 20\tanh[(x-150)/70], x \in R, i = 1,2,...,n \tag{5}$$

Further, the pixel valuc matrix must be transformed into vector, we define the conversion between the matrix and vector as:

$$\overline{P} = \begin{bmatrix} \overline{p}_1 \\ \overline{p}_2 \\ \vdots \\ \overline{p}_{n_p} \end{bmatrix} \underset{vcctor\ \ to\ \ matrix}{\overset{matrix\ \ to\ \ vcctor}{\rightleftharpoons}} [\overline{p}_1, \overline{p}_2, ..., \overline{p}_{n_p}]^T \tag{6}$$

Here $\overline{p}_i (i = 1,2,...,n_p)$ is i-th row of matrix $\overline{p}_i$.

In order to tronsform the output of the network into a matrix, and further tronsform them into an image.First, took integer according to the output of the network,and tronsform the integer vector into a matrix,then tronsformed them into an image,the formula $Y_o^j = [y_{o1}^j, y_{o2}^j, ..., y_{on}^j]^T$ is the final output for the network, X^j is the the corresponding desired output, its specific form is as follows:

$$NMSE = \frac{\sum_{i=1}^{n}(x_i^j - y_{oi}^j)^2}{\sum_{i=1}^{n}(x_i^j)^2} \tag{7}$$

The network performance, namely the reliability of recovered memories, depend on many factors, such as the pseudo-state problem, storage capability of the network, memory retrieving time,etc. Because the network is non-linear coupling, it is difficult to fully explain these questions with analytical methods, but the performance of the Hopfield network can be easily observed from some simulation examples, and the effectiveness of the method can be determined.

4 System Simulation

There are 8 256-color grayscale in Figure 1, each image is 40 pixels wide and 33 pixels high. Usually gray-scale image can be represented with a matrix, the data type of the matrix is unit8,these pixel values are positive integers in the interval of [0,255]. Hopfield network transfer function is an odd function,it is symmetrical about the origin, in order to fully transfer the odd function,the image pixel values are converted to $\tilde{\phi}_n$:

$$\tilde{\phi}_n = \begin{cases} \tilde{\phi}_o - 200, & \tilde{\phi}_o \le 125 \\ \tilde{\phi}_o - 55, & \tilde{\phi}_o > 125 \end{cases} \tag{8}$$

$\tilde{\phi}_o$ are the original pixel values, $\tilde{\phi}_n$ are the transformed pixel values.

Next, the pixel value matrix can be transformed into a 1320-dimensional vector through formula (6), we use single-layer feedforward neural network (SFN) method to design the Hopfield network, the specific designing parameters were:n=1320, m= 8, A=diag[10, 10, . . . , 10], the initial Settings was: $\omega_{jj} = 10, j = 1,2,...,1320$,set $f_j(\cdot)$ as $f_j(x) = 200\tanh(x/100), x \in R$.

Fig. 1. 8 original 256-color grayscale images

Set $Y^{(i)} = F(X^{(i)})$,then training samples of the network ($Y^{(i)}, A(X^{(i)})$) was obtained,here $x^{(i)}$ was the transformed pixel value vector. Five different networks are designed, a network with the smallest value was chosed as the final network.

In order to test the performance of the network , Figure 2(a) shows, the damaged images(Gaussian white noise with mean 0 and variance 0.05)were provided to the network as the initial state. With the operation of the network, the network eventually converged to some stored patterns. The average NMSE was calculated, $NMSE = 2.1\times10^{-5}$. The output of the network was rounded,then it was transformed into a matrix from formula (6). Through the image's color table, the value matrix can be converted into grayscale image. It is noted that, after the output of the network is rounded,NMSE = 0.

Each damaged image corresponds to a restored image.Figure 2(b) shows, these stored images are exactly the same as the stored image of the net work. The reliability of the network is high.

(a) Damaged images(Gaussian white noise with mean 0 and variance 0.05)

(b) Restored image

Fig. 2. Damaged images and restored images in Hopfield Network

Figure 2(a) shows, 4 damaged images(Salt and pepper noise with density of 0.4) were provided to the network as the initial state. With the operation of the network, the network converged to the stored patterns. Eventually the restored image shown in Figure 3 (b). The average NMSE was rounded, NMSE = 0.Figure 2 (a) and Figure 3(a) shows,the images used for network performance testing have been seriously polluted, even the human brain is difficult to recognize these images, however, the network can restore the polluted image accurately.

(a) Damaged images (salt and pepper noise with density of 0.4)

(b) Restored image

Fig. 3. Damaged images(Salt and pepper noise with density of 0.4) and Restored image Restored image

5 Conclusions

The simulation results show that Hopfield network can be used as an effective image memory.The operating cost of the neural memory is positive related with the dimension of the network.It is noted that, neural memory designed in this work can remember every pixel of the image, in fact, the human brain can only remember some important parts of the image.Therefore, the method of feature extraction and principal component analysising can be used to design the network, this will greatly reduces the dimensions of Hopfield network.

Acknowledgment. This work was supported by Scientific Research Fund of Hunan Provincial Science Department under Grant (No: 2011FJ3022, 2011FJ3016).

References

1. Lu, W., Chen, T.: New conditions on global stability of Cohen-Grossberg neural networks. J. Neural Comput. 15(5), 1173–1189 (2003)
2. Valle, M.E.: A class of sparsely connected autoassociative morphological memories for large color images. IEEE Trans. Neural Networks 20(6), 1045–1050 (2009)
3. Huang, W.Z., Huang, Y.: Chaos of a new class of Hopfield neural networks. Appl. Math. C. 206, 1–11 (2008)
4. Huang, Y., Yang, X.S.: Hyperchaos and bifurcation in a new class of four-dimensional Hopfield neural networks. Neurocomputing 69, 1787–1795 (2006)

[illegible] (1)

[illegible]

Design of Mobile Learning-Recommendation Services Supporting Small-Screen Display Interfaces

Hong-Ren Chen

Department of Digital Content and Technology
National Taichung University of Education
Taichung 403, Taiwan

Abstract. Increasing numbers of e-learning websites are being established. However, with the overload of learning material, learners may not be able to efficiently locate the required study materials. Furthermore, with the evolution of technology for constructing information-rich websites, learners are likely to feel confused about the information they receive. Therefore, the concept of learning-recommendation services was proposed to help learners obtain suitable study resources. With the rapid development of wireless networking technology and innovation in mobile devices, mobile learning has become another learning area. Known for its convenience, expediency, and immediacy, mobile learning enables learners to access information more easily and learn at any time. Their motivation is stronger, and the learning effects are also greater. Based on the structure of learning-recommendation websites, the aim of this study was to integrate and establish a mobile content-oriented learning-recommendation system supporting small-screen display interfaces. Learners can access the learning materials conveniently at any time from handheld devices and wireless communication devices. The navigator bar and browsing page are stacked together so that the learners can clearly see which section they are reading and obtain the relevant information at the same time to meet their individual needs.

1 Introduction

With the development of e-learning, improvements in wireless Internet technologies and the increasing prevalence of mobile devices, mobile learning has emerged as a novel education model [1, 2]. Neither the services nor the equipment used by mobile learning are limited by time or space. Mobile learning can help learners acquire knowledge by providing them with digital information and learning materials [2, 3]. Hence, mobile learning has the advantages of convenience, expediency, and immediacy. Users of mobile learning systems will be able to go online for lessons regardless of time and location constraints. The system will also effectively promote learning motivation and increase the effectiveness of education [4].

Current research on mobile learning can be divided into five categories classified by the practicality of mobile devices and the design of course contents. The first type of research is on content perception and how the system can automatically detect the type of mobile device used by the learner, with appropriate adjustment of the content to the most suitable visual format [5]. The second type of research revolves around

A. Xie & X. Huang (Eds.): Advances in Computer Science and Education, AISC 140, pp. 177–182.
springerlink.com

the utilization by mobile devices of wireless networks for the purpose of education, and how learners use the technologies of wireless Internet for their personal learning needs to acquire the newest information online at any time. This demonstrates three characteristics of mobile learning: rapid satisfaction of urgent need for information, encouragement of the learners' initiative in acquiring knowledge, and mobility in terms of learning locations [6]. The third type of research focuses on how mobile devices supplement course instruction. This research investigates supplemental learning programs designed to assist course instruction and their adaptation for use in mobile devices. Learners using such supplemental programs would have an improved understanding and retention of course contents, as well as enhanced learning mobility with the option of bringing course materials outside the classroom [7]. The fourth type of research involves the use of mobile devices and measuring equipment as tools for measurement methods. Mobile devices can be connected with any type of measurement or recording equipment, such as microphones, cameras, or water quality meters. Learners would be able to understand the situation of their surroundings, while conducting long-term environmental measurements and recordings [6]. The fifth type of research is the use of situational perception in education. Using digital course contents with supplemental learning software and a mobile device with perception technologies, the most suitable learning material can be provided based on students' perceptions [8].

2 Learning-Recommendation Service and Its Application

Currently, information regarding learners is commonly gathered using two methods, explicit rating and implicit rating, which are implemented during learning-recommendation research [9]. Explicit rating, in its simplest form, involves the acquisition of information by giving learners questionnaires. The advantage of questionnaires is that they provide a high degree of accuracy; however, they can be inconvenient for the users. Implicit rating, on the other hand, is the opposite of explicit rating. Information that had been browsed by learners is recorded without their input. The system then rates the preferences of learners for various types of information. This method would greatly reduce the users' burden, as they do not need to know when they are being recorded [10].

Previous research regarding recommendation technologies tended to focus on its use in e-commerce. However, this technology has recently spread into the field of education [11]. When the technology is applied to learning recommendation, a learner's preferences, basic information, learning style, and learning experience can be compiled and processed to produce general rules. The system can then recommend relevant books or educational materials suitable for the learner's perusal. Learners can also find other learners with similar learning styles or interests, giving them recommendations on books or other supplemental reading materials. For example, Qiu [12] designed a course recommendation system suite that uses data-mining technologies. The system recommends online learning courses based on various channels, such as social groups, interests, and choices. Shau [13] designed a novel categorization of learners based on browsing behaviors and characteristics of online courses. The system will then provide personalized recommendations for online learning materials, utilizing a more accurate and suitable set of recommended browsing of course materials for the student's needs. Kuan [14] investigated learning style as the basis for a learning-resource recommendation system. Browsing histories of learners with similar learning styles were analyzed for association rules to understand their browsing behaviors. This information, in turn, could be used to provide recommendations for new articles as well as relevant literature as learning materials for the learners.

3 Design of Mobile Learning-Recommendation System

From analysis of the relevant literature mentioned above regarding mobile learning and learning recommendation, it is clear that the advancement of digital technology is driving changes in learning styles and methodologies in traditional and mobile learning systems as well as applications to learning. Learning recommendation can achieve personalization of learning such that the educator is no longer the sole focus. as personalization now involves learning interests, suitability, and two-way interaction. Although there have been many important studies regarding mobile learning and learning recommendation [15,16], this study combined the advantages of these two learning models and developed a mobile learning-recommendation environment where learners would be able to learn regardless of time and location constraints, and their learning experience would be personalized with real-time recommendations. For example, medical school students undergoing practical training would be able to quickly access necessary learning materials or case studies before making a diagnosis. Another example would be a course identifying outdoor plants on campus grounds, in which the learners would be able to identify plants based on photographs presented in the online learning materials. They would also be able to acquire recommended knowledge relevant to their course unit while their memories of course materials were the clearest, thus enhancing learning efficacy [17].

As shown in Figure 1, this study established a learning-recommendation service system, as extended by Chen and Huang [18], as the basis for determining the materials to be recommended as appropriate for the learner. Recommendations of course materials in different subjects would be assessed and quantified mainly through the expansion and development of information-filtering techniques. Explicit and implicit ratings deriving from data-mining techniques were also used to determine the order of materials recommended. Calculation speed is a critical point requiring assessment especially on the implementation of mobile learning environments. The methods designed here would satisfy the learners and allow them to carry out learning-recommendation activities regardless of time and location, acquiring suitable or preferred learning materials on a real-time basis. For an easier understanding of the operational concepts, the core idea can be explained as follows. Learners would first use their handheld devices or wireless network to connect and login to the system whenever and from wherever they wished to peruse relevant learning materials. The selected recommendation algorithms entail information-filtering techniques and association-rule learning based on data mining to calculate the suitability of recommended learning materials. Hence, data based on the settings of learners' preferences and relevance to their learning experiences must be taken into consideration. After running the recommendation algorithms, the system ranks the learning materials based on recommendation scores, recommending the highest scoring material for the learners' use.

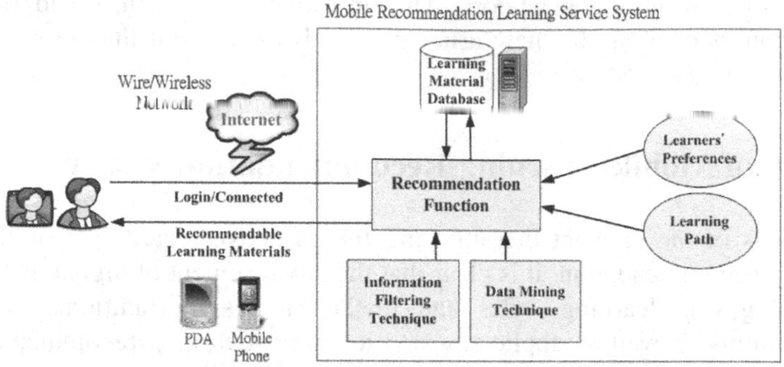

Fig. 1. Framework of the mobile learning-recommendation service system

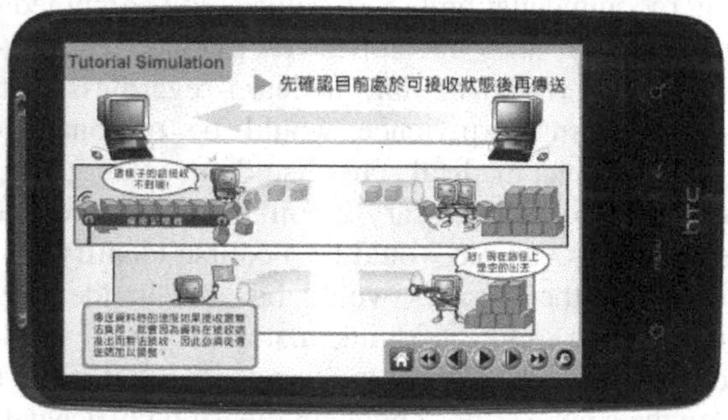

Fig. 2. Interactive interface design for one-window page browsing

Although mobile learning has been shown to be valuable in application, there have also been studies exploring the problems of learning on a small-screen display, which are closely correlated with the convenience of interface design. In the design of mobile learning-recommendation service systems, to allow learners to browse the learning contents easily and conveniently, it is necessary to take into consideration the slower transfer speeds of wireless internet. The interactive interface design uses one-window page browsing for the instructional materials, as shown in Figure 2. The guidance row and pages are combined in a cascading format, so that when learners engage in mobile learning-recommendation services, they can clearly understand the reading material units and the information shown by the system

4 Conclusions

The mobile learning-recommendation service system implemented by this study can support learning by browsing small-screen display interfaces, providing learners with the recommended instructional materials on handheld devices at any time and place, thereby increasing learning effects. To effectively conform to the browsing limitations of small-screen displays for instructional content and to improve upon the inconvenience of reading on handheld devices, the interactive interface uses the design principles of one-window paging, which can decrease extraneous clicking and switching between different learning windows, allowing for convenient learning of recommended instructional materials as well as clear understanding of the current learning unit and personal learning progress.

Acknowledgement. This research was partially supported by the National Science Council in Taiwan through Grant NSC99-2511-S-142-010-MY3.

References

1. Harris, P.: Goin Mobile, from the World Wide Web (2001), `http://www.learningcircuit.org/2001/jul2001/harris.html` (retrieved December 18, 2005)
2. Chabra, T., Figueiredo, J.: How To Design and Deploy Handheld Learning (2002), `http://www.empoweringtechnologies.net/eLearning/eLearning/_expov5/_files/frame.htm` (retrieved September 20, 2005)
3. Lehner, F., Nösekabel, H., Lehmann, H.: Wireless E-Learning and Communication Environment: Welcome at the University of Regensburg. E-Service Journal 2(3), 23–41 (2003)
4. Kynaslahti, H.: In search of elements of mobility in the context of education. In: Kynaslahti, H., Seppala, P. (eds.) Mobile learning, pp. 41–48 (2003)
5. Lin, H.-X.: Context-Sensitive Content Representation for Mobile Learning. Master Thesis, TungHai University, Taichung, Taiwan (2005)
6. Wu, L.-J.: A Study of Mathematics Path and Interactive Problem Solving Discussion System in Mobile Learning Environment. Master Thesis, National Taiwan Normal University, Taipei, Taiwan (2005)
7. Fang, T.-L.: A Study on Mobile Learning Device Applied to the Nature Science Class in Elementary School. Master Thesis, Taipei Municipal University of Education, Taipei, Taiwan (2005)
8. Wang, C.-C.: Mobile Engilsh Situated Assistant Learning Systemabout Campus Living Environment. Master Thesis, National Pingtung University of Education, Pingtung, Taiwan (2007)
9. Morita, J., Shinoda, Y.: Information filtering based on user behavioranalysis and best match text retrieval. Paper Presented at the 7th Annual ACM-SIGIR Conference on Research and Development in Information Retrieval (1994)
10. Chen, H.R., Huang, J.G.: A Practice Framework of Recommendation Learning Service System for Interdisciplinary Learning. Paper presented at the International Conference Interactive Computer Aided Learning, Austria (2009)
11. Yang, H.-L., Huang, R.-Z.: Consideration with the recommendation of the overall system point of view: an example of family child. Web Journal of Chinese Management Review 11(3), 1–26 (2008)
12. Qiu, Y.-H.: A Study of Web-Based Curriculum Recommended by using neural network and data mining technique. Master Thesis, Ming Chuan University, Taipei, Taiwan (2003)

13. Shau, S.-M.: A Study on Personalized Web-based Learning Recommendation based on Data Mining Technology. Master Thesis, Ming Chuan University, Taipei, Taiwan (2003)
14. Kuan, Y.-T.: The Study of Learning Resource Recommandation Mechanism Based on Learning Style. Master Thesis, Chung Yuan Christian University, ChungLi, Taiwan (2004)
15. Ghauth, K.I., Abdullah, N.A.: The Effect of Incorporating Good Learners' Ratings in e-Learning Content-based Recommender System. Educational Technology & Society 14(2), 248–257 (2011)
16. Tai, D.W., Wu, H., Li, P.: Effective E-Learning Recommendation System Based on Self-Organizing Maps and Association Mining. The Electronic Library 26(3), 329–344 (2008)
17. Chen, H.-R., Huang, H.-L.: User Acceptance of Mobile Knowledge Management Learning System: Design and Analysis. Educational Technology & Society 13(3), 70–77 (2010)
18. Chen, H.R., Huang, J.G.: Exploring learner attitudes toward web-based recommendation learning service system for interdisciplinary application. Journal of Educational Technology & Society (accepted 2011)

The Effect of Perceived Organizational Support on Employee Work Engagement: A Case Study Analysis

TingJun Kou

Zhejiang University, Zhejiang Hangzhou 310058

Abstract. Employees' work engagement level will directly affect company's performance. And among these effect factors, perceived organizational support is the vital antecedent. The perception level of organizational support will affect employees' commitment on their organizations and their work engagement. This study discussed the concept of perceived organizational support and the effect system toward on work engagement. A case study is also introduced for further discussion.

Keywords: Work engagement, Perceived Organizational Support, Employee Initiative.

1 Introduction

As the increasing competition in business, employees who directly affect the performance of company, already attains more and more attention from entrepreneurs and scholars. And how to effective raise employees' initiative and involvement to further improve company's performance, is becoming the key question for companies.

Traditional, human resource management is in terms of recruitment, selection, training, motivation and performance management. These HR practices are designed to help employees improve their work skills, enhance their agreement on organizations and motivate them based on their performance. So the support from organizational level is necessary for employees to self-improvement and work engagement.

But during the launching of HR practices, only few will do focus on these practices are effectively perceived and satisfied by employees. In other way, from employees' perceptive, whether they can realize these supports from organizations will be helpful for their work and performance needs to be discussed.

As a conclusion, the perceived organizational supportive will better help employees to enhance their work engagement and gain excellent performance. And a well-designed HR strategy should also focus on the acceptable of employees.

2 Perceived Organizational Support

Homans(1958) gives the concept of social exchange. This theory concerns that every behaviors of human beings can be attributed to the exchange for rewards and

A. Xie & X. Huang (Eds.): Advances in Computer Science and Education, AISC 140, pp. 183–187.
springerlink.com

compensation. Based on this concept, people's social behaviors can also be concluded as kind of exchange, so as the social relationships. To some extent, individual'scontribution to organization is also for the exchange of the support from organizations. And this exchange can form Mutually beneficial relationship.

This theory believes during the relationship between employees and organizations, the most important is perceived reward from organizations should equal to the contribution employees made to organizations. If employees can realize that, they will work hard and commit to organizations to gain the support in terms of spiritual and material rewards (Harris, Kenneth, Harvey, 2007).

Blau(1964) further develops this theory. He defines the social exchange into internal exchange and external exchange. In management practical, external exchange is more as the beneficial exchange between individual and organizations, such as salary and compensation. Internal exchange here is the emotional contract and commitment between the two (Loi, Ngo, Foley, 2006).

Based on social exchange theory and compensation theory, Eisenberger(1986) proposed organizational support theory, which is about physiological perception of employees for the support and help from their organizations. This is mainly included whether the feedback of employees' contribution can be effective informed and whether they perceived positive support during their work process.

Besides, McMillin(1997) adds that successful organizational support should also includes instrumental support in terms of training and others, which can better help employees realize those positive support from organizations.

As a conclusion, effective organizational support should include emotional organizational support and instrumental support. And when employees have a positive perceived organizational support, they will intend to know their contribution is accepted by organizations, which will enhance their commitment, citizenship behaviors and performance.

3 The Effect System of Perceived Organizational Support on Work Engagement

Kahn (1990) defines work engagement as organizational members control themselves to fit with the job roles. Based on this point, Kahn (1991) further divided work engagement into three dimensions as physical, cognitive and emotional. These three dimensions are relatively independent, but the total work engagement will be higher if one dimension becomes higher.

Besides, Maslach' (1997) research shows that as the opponent side of job burnout, three dimensions of work engagement are energy, involvement and efficacy, which are compared with job burnout' dimensions as exhaustion,cynicism and lack of professional efficacy.

Based on the triangular model of responsibility raised by Schlenker(1994), Britt(2001)redefined work engagement as perceived responsibility, commitment and perceived influence of job performance. For this definition, work engagement is about individual's commitment and responsibility to their performance and the relative with themselves.

According to above research and analysis, one important antecedent of employees' work engagement is perceived organizational support. And scholars believe during practical management and empirical study, effective organizational support can positively affect work engagement through following ways (Kühnel, Sonnentag, Westman, 2009).

Perceived organizational support can enhance employees' commitment. Organizational commitment means physiological contract made between employees and organizations. Higher commitment can obviously improve employees' work engagement (Stinghamber, Vandenerge, 2003)

Perceived organizational support can influence employees' job satisfaction, which will also effect the engagement to the works. Muse and Stamper (2007) realized perceived organizational support will positively affect employees' job satisfaction, which will indirectly influence their work engagement.

Perceived organizational support will obviously improve employees' organizational citizenship behaviors. Podsakof, Aheame and Mackenzie (1997) proposed citizenship behaviors could significantly help reduce organizational exchange cost and improve job performance. And in the other words, it can help improve work engagement.

Above all, perceived organizational support will influence employees' citizenship behaviors and their commitment to organizations, which will finally effect their engagement to jobs.

4 Case Study of Effect System in Practical Management

In Chinese realistic management behaviors, organizations provide different support to their employees, so the perceived situations are diverse according to different companies. This case will further discuss the effect system of perceived organizational support on work engagement, which happens on a chemical plant in Zhenghai, Ningbo.

This Chemical plant, named as Jinliu Chemical plant, was established in 1997, one of famous private-owned chemical plant. The main product is Synthetic Fatliquors. Right now there are more than 150 employees worked in this plant. And it is a typical Chinese family business.

4.1 Human Resource Management System

During the conversation with the entrepreneur Mr. Liu, it can be found out that company has already established the HRM system, but Liu mainly controls this system instead of professional manager. Compared with modern HRM system, it is still a long way to go for a better HRM in this plant.

For instance, HR department is in charge of employees' compensation management and performance appraisal. But lack of employee promotion and training management will make the HR department cannot well take its responsibility. And above all, in Jinliu Chemical plant, task and performance-oriented HR management is the main strategy.

Besides, it can also be seen that Mr. Liu shows strong responsibility and obligation to his employees, which can be attributed as the role of traditional family culture. When discussed about employee layoff, Liu says company respect every employee and will not layoff employees to cut down the cost, unless employees cannot fit the demand of company.

4.2 Support Provided to Employees

During the conversation, we also find out between instructional support and emotional support, company has strong intention for the later one. For the instructional support part, because of the obvious paternalistic leadership in Chinese family business, company used to control all the information about company, and it also difficult for ordinary employees to get relevant information and materials.

As mentioned by Liu, this also happens in Jinliu company, and company do not do well for providing enough for their employees such as training and employee career development management. Besides, few employees can involve in this process, and Liu mainly decides who can have the promotion opportunity.

But strong emotional organizational support can also be observed in Jinliu. It may be traditional harmonious culture also works in this company, that company should provide emotional support for employees.

During the conversation, Liu emphasizes company and manager should concern common employees' life and family situation. A tradition in Jinliu is in most festival Liu will visit outstanding or life-difficult employees, to better connect employees and company. Besides, many managers are from the founding team, so the atmosphere inside the company is well developed.

4.3 Conclusion

From above corporate case we can know that in the management practice of Chinese private-owned company, HR system is not well established. But because of traditional culture and other factors, managers show strong responsibility and emotional support to employees.

Managers will positively concern employees' life situation and work condition, which will help company form better commitment and agreement on company. For example, in Jinliu, many employees stay more than 5 years.

But company usually controlled by the manager, in other words, the entrepreneur have all resources and information about and strongly influence the company. But employees are hard to get those necessary information and resources. Besides, because of the attitude of entrepreneurs, not enough instrumental support is given to employees.

5 The Way to Improve Perceived Organizational Support

Based on above discussion, we can get following suggestions and methods to better improve perceived organizational support in Chinese situation, and to enhance their work engagement.

Establish and perfect well work condition and in-time feedback. Well-designed work condition can help provide employees' necessary resources to complete their job, and effective feedback can also motivate employees to better engaged into their jobs.

Enhance employee involvement. During the HR management system re-designing process, employees should be encouraged to involve into this process. This can help employees realize the strong connect between company and themselves, which will also solid their physiological contract and commitment to company.

Improve perceived support from management. In most Chinese middle and small companies, top management are also seen as the symbol of company, and their behaviors will directly influence the perception level of organizational support.

So fluent communication between management and employees can positively affect their engagement to their jobs.

Build the trust and supportive relationship between employees and middle management. In practical management, middle management is in charge of launching the strategic decision and giving first-line feedback between employees and top management. It is usual for employees to take the support extent from middle management as organizational support.

Insist procedural fairness. Procedural fairness is vital for performance appraisal design and can bring lots benefits to companies, such as reduce cost and turnover rate, enhance the organizational commitment, and improve the work engagement.

Above all, well designed management and reassignment system, effective feedback, employee involvement and procedural fairness will help enhance employees' work engagement, which finally will improve employees' performance.

References

1. Bakker, A.B., Hakanen, J.J., Demerouti, E., Xanthopoulou, D.: Job Resources Boost Work Engagement, Particularly When Job Demands Are High. Journal of Educational Psychology 99(2), 274–284 (2007)
2. Bateman, T.S., Organ, D.W.: Job satisfaction and the good soldier:The relationship between affect and employee "citizenship". Academy of Management Journal 26, 587–595 (1983)
3. Eisenberger, R., Armeli, S., Rexwinkel, B., et al.: Reciprocation of perceived organizational support. Journal of Applied Psychology 86, 42–51 (2001)
4. Harris, R.B., Harris, K.J., Harvey, P.: A Test of Competing Models of the Relationships Among Perceptions of Organizational Politics, Perceived Organizational Support, and Individual Outcomes. Journal of Social Psychology 147(6), 631–656 (2007)
5. Loi, R., Ngo, H.-Y., Foley, S.: Linking employees' justice perceptions to organizational commitment and intention to leave: The mediating role of perceived organizational support. Journal of Occupational & Organizational Psychology 79(1), 101–120 (2006)
6. Stinglhamber, F., Vandenberghe, C.: Organizations and supervises as sources of support and targets of commitment: a longitudinal study. Journal of Organizational Behavior 24(3), 251–270 (2003)
7. Kühnel, J., Sonnentag, S., Westman, M.: Does work engagement increase after a short respite? The role of job involvement as a double-edged sword. Journal of Occupational & Organizational Psychology 82(3), 575–594 (2009)

References

Web-Based Data Mining in Computer Culture Basic Course in College Using an Improved GA

Jianxin Bi

Computer Science and Information Technology School,
Zhejiang Wanli University, Ningbo, 315100, China
bijianxinnb@163.com

Abstract. In the paper, how to apply an improved Genetic Algorithm (GA), which introduces a diversity control method in GA, is presented for data mining of student information obtained in a Web-based Computer Culture Basic Course in College. It aims at obtaining interesting association rules so that teachers can improve the performance of the system. To check the proposed algorithm, a Web-based Computer Culture Basic Course developed for use by students is used. First, the proposed methodology and the specific characteristics of the course is described, and then the information obtained about the students is explained. We continue on with the implemented Genetic Algorithm (GA) and finally with the rules discovered and the conclusions.

Keywords: Data Mining, Computer Culture Basic Course in College, Genetic Algorithm.

1 Introduction

Nowadays, there is a growing trend of web-based technology applied for distance education [1]. But usually, the methodology used to elaborate them is static, that is, when the course elaboration is finished and published on the Internet it is never modified again. The teacher only accesses the student evaluation information obtained from the course to analyze the student's progress. We present, a dynamic elaboration methodology, where the evaluation information is used to modify the course and to improve its performance for better student's learning [2]. The approach is to use a knowledge acquisition method (machine learning and data mining) to discover useful information that might help the teacher to improve the course. We propose an improved Genetic Algorithm for data mining to evaluate the student information obtained from a Web-based Computer Culture Basic Course in College. We have used a Web-based Computer Culture Basic Course in College that was designed to be used by student as an example to evaluate our algorithm and to obtain association rules [3] [4]. These rules could then be shown to the teacher in order to help him decide how the course could be modified to obtain best performance [5].

A. Xie & X. Huang (Eds.): Advances in Computer Science and Education, AISC 140, pp. 189–194.
springerlink.com

2 Methodologies

The dynamic construction methodology of Web-based Computer Culture Basic Course in College that we propose is recurrent and evolutionary and while the number of students who use the system increases, more information is available to the teacher to improve it. We can divide it into four main steps:

Construction of the Course. The teacher builds the Computer Culture Basic Course in College providing information of the domain model, the pedagogic model and the interface module. An authoring tool is usually used to facilitate this task. The remaining information, tutor model and the student model usually is given or acquired by the system itself. Once the teacher and the authoring tool finish the elaboration of the course, then, the full course's content may be published on a web server.

Execution of the Course. The students execute the course using a web navigator and in a transparent way the usage information is picked up and stored in the server in a huge database of all the students.

Application of Data Mining. The teacher applies data mining algorithms [6] to the database to obtain important relationships among the data picked up. For this, he uses a graphical data-mining tool.

Improving the Course. The teacher using the discovered relationships carries out the modifications that he believes more appropriate to improve the performance of the course. To do it, he again uses the authoring tool. The process of execution, application and improvement can be repeated as many times as the teacher wants to do so, although it is recommendable to have a significant amount of new students' usage information before repeating it.

3 Web-Based Computer Culture Basic Course in College

We have used the data obtained from the evaluation of a Web-based Computer Culture Basic Course in College in the study of Data Culture. The system used to develop the course is an adaptive Data Culture system [7], but the evaluation data used for this paper was obtained from a usability study and there was no attempt to use the result to redesign the Course. As part of this evaluation, the system was used by 50 users, of which 30 were computer majors and 30 were non-computer majors. All the information was stored in a single database in the following tables: USER: String value that represents a system user, in our case they are 50.

PERFORMANCE: Real value that represents the user's performance in the 15 case studies in this application.

AVEP_AH: Real value that represents the average performance of the users in the 15 case studies, adaptive application version.

AVEP_NOAH: Real value that represents the average performance of the users in the 15 case studies, but in the version of the application without adaptation.

CASETIME: Integer value that represents the time that a user takes in visualizing a complete case study.

CASESCORE: Integer value that represents the score that a user has obtained when undertaking a case study.

ACCESSTIME: Real value that represents the number of times a user has accessed the application.

CONCEPT: Real value that represents the user's effort spent in the different concepts.

QUESTIONSCORES: Integer value that represents the score obtained by the users in the relating questions to the case studies. The data was preprocessed so that it will be easier to obtain relationship rules from them. This transformation consisted of a discretization, which mapped from continuous values (usually real values) to discrete values (strings that represent values groups) and integer values only needed to be labeled. In this way the modifications made to the tables are as follows: PERFORMANCE, AVEP_AH, AVEP_NOAH, CASETIME, ACCESSTIME and CONCEPT have been discretized to the labels VERYHIGH, HIGH, MEDIUM, LOW and VERYLOW. The values of CASESCORE and QUESTIONSSCORES have been named with the labels SUCCESSFIRST, SUCCESSSECOND, SUCCESSTHIRD and SUCCESSFOUR, which means getting the answer correct at the first attempt, second attempt and so forth. USER does not need modification.

4 An Improved Genetic Algorithm for Data Mining

Some of the main data mining tasks are [8]: classification, clustering, discovery of association rules, etc. We have used an improved Genetic Algorithm to obtain association rules from the user evaluation data. The association rules relate variable values. They are more general than classification rules due to the fact that in association rules any variable may be in the consequent or antecedent part of the rule. The classical problem of discovering association rules is defined as the acquisition of all the association rules between the variables. Genetic algorithms are a paradigm based on the Darwin evolution process, where each individual codifies a solution and evolves to a better individual by means of genetic operators (mutation and crossover). In this paper we use an improved Genetic Algorithm with the diversity controlled for data mining. The improved GA Process we have used consists of the following steps:

4.1 Initialization

Initialization consists of generating a group of initial rules specified by the user (50 - 500 rules). Half of them are generated randomly and the other half starting from the most frequent values in the database. We use a Michigan approach in which each individual (chromosomes) encodes a single rule. The format of the rules we are going to discover is:

```
  IF Variable1 = Value1 (AND Variable2 = Value2)
 THEN
 VariableX =ValueX
 Where:
 Variable1, Variable2, VariableX: Are the database's
field names. (P0...P6, AVEPH, AVEPNOH,
```

CASETIME0...CASETIME6,
CASESCORE0.CASESCORE6,
ACCESSTIME0...ACCESSTIME6, C0...C37, MCQ0..
MCQ6).
Value1, Value2, ValueX: Are the possible values of the previous database fields (VERYLOW, LOW, MEDIUM, HIGH, VERYHIGH, SUCCESSFIRST, SUCCESSSECOND, SUCCESSTHIRD, and SUCCESSFOUR).

We use value encoding in which a rule is a linear string of conditions, where each condition is a variable-value pair. The size of the rules is dynamic depend of the number of elements in antecedent and the last element always represents the consequent.

4.2 Evaluation

Evaluation consists of calculating the fitness of the current rules and keeping with the best ones. To calculate the fitness we count the precision of the rule, the number of patterns in the database that fulfill both antecedent and consequent and do not fulfill both antecedent and consequent. That is, we obtain very strong association rules [2] that fulfill [A=a] _ [C=c] and [C≠c] _ [A≠a]. So a rule is very strong if the previous two rules are strong, that is, both rules have greater support and confidence than a minimum values set by the user (0.5-1). Our formula detects both statistical negative dependence and independence between antecedent and consequent.

4.3 Selection

The selection chooses rules from the population to be parents to crossover or mutate. We use rank-based selection that first ranks the population and then every rule receives fitness from its ranking. The worst will have fitness 1, second worst 2, etc. and the best will have fitness N (number of rules in population). Parents are selected according to their fitness. With this method all the rules have a chance to be selected.

4.4 Reproduction

Reproduction consists of creating new rules, mutating and crossing current rules (rules obtained in the previous evolution step). The user sets the crossover and mutation probabilities. A higher crossover rate (50-95%) and a lower mutation rate (0.5-2%) are recommended. Additionally it is good to leave some part of population survive up to next generation. Mutation consists of the creation of a new rule, starting from an older rule where we change a variable or value. We randomly mutate a variable or values in the consequent or antecedent. Crossover consists of making two new rules, starting from the crossing of two existent rules. In crossing the antecedent of a rule is joined to the consequent of the other rule in order to form a new rule and vice versa (the consequent of the first rule is joined to an antecedent of the second). So it is necessary to have two rules to do the crossover.

4.5 Diversity Controlling

After the above genetic operations, the diversity controlling procedure is executed. The diversity of the population is measure by the average Hamming distance among any pair of individuals. If the diversity decline to below a pre-specified low bound (the value is set to 1 in this work), the mutation rate will be set to 0.1 until the diversity return to above the low bound. This diversity control method could enhance the global search ability of the GA and result in a good performance.

4.6 Finalization

Finalization is the number of steps or generations that will be applied to thc genetic process. The user chooses this value (10-500 steps). We could also have chosen to stop when a certain number of rules are obtained.

5 Results and Conclusions

We have applied the algorithm to the whole data, only to the medical students and only to the other users. For each case, we have obtained different rules both in the content and in number and fit. For example, one of the rules obtained when using 100 initial rules and 100 steps is:

```
  IF CASESCORE2=SUCCESSFIRT AND
  CASESCORE4=SUCCESSFIRT THEN
  CASESCORE1=SUCCESSFIRST   (with   support   =   0.73   and
confidence=1).
```

This can be interpreted as if a user gets the answer for case 2 and case 4 correct, he/she is likely to do well in case 1. The support of a rule gives the importance of a rule and the confidence of a rule gives its predictability power. All the rules discovered are showed to the teacher in order he can obtain conclusion about the course functionality. The teacher has to analyze them and he has to decide the best modifications that can improve the performance of the course. Summarizing the main conclusions that we obtained starting from the discovered rules are: We obtained expected relations, for example: between CASESCORE and CONCEPT, and between MCQ and CONCEPT, due to the fact that the questions are about the concepts. We obtained useful relations, for example: between CASESCORE and MCQ. This could be because the questions are the same ones, or they refer to the same concepts, or they have equal difficulty. We obtained strange relations, for example between AVEP_AH and P. it is probable that this relation takes place by chance. And we didn't find any other relation, for example with ACCESSTIME or CASETIME. This could be because user access times were completely random and they did not determine any other variable as it might be expected.

References

1. Srikant, R., Agrawal, R.: Mining Sequential Patterns: Generalizations and performance improvements. In: Proceedings of the 5th International Conference on Extending Database Technology Avignon, France, pp. 412–421 (1996)
2. Delgado, M., Sánchez, D., et al.: Mining association rules with improved semantics in medical databases. Artificial Intelligence in Medicine (2001)
3. Liu, X., Wang, L.: Application of Web Data Mining in Web-Based Education. Computer Engineering (21), 34–38 (2006)
4. Freitas, A.A.: A survey of evolutionary algorithms for data mining and knowledge discovery. To appear in: Advances in Evolutionary Computation. Springer (2002)
5. Chen, L.D., Frolick, M.N.: Web-based Data Warehousing Fundamentals, Challenges, and Solutions. Information System Management (1) (2000)
6. Chen, M.S., Park, J.S., Yu, P.S.: Data mining for path traversal patterns in a web environment. In: Proceedings of the 16th International Conference on Distributed Computing System, Hong Kong, pp. 385–392 (1996)
7. Osmar, R.Z.: Web Usage Mining for a Better Web-Based Learning Environment. Technical Report (2001)
8. Han, J., Kamber, M.: Data Mining Concepts and Techniques. Morgan Kaufmann Publishers, Harcourt (2006)
9. Inokuchi, A., Washio, T., Motoda, H.: Complete Mining of Frequent Patterns from Graphs: Mining Graph Data. Machine Learning (50), 13–19 (2003)

Teaching Mode Innovation and Practice of "Foundation of Computer Programming"

Yaqing Shi*, Meijuan Wang, ChengHu, Hongyan Wei, and Jinyao Jiang

Lecturer, Foundation Electron Department Institute of Science,
PLA Univ. of Sci. & Tech., 211101 Nanjing JiangSu, China
qingyashi_blue@sina.com, wangmeijuan1984@hotmail.com

Abstract. "Foundation of computer programming" is an important public implemental core curriculum of non-computer science and engineering undergraduate specialty. Taking the teaching mode innovation of basic course as an opportunity, the author has proposed adopting "Four Actions" teaching mode based on network environment in "Foundation of computer programming".

Keywords: network environment, Four Actions, 3 any principle.

1 Introduction

"Foundation of computer programming" is another important computer course setting up for the non-computer science and engineering students after "College Basic Computer". It not only builds the foundation of program design for other professional courses, but also is the programming tool for other professional courses, and is an important public implemental core curriculum of non-computer science and engineering undergraduate specialty. The course is computer language teaching and requires students to be able to skillfully use computers for simple programming. For non-computer science and engineering undergraduates, the computer is used as a tool. We should take application as the purpose and a starting point, and start from the actual, and focus on mastering the methods and skills of the application, and select learning content according to the specialty's need, and learn around the application.

Based on the actual teaching experience, the paper discusses the existing problems of "Foundation of computer programming" course teaching mode, and introduces the "Four Actions" teaching mode basis of network environment to better adapt to the transformation of eduction and to train more talents for society.

2 The Shortcomings of Traditional Teaching Mode

The traditional teaching mode of "Foundation of computer programming" is the "theory + practice" teaching mode. The total hour is 60 hours, and is divided into theory classes and practical classes and is basically a 2:1 ratio. Theory courses is completed by the form of classroom teaching, and the primary contents include: learning the basic concepts and general methods of computer programming; learning the C language

---

* Yaqing Shi, lecturer, foundation electron department, Institute of Science, PLA Univ. of Sci.& Tech. Jiangsu Nanjing 211101.

A. Xie & X. Huang (Eds.): Advances in Computer Science and Education, AISC 140, pp. 195–198.
springerlink.com

grammar and use of various control statements; learning the concept and use of a variety of program structures and data types of C language. Usually after completing theory classes, and then we let students exercise on computer to verify and consolidate the knowledge. Thus practical classes are carried out in the form of teachers guiding and students exercising on computers. Taking Knowledge validation and knowledge integration as the principle, we design test contents and further understand and digest the classroom teaching and train program design and improve students' ability to analyze and solve problems.

This teaching mode advocates that teachers teach clearly and students learn plainly. Teacher's primary task is to teaching, summarize, and subtly teach throughout the teaching process. The key task of students is to listen, know, and understand to weigh the effect of the whole teaching. At the same time "Foundation of computer programming" is a computer language teaching and its operation is obvious. Proving the effect of teaching by exercising on computer after classroom teaching completed is the common way. But this teaching method strips the immediate response between the theory teaching and the practice teaching. And so a "delay" is engendered between the two which induces bad amalgamation of theory and practice. Furthermore, based on the above teaching model, students will only mechanically listen and exercise and real "personality education" and "talent creation" is out of the question.

3 "Four Actions" Teaching Mode Basis of Network Environment

Given the shortcomings of the traditional teaching mode, we propose "Four Actions" teaching mode basis of network environment in the "basis of computer programming" course.

"Four Actions" teaching mode basis of network environment is under the guidance of teaching thought which emphasizes 'learning' is the basis. Fully taking advantage of the immediacy and spread of network environment, it applies the "issues-driven, teachers- motivated, students' initiative, multiple interactions" such "Four Actions" strategy to the teaching process, and establishes a more stable framework for teaching activities and activities program, and ultimately realizes the teaching objectives of developing students' application ability.

3.1 What Is "Four Actions"

The so-called "Four Actions" is "issues-driven, teachers-motivated, students' initiative, multiple interactions". The whole process of teaching progresses under the "teacher-motivated" and then implements "issues-driven", thereby fully mobilizes students' enthusiasm and initiative to learn programming language to achieve a real sense of the "students' initiative", and ultimately realizes "multiple interactions" among faculties, students and course contents. The "issue-driven" is the theme throughout the teaching process, and "students' initiative" is the key point. Only the "students' initiative" achieved, the other three can really move up to serve for and fully mobilize the students' initiative to stimulate students' learning interest and potential, so that students master the basic method of computer programming and form a correct idea of program design and skillful use C language to program and learn how to use computer applications with relevant expertise to solve practical problems, and lay a solid foundation for the follow-up courses.

"Issues-driven" of "Four actions" strategy is throughout the main line of teaching process. It mainly includes the following four steps: "ask questions - analyze and solve - summary - new issues verified". Compare with traditional "issues-driven method", "new issues verify" is added. And it verifies the effectiveness of knowledge points and further mobilizes students' enthusiasm to participate in and enhances the urgency of faculty issues designed and highlight the application of course contents.

3.2 How to Base on Network Environment

Traditional teaching progresses in the classroom. Classroom is the basic unit of teaching. The ultimate purpose of what we called "base on network environment" is to make network environment as a large "classroom". The "classroom" is beyond restriction of time and space and can cover all the point which has network environment. In the classroom, the boundary between teachers and students is gradually blurred and interval between region and time cannot determine by the usual way. We try to explain the application of network environment following the "3 any principle", namely "anywhere, anytime, anybody". At any place, any time, anyone can learn from others. Other people can be your teacher, as you can be other people's teacher.

4 Innovations of "Four Actions" Teaching Mode Basis of Network Environment

"Four Actions" teaching mode basis of network environment emphasizes that student is the subject of teaching activities and emphasizes students' participation into teaching. According to the needs of teaching, it rationally designs "teaching" and "learning" activities. The teaching mode introduces "network environment" such advanced teaching resources into the "Four Actions" mentioned before. This teaching mode changes the traditional classroom teaching to the interactive teaching basis of network environment with the characteristics of extension of time and space, and its main innovations are the following five: areas:

1. The teaching body: realizing the conversion from teacher-centered to students, faculties, and course contents integrated.
2. The teaching contents: realizing the conversion from focusing on imparting knowledge to training strong innovation and the comprehensive ability to analyze and solve problems.
3. Teaching resources: realizing the conversion from blindly increasing amount of teaching resources to putting emphasis on efficiency, purpose and selectivity.
4. Teaching forms: realizing the conversion from a single mode to diverse modes.
5. Teaching methods: realizing the extension from multi-media technology to the network technology.
6. Compared with the traditional teaching mode, "Four Actions" teaching mode basis of network environment has its obvious advantages: First, introduction of "network platform" brings breakthrough to the traditional teaching methods of "blackboard + multimedia". Second, this teaching mode abandons the former teaching method of

"teaching-oriented" and "teacher teaching all the class" and boldly adopts the method of "teaching for the lead and learning for the head" to fully mobilize students' initiative and achieves the real meaning of "students' initiative". Third, this teaching mode doesn't simply consider any party of the teaching, but from macroscopically perspective, synthetically considers the "teacher" "student", "curriculum" and other factors to realize "multiple integration". Fourth, this teaching mode makes use of all-round and multi-angle "guide", "teaching", "learning", "practice" instead of the traditional indoctrinated "teaching" and "learning" to maximize excavate the real meaning of classroom teaching.

5 Conclusion

Teaching practice has proved that "Foundation of computer programming" course changing from the traditional teaching mode to innovative "Four Actions" teaching mode basis of network environment, the substantial adjustment has happened in teaching subject, teaching contents, teaching resources, teaching forms, teaching methods and other aspect. And this adjustment will better reflect the course's characteristic of theory and practice combination, and emphasize the practical operability of the course and highlight the pertinence of the purpose of application. It fully mobilizes students' enthusiasm of participation and makes students master the idea of programming design to solve practical problems and possess of good programming habits and the consciousness and ability to solve practical problems with computers. And then it makes the students use computers as a tool to solve practical problems in the follow-up professional courses study and improve students' computer application ability.

References

1. Xu, Y.: C Programming Language Application Tutorial. Science Press (2011)
2. Xu, Y.: On the computer network environment of online teaching. China-School Education (Theory) (16) (2010)

MBR Phrase Scoring and Pruning for SMT

Nan Duan

School of Computer Science and Technology
Tianjin University
nannanduan@hotmail.com

Abstract. One of the major reasons for translation errors in phrase-based SMT systems is the incorrect phrases induced from inaccuracy word-aligned parallel data. In this paper, we propose a novel approach that uses the minimum Bayes-risk (MBR) principle to improve the accuracy of phrase extraction. Our approach performs as a four-stage pipeline: first, bilingual phrases are extracted from parallel corpus using a standard phrase induction method; then, phrases are separated into groups under specific constraints and scored using an MBR model; next, word alignment links contained in phrases with their MBR scores lower than a certain threshold are pruned in the parallel data; last, a new phrase table is learned from the link-pruned parallel data and used in SMT decoding. We evaluate our approach on the SMT Chinese-English MT tasks, and show significant improvements on parallel data sets of different scales.

Keywords: statistical machine translation, MBR, phrase extraction.

1 Introduction

The common practice of extracting bilingual phrases from the parallel data usually consists of three steps: first, words in bilingual sentence pairs are aligned using state-of-the-art automatic word alignment tools, such as GIZA++ [1], in both directions; second, word alignment links are refined using heuristics, such as Grow-Diagonal-Final (GDF) method; third, bilingual phrases are extracted from the parallel data based on the refined word alignments with predefined constraints [1]. Such phrase extraction methods, however, are not performed in a clean room. They are usually subject to various kinds of errors, such as noises in training corpus or mistakes caused by word alignment models. These errors could produce low-quality bilingual phrases in the final phrase table. Incorrect phrase entries fed into SMT decoder are one of the major reasons for translation errors in phrase-based SMT systems.

Motivated by the success of consensus-based methods in SMT research [2] [3] [4], this paper proposes a novel approach that makes use of MBR principle to improve the accuracy of phrase extraction. Our approach operates as a four-stage pipeline: first, bilingual phrases are extracted from parallel corpus using a standard phrase induction method; then, phrases are separated into groups under specific constraints and scored using an MBR model; next, word alignment links contained in phrases with their MBR scores lower than a certain threshold are pruned in the parallel data; last, a new phrase table is learned from the link-pruned parallel data and used in SMT decoding.

A. Xie & X. Huang (Eds.): Advances in Computer Science and Education, AISC 140, pp. 199–204.
springerlink.com

2 A New MBR-Based Phrase Extraction Pipeline

In Step 1, all potential phrase pairs that are consistent with word alignments are extracted from a given training corpus $\mathcal{C}$ using the standard phrase extraction method. Furthermore, inspired by several studies [5] [6], in which *n*-best alternatives of annotations to SMT systems are leveraged to improve translation quality, we allow our proposed phrase extraction method to operate on *n*-best word alignments as well: given a sentence pair with *n*-best alignment candidates, we use alignments in the *n*-best list one at a time with the same sentence pair to form a new word-aligned sentence pair, and annotate it with the posterior probability of the alignment it used.

In Step 2, the objective of this step is to score phrase pairs extracted in Step 1 based on an MBR model. Similar to those bi-direction translation features, we compute two MBR scores for each phrase pair from two different language sides as well. We first consider scoring phrase pairs based on their source phrases. Given all phrase pairs $\{(a', e', f)\}$ with the same source side f, we define a score $S_f(\rho)$ that is assigned to each phrase pair $\rho = (a', e', f)$ based on an MBR model as:

$$S_f(\rho) = \sum_{\rho' \in \mathcal{H}(f)} G(\rho, \rho') P(\rho' | f) \tag{1}$$

- $\mathcal{H}(f)$ is the *hypothesis space* that contains all phrase pairs $\{(a', e', f)\}$ extracted in Step 1 sharing the same source phrase f.
- $G(\rho, \rho')$ is the *gain function.* We define $G(\rho, \rho')$ as the similarity measure between two hypotheses ρ and ρ'. In this sense, $S_f(\rho)$ can be viewed as the expected similarity between ρ and all hypotheses in $\mathcal{H}(f)$. W formulate $G(\rho, \rho')$ as a weighted combination of a set of similarity features:

$$G(\rho, \rho') = \sum_i \lambda_i \theta_i(\rho, \rho') \tag{2}$$

- $P(\rho|f)$ is the *hypothesis distribution* over all hypotheses contained in $\mathcal{H}(f)$:

$$P((\rho|f)) = \frac{\sum_{(E,F)\in\mathcal{C}} \sum_{\mathcal{A}} \delta_{(\mathcal{A},E,F)}(\rho) P(\mathcal{A}|E,F)}{\sum_{\rho' \in H(f)} \sum_{(E,F)\in\mathcal{C}} \sum_{\mathcal{A}} \delta_{(\mathcal{A},E,F)}(\rho') P(\mathcal{A}|E,F)}$$

where $\delta_{(\mathcal{A},E,F)}(\rho)$ equals to 1 when ρ can be extracted from $(\mathcal{A}, E, F)$, and 0 otherwise, $P(\mathcal{A}|E, F)$ is the posterior probability of $\mathcal{A}$:

$$P(\mathcal{A}|E,F) = \frac{exp\,\{\alpha \cdot \varphi(\mathcal{A}, E, F)\}}{\sum_{\mathcal{A}'} exp\,\{\alpha \cdot \varphi(\mathcal{A}', E, F)\}}$$

where $\varphi(\mathcal{A}, E, F)$ is the score predicted by the alignment model for an alignment $\mathcal{A}$, α controls the entropy of resulting distribution.

We rewrite Equation (1) by replacing $G(\rho, \rho')$ using Equation (2) as:

$$S_f(\rho) = \sum_i \lambda_i \{ \sum_{\rho' \in \mathcal{H}(f)} \theta_i(\rho, \rho') P(\rho'|f) \} \\ = \sum_i \lambda_i Sim_i(\rho, \mathcal{H}(f)) \tag{3}$$

$Sim_i(\rho, \mathcal{H}(f)) = \sum_{\rho' \in \mathcal{H}(f)} \theta_i(\rho, \rho') P(\rho'|f)$ is defined as the expected value of the ith similarity feature θ_i for ρ based on the entire $\mathcal{H}(f)$.

We then consider scoring phrase pairs based on their target phrases. Given all phrase pairs $\{(a', e, f')\}$ with the same target side e, we score each of them in a similar way as in Equation (3):

$$S_e(\rho) = \sum_j \lambda_j \sum_{\rho' \in \mathcal{H}(e)} \theta_j(\rho, \rho') P(\rho'|e) \\ = \sum_j \lambda_j Sim_j(\rho, \mathcal{H}(e)) \tag{4}$$

$Sim_j(\rho, \mathcal{H}(e)) = \sum_{\rho' \in \mathcal{H}(e)} \theta_j(\rho, \rho') P(\rho'|e)$ is defined as the expected value of the jth similarity feature θ_j for ρ based on the entire $\mathcal{H}(e)$.

In Step 3, in order to discard low-quality phrase pairs from the final phrase table to alleviate decoding errors, an alignment pruning strategy is proposed.

Given each phrase pair $\rho = \{a, e, f\}$, we first find out two maximum MBR scores, $\hat{S}_f(\rho')$ and $\hat{S}_e(\rho')$, from its corresponding hypothesis spaces $\mathcal{H}(f)$ and $\mathcal{H}(e)$ respectively and multiply them to obtain a reference value as $S_{max}(\rho)$; we then prune all alignment links contained in ρ from C, if the product of $S_f(\rho)$ and $S_e(\rho)$ is lower than $S_{max}(\rho)$ by a certain threshold.

In Step 4, we re-extract bilingual phrases based on the link-pruned training corpus to learn a new phrase table. For each phrase pair in this phrase table, we also compute two sets of expected similarity scores based on the MBR model, and use them as extra phrasal features. We will show that besides alignment pruning, using similarity scores as additional features can provide further improvements as well. When using n-best alignment results instead of 1-best ones, translation probabilities and lexical weights are estimated based on fractional counts instead of absolute frequencies of phrases.

3 Similarity Features

This section presents all similarity features that are used in computing $G(\rho, \rho')$. We summarize them into two categories as follows.

- ***Alignment-Based Features***

1) $\theta_{W2W}(\rho, \rho')$. A feature that counts how many (source word)-to-(target word) link pairs in ρ co-occur in ρ'.
2) $\theta_{W2P}(\rho, \rho')$. A feature that counts how many (source word)-to-(target word's POS) link pairs in ρ co-occur in ρ'. Two MaxEnt-based POS taggers are used to tag Chinese and English words contained in the bilingual corpus respectively.

3) $\theta_{W2C}(\rho, \rho')$. A feature that counts how many (source word)-to-(target word's class) link pairs in ρ co-occur in ρ'. Word clusters are obtained by using *mkcls* toolkit [7] that trains word classes based on the maximum-likelihood criterion. The total numbers of word classes are set to be 80 for both Chinese and English.
4) $\theta_{W2S}(\rho, \rho')$. A feature that counts how many (source word)-to-(target word's stem) link pairs in ρ co-occur in ρ'. A stem dictionary that contains 22,660 entries is used to convert English words into their stem forms. We consider the stem for each Chinese word as the Chinese word itself.
5) $\theta_{Fert}(\rho, \rho')$. A feature that reflects agreement on word fertilities for ρ and ρ':
$$\theta_{Fert}(\rho, \rho') = \#(a)\delta_{\#(a)}(\#(a'))$$
$\#(a)$ *is the number of word link pairs in* a, *and* $\delta_{\#(a)}(\#(a'))$ *equals to 1 when* $\#(a) = \#(a')$, *and 0 otherwise.*
6) $\theta_{Ratio}(\rho, \rho')$. A feature that reflects the agreement on link ratio defined as $r(\rho) = \#(\rho)/\{|\rho_s| + |\rho_t|\}$ between ρ and ρ', where $\#(\rho)$ is the total number of linked words contained in both sides of ρ, ρ_s and ρ_t are source and target phrases of ρ respectively:
$$\theta_{Ratio}(\rho, \rho') = r(\rho)\delta_{r(\rho)}(r(\rho'))$$

We use alignment-based features due to the fact that alignment quality usually determines the quality of phrase pairs. Besides words, we also incorporate the knowledge of POS tags, word classes and word stems to alleviate data sparseness and to make our model to be more general.

- ***N-Gram-Based Features***

7) $\theta_n(\rho, \rho')$. A feature that counts how many *n*-grams in ρ_t co-occur in $\rho_t{}'$:
$$\theta_n(\rho, \rho') = \sum_{\omega \in \rho_t} \#_\omega(\rho_t)\delta_{\rho_t'}(\omega)$$
$\#_\omega(\rho_t)$ is the number of times that ω occurs in ρ_t, $\delta_{\rho_t'}(\omega)$ equals to 1 when ω occurs in ρ_t', and 0 otherwise.
8) $\theta_{Len}(\rho, \rho')$. A feature that reflects the agreement on word lengths for ρ_t and ρ_t':
$$\theta_{Len}(\rho, \rho') = |\rho_t|\delta_{|\rho_t|}(|\rho_t'|)$$
$\delta_{|\rho_t|}(|\rho_t'|)$ equals to 1 when $|\rho_t| = |\rho_t'|$, and 0 otherwise.

4 Experiments

We evaluate on the NIST Chinese-to-English MT tasks. The NIST 2003 (MT03) data set is used as the development set to tune model parameters, and evaluation results are reported on the NIST 2005 (MT05) and 2008 (MT08) data sets.

Two parallel data sets with different scales are used as training corpus: the first data set includes FBIS only, which contains 128K sentence pairs after pre-processing; the second data set includes 4 data sets, with 354K sentence pairs contained after pre-processing. We confirm the effectiveness of our method on it using the optimal parameter setting determined on the first data set.

Translation quality is measured in terms of the case-insensitive *IBM-BLEU* scores. A re-implemented phrase-based SMT decoder [8] is used to generate translation outputs. The standard phrase extraction method (*Base-PE*) proposed by [1] is utilized to generate the baseline phrase table. The length limitations are set to be 5 and 10 for source and target phrases respectively. We denote our MBR-based phrase extraction method as *MBR-PE* and compare it to Base-PE on the first data set (Table 1). Disc-Aligner is used to predict word alignments. We first use Base-PE to generate two baseline phrase tables, $\textit{Base-PE}^{\textit{1-best}}$ and $\textit{Base-PE}^{\textit{15-best}}$, using 1-best and 15-best alignments respectively; we then use MBR-PE to generate two improved phrase tables, $\textit{MBR-PE}^{\textit{1-best}}$ and $\textit{MBR-PE}^{\textit{15-best}}$, using link-pruned 1-best and 15-best alignments. The pruning threshold is set to be 10^{-5}. Similarity features computed in our MBR model for phrase pairs are also used as features in decoding.

Table 1. MBR-PE vs. Base-PE on the first data set

	IBM-BLEU%		
	MT03	MT05	MT08
Base-PE$^{1\text{-best}}$	32.87	31.40	21.09
Base-PE$^{15\text{-best}}$	33.38	31.83	21.57
MBR-PE$^{1\text{-best}}$	**33.50**	**32.18**	**21.77**
MBR-PE$^{15\text{-best}}$	**34.47**	**32.85**	**22.41**

Table 2. MBR-PE vs. Base-PE on the second data set

	IBM-BLEU%		
	MT03	MT05	MT08
Base-PE$^{1\text{-best}}$	36.13	34.19	22.50
Base-PE$^{15\text{-best}}$	36.54	34.50	22.94
MBR-PE$^{1\text{-best}}$	**36.97**	**34.80**	**23.26**
MBR-PE$^{15\text{-best}}$	**37.21**	**35.07**	**23.85**

We can draw several conclusions from Table 1: (i) by using *n*-best instead of 1-best alignments in phrase extraction, significant improvements are obtained (+0.43/+0.48 BLEU on MT05/MT08), which confirm that SMT systems can benefit from making the annotation pipeline wider; (ii) both MBR-PE$^{1\text{-best}}$ and MBR-PE$^{15\text{-best}}$ outperform two baseline phrase tables. These improvements, we think, mainly come from using all similarity features as additional features in decoding, which brings more discriminative power to the SMT model, and using alignment pruning to remove those low-quality phrase pairs, which reduces the possibility of making decoding errors; (iii) MBR-PE$^{15\text{-best}}$ performs significantly better than MBR-PE$^{1\text{-best}}$. We think that the reasons are two-fold: 1) more phrase pairs included in hypothesis spaces makes the computation of consensus statistics to be more accurate; 2) more alignments involved makes the hypothesis distributions to be more accurate. Compared to Base-PE$^{1\text{-best}}$, MBR-PE$^{15\text{-best}}$ obtains +1.45/+1.32 BLEU on MT05/MT08. From Table 2 we can see that, compared to

Base-PE$^{1\text{-best}}$ and Base-PE$^{15\text{-best}}$, MBR-PE$^{1\text{-best}}$ and MBR-PE$^{15\text{-best}}$ achieve significant improvements on both dev and test data sets as well, which solidly demonstrate the effectiveness of our approach again.

5 Conclusions

We have presented a novel MBR-based phrase extraction pipeline for SMT training. Under this pipeline, the quality of phrase pairs are measured by their internal similarities, and phrase pairs with low MBR model scores are pruned based on an alignment pruning strategy. One future work to do is to investigate more features to measure similarities between phrase pairs. The other is to compare our approach with both discriminative learning-based methods and significance test-based methods for SMT phrase extraction.

References

1. Och, F., Ney, H.: The Alignment template approach to Statistical Machine Translation. Computational Linguistics 30(4), 417–449 (2004)
2. Kumar, S., Byrne, W.: Minimum Bayes-Risk Word Alignment of Bilingual Texts. In: Proc. EMNLP, pp. 140–147 (2002)
3. Kumar, S., Byrne, W.: Minimum Bayes-Risk Decoding for Statistical Machine Translation. In: Proc. HLT-NAACL, pp. 169–176 (2004)
4. Kumar, S., Macherey, W., Dyer, C., Och, F.: Efficient Minimum Error Rate Training and Minimum Bayes-Risk Decoding for Translation Hypergraphs and Lattices. In: Proc. ACL, pp. 163–171 (2009)
5. Mi, H., Huang, L., Liu, Q.: Forest-based Translation. In: Proc. ACL, pp. 192–199 (2008)
6. Liu, Y., Xia, T., Xiao, X., Liu, Q.: Weighted Alignment Matrices for Statistical Machine Translation. In: Proc. EMNLP, pp. 1017–1026 (2009)
7. Och, F.: An Efficient Method for Determining Bilingual Word Classes. In: Proc. EACL, pp. 160–167 (1999)
8. Xiong, D., Liu, Q., Lin, S.: Maximum Entropy based Phrase Reordering Model for Statistical Machine Translation. In: Proc. ACL, pp. 521–528 (2006)

Application of Computer Aided Instruction (CAI) in English Teaching for Non-English-Major Postgraduates Based on Constructivism

Xueyao Chen

School of Foreign Languages, Wuhan University of Technology,
430070 Wuhan, P.R. China
cherubinchen@163.com

Abstract. To merge with multimedia technology, IT and educational science, Computer Aided Instruction (CAI) applied in English teaching has displayed its incomparable advantages as a modern teaching technology. This paper analyzes the current situation of English teaching for non-English-major postgraduates, illustrates the functions of CAI in English teaching for non-English-major postgraduates based on the theory of constructivism, and puts forward a new teaching model with the application of CAI, for the purpose of transforming thoroughly the traditional cramming teaching so as to improve the quality education. Some problems involved in the application of CAI are also discussed.

Keywords: Computer Aided Teaching (CAI), English teaching, Non-English-Major postgraduates, Constructivism.

1 Introduction

The purpose of English teaching for non-English-major postgraduates is to help strengthen the students' ability of applying English to both academic field and practical life like how to communicate, acquire information, and solve problems independently by use of English. But it has been test-oriented for a long time, being influenced by traditional teaching methods and ideas. Therefore, it is very hard to develop the students' initiative, activeness, creativity and independence of self-learning. Computer Aided Instruction (CAI) as a modern teaching technology characterizes not only large capacity, high speed and information processing in a wide range, but also the interaction and intellectual faculties[1]. This paper is going to discuss how to apply modern computer technology to establish a teacher-guided and student-centered teaching model, based on the theory of constructivism.

[1] Gang Deng: CAI in China's English Education Reform, the Mode of Instruction and Teaching Theories (2007).

A. Xie & X. Huang (Eds.): Advances in Computer Science and Education, AISC 140, pp. 205–210.
springerlink.com

2 Current Situation

Postgraduate English teaching in China is now undergoing a reform, trying to transform the traditional teacher-centered model to the student-centered teaching model[2].

2.1 Curriculum

Take Wuhan University of Technology in China for example, the proportion of different types of courses and application of CAI are shown in table 1.

Table 1. Curriculum of English Course for Non-English-Major Postgraduates

Type of Course		Proportion比例	Application of CAI
Listening		15%	Multimedia-system
Intensive Reading	Reading	25%	None
	Writing	25%	None
	Translating	20%	None
Speaking		15%	None

From table 1, we can see that although the curriculum is rather comprehensive, the emphasis is mainly laid on traditional classroom teaching where there is no application of CAI at all. The only course where CAI is applied is listening class that only accounts for 15% in the whole curriculum. It is also the most welcomed class.

2.2 Teaching Model

In a traditional class, teachers transmit knowledge to students by utilizing, analyzing and explaining the content of textbooks[3]. The teaching model is shown in figure 1.

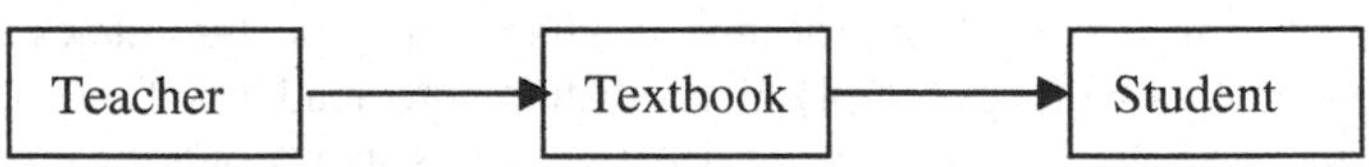

Fig. 1. Teaching Model Based on Textbook

A survey has been carried out among both teachers and students about their appraisal on the traditional teaching model. They are required to score the current teaching model from 1 (the least satisfied) to 5 (the most satisfied). The results are shown in table 2 and compared in figure 2.

[2] Weining Ji: Reflections on the Reform of Postgraduate English Teaching. (2007)

[3] Jianlin Chen: The Integration of Computer and Networks into Foreign Language Curriculum – A Research Based on College English Reform. (2010)

Table 2. Appraisal on Current Teaching Model

Teaching Model		Students' Average Score	Teachers' Average Score
Class Teaching	Teachers' teaching is based on the content of textbook.	2.3	3.1
Teaching Organization	Teachers give lecture mostly with a few student activities.	2.7	2.5
Teaching Design	Emphasis is put on the design and development of teaching content.	2.3	3.4
Teaching Procedure	Teachers analyze and explain the text, illustrate examples and give exercises.	2.4	2.1
Teaching Assessment	Students are assessed in the end by being given a final score.	1.9	1.8

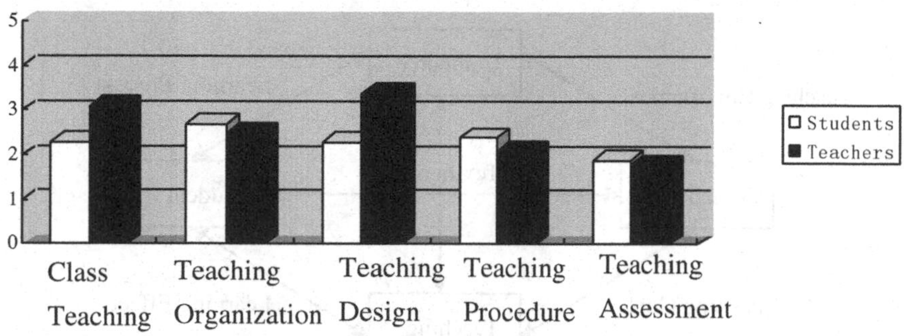

Appraisal on Current Teaching Model

Fig. 2. The survey shows that most of the students and teachers' scores are below 3, which proves that they are unsatisfied with the current teaching model of English teaching for non-English-major postgraduates

3 Application of CAI in English Teaching for Non-English-Major Postgraduates Based on Constructivism

Constructivism has a long history, but it is modern computer technology that makes it possible to create the learning environment of constructivism in a real sense.

3.1 Theory of Constructivism

The theory of constructivism was first put forward by Jean Piaget, a Swiss psychologist, in 1960's. According to constructivism, learning is a kind of active process of knowledge construction and understanding where learners adjust their knowledge structure through assimilation and adaptation and acquire knowledge when interacting with the objective environment[4].

3.2 Technical Support of CAI

Traditional English teaching model overemphasizes the function of teachers while neglecting and even depressing the individuality and initiative of students. Thanks to computers, these problems can be eliminated. The interaction of computers and human can help to increase the students' motivation and initiative. Computers can also build vivid and visualized learning environment by creating multiple sensory stimulation through pictures, words, sounds and images to arouse the students' interest so as to realize their active language practice. Thus the traditional textbook-based teaching model will turn into computer-based model, as indicated in figure 3 (Jianlin Chen: 2010).

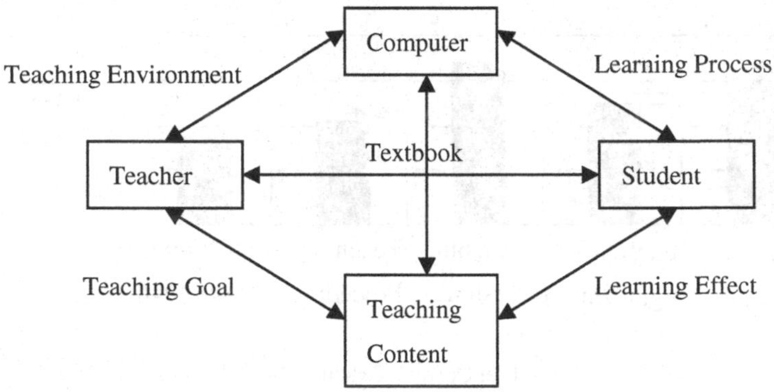

Fig. 3. Teaching Model Based on Computer

Fig. 3. Obviously, CAI provides technical support for the realization of the core ideas of constructivism, which will eventually boost the innovation of English teaching for non-English-major postgraduates.

3.3 Construction of a New Teaching Model

Task Presentation. The primary work of the teachers is to consider how to present the learning tasks to stimulate the students' enthusiasm and impetus to learn. If teaching information can be conveyed to the students through computers and multimedia technology, students will be easily attracted and actively get involved. Their imagination, impetus and interest will be fully developed. Motivated by tasks, the students will also take initiatives to accomplish their own learning tasks.

[4] Yuanyuan Liu: Advantages and Disadvantages of English Teaching via Multimedia and Teacher Roles. (2009).

Situation Setup. According to constructivism, it is in a certain situation that knowledge will be acquired in a way of meaning construction through the help of teachers and collaborators by use of study resources. Therefore, situation setup is the premise of meaning construction. We can create an ideal English acquisition environment through multimedia integration, individualized study, interacted and remote teaching and learning. In this environment, students can experience various creative and cooperative activities in person so as to meet the needs of meaning construction to the greatest extent.

Interactive Learning. Teaching is actually an interacted process. The most effective study lies in the mutual interaction between learners and teachers, learners and learners, learners and teaching content, and teaching and multimedia. Modern computer technology helps to extend teaching activities beyond classroom and break the limitation in time and space.

4 Problems Involved in the Application of CIA

Like other teaching approaches, CIA is not almighty and has its own shortcomings. So when applying CIA in English teaching, we also need to pay attention to the following problems.

First, teachers should not completely rely on the computers to organize teaching activities without interfering[5]. The role of teachers can never be replaced by computers.

Second, computers and multimedia also have some side effects. Since they provide learning resources with more fun, students' attention are easily distracted. Thus the autonomous learning should be under close and strict management and supervision.

Last, although it is an inevitable trend that CAI will be widely applied in teaching, it doesn't mean we can completely abandon traditional teaching methods. Actually, we need to take advantage of the strength of both and avoid their shortcomings so as promote teaching efficiency.

5 Conclusions

The application of CAI is an important way for the reform and development of English teaching for non-English-major postgraduates by providing comparatively ideal approach to create a healthy environment for quality education. CAI embodies the core ideas of constructivism, so it is also the breakthrough point for establishing a whole new English teaching system to transform traditional cramming teaching to student-centeredness. CAI not only optimizes English teaching resource, environment, process, and objective, but also boosts teaching effect and enhances students' learning efficiency. Computers can/will never substitute teachers but they offer new opportunities for better language practice. We need to take full advantage of CAI to improve English teaching for non-English-major postgraduates.

---

[5] Shouren Wang: Toward Web-Based English Education. (2002).

References

1. Deng, G.: CAI in China's English Education Reform, the Mode of Instruction and Teaching Theories. Foreign Language of China 4(3), 10–14 (2007) (in Chinese)
2. Chen, J.: The Integration of Computer and Networks into Foreign Language Curriculum – A Research Based on College English Reform. Shanghai Foreign Language Education Press, Shanghai (2010) (in Chinese)
3. Wang, S., Zhao, W.: Toward Web-Based English Education. Foreign Language Research 5, 62–65 (2002) (in Chinese)
4. Ji, W., Zhao, Y.: Reflections on the Reform of Postgraduate English Teaching. Journal of Anhui University of Technology (Social Science) 24(6), 95–96 (2007) (in Chinese)
5. Liu, Y.: Advantages and Disadvantages of English Teaching via Multimedia and Teacher Roles. In: Research in Digital Foreign Language Teaching, Beihang University Press, Beijing (2009) (in Chinese)

Discussion on Digital Storytelling in Comprehensive Practice Courses Application

JiangYan Zheng, Hao Zhang, Liang Hong, HongHui Wang, and ChunPing Zhang

Qinggong College, Hebei United University
zjymm2005@126.com

Abstract. The increasing development of computer and information technology has brought new challenges as well as opportunities to education. As a way of personal multimedia storytelling, digital storytelling is very popular in foreign countries and begins to be widely used in the field of education, and have achieved some results. Firstly, digital storytelling is given detailed explanation. Then functions and advantages in applying digital storytelling in comprehensive practice courses are put forward. Finally, the feasibility of applying digital storytelling in comprehensive practice courses is confirmed by a teaching case.

Keywords: Digital storytelling, Comprehensive Practice Course, Teaching cases.

1 Introduction

1.1 An Overview of Digital Storytelling

Since people learn to talk," tell me a story" has become child and adult mutually requirement. The story to tell another person something, can be a real thing or a fabricated things, the story has become our daily conversation part. Story is the most primitive way of teaching, not only to attract children's attention, to stimulate people's imagination, also confusing to explain ideas and concepts. Prior to the formal establishment of the school, parents teach their children by telling the story of knowledge and experience to build their knowledge, It is a important method to the first human to share knowledge, ethics and values. Thus, the story and education are inextricably linked; the story in education plays a significant role.

With the development of information technology and the media development, the traditional story of application in teaching is increasingly reduced, technology during the presentation of the story began to play a role, so that the way of telling the story is more and more abundant, spreading more and more widely. A use of multimedia technology and network technology have a way of telling stories, people call it the digital storytelling.

Digital storytelling is through a series of pictures, video and the narrator 's voice to tell the story, which combines sound and vision two elements, so that students from the sound and picture at the same time getting the story content, and can obtain the intuitive feeling and experience [1]. Digital storytelling as a special way of expression, to expand the

A. Xie & X. Huang (Eds.): Advances in Computer Science and Education, AISC 140, pp. 211–216.
springerlink.com

sound, especially for those expressing ability of poor students provide a stage to show themselves. In addition, Digital storytelling as a new teaching and learning mode, reflected a kind of new course concept. It emphasizes student-centered and the group cooperation learning, students are required to explore the real life, and through digital storytelling to show knowledge and ideas about the problem. In this process, students are both as readers and writers, as well as writers, artists, designers and directors.

1.2 The Characteristics and Production Process of Digital Storytelling

Upon completion of the task of digital storytelling, students under the guidance of teachers, search for relevant information about the subject through interviews, surveys, collect information from life or online stories, write the script, and tell the story with the help of multimedia. We can see that the production of digital storytelling does not only focus on the production of student work, and more emphasis is on cognitive, emotional and improvement of information literacy. Students can master the textbook knowledge, but more importantly let students be able to process through the acquisition of knowledge, handle the process of social relationships, and guide students to calmly face the complex relationship between the living worlds. In addition, promote student visual, musical, aesthetic and other abilities. In the course, practice is an essential part. In this process, teachers guide students to use various means and methods, from different angles to carry out inquiry learning, through a series of events, students acquire a certain amount of learning materials, the accumulation of certain knowledge can write a story after the script, through the forms of multimedia can make their knowledge visualized.

Digital storytelling as a teaching method are applied and developed continuously in the classroom teaching, but in the process two major restricted factors appear: the first one is the limited class time, and the second one is the hardware resources. As for these two issues, foreign experts made the following six steps in the production of digital storytelling [2]:

Write a script draft.

Design a story and script series corresponding storyboard. Storyboard shown in Figure:

Screen number	Diagrams or pictures	Narrative / sound track
1		
2		
…		

Discussion, modify the script.

Layout photos in the video time line of video editing software.

Add narrative content to audio track.

Add photo effects and transition effects.

When teachers design for students to create digital storytelling of these six steps, there are two very important benefits: first, to specify tasks, the extension of the student extra-curricular time to learn and experiment, while in video editing software, students can more clearly see senses of participation. Followed by the production of teachers into the blackboard or whiteboard on the steps, teachers can understand the progress of students easily and provide timely study guide.

2 The Role and Advantages of Stories in the Comprehensive Practice Course [3][4]

Digital storytelling as a novel application of the comprehensive practical learning activities, have had a great impact on multiple intelligences of students, as well as cognitive, emotional and behavioral inputs, but also provide a form for curriculum integration of information technology.

2.1 To Make Up for Students in the Course of Comprehensive Practical Activities of Uneven Development of Multiple Intelligences

Comprehensive Practice Course is the existing knowledge and experience of students and based on real life. The curriculum design reflects the comprehensive application of knowledge, emphasizes the practice, students focus on analysis of issues and problem-solving skills development. In this process, teachers train students to develop multiple intelligences, but due to students' intellectual development vary, each other's strengths and weaknesses are different. So in the event there will be some students who are often actively involved in the activities while some students have cynicism problems. The digital storytelling is to let students use digital technology and multimedia tools learn. Digital storytelling consists of video, audio and other digital resources. These visual and auditory sensory involvement, provide a broad space for students' balance development of multiple intelligences, especially for introverted students to provide the advantages of intelligent stage show, enhancing their confidence and participation initiatives.

2.2 A Form of Information Technology and Curriculum Integration

In digital storytelling comprehensive practical activities, students can not only collect text, pictures, video and other data, but also find useful information from the network. Then these information and information technology serve as a platform to create creative and unique works. In the process, digital storytelling provides information processing support for the Comprehensive Practice Course, as well as support for the exchange of students' collaboration. In addition, digital storytelling provide students with creative thinking space, and realize the function of information technology in the curriculum resources collection, information processing, collaboration, communication and research, which truly reflect the organic information technology and curriculum integration.

2.3 Promote Students' Behavior Investment, Emotional and Cognitive Inputs

These are mainly demonstrated in the following three aspects: **Involvement of digital technology and media to promote the development of deep-level cognitive engagement.** photos, videos and audios visualize and specify the abstract contents and give significance to the story's pictures, videos and audios. This is conducive to the promotion of student learning activities in deep thought, which also shows the degree of cognitive involvement of students has increased.

Encourage Students' Emotional Investment by Storytelling. In the process of making digital stories, students' telling the emotional story allows them to experience stories and stimulate students' interest and curiosity. The story can enhance communication and emotion among students, between students and teachers.

Digital Storytelling Centered around Comprehensive Practice Activities Promote the Investment Behavior. When digital storytelling are used in the Comprehensive Practice Course, students are motivated to be more conscious to review the learnt knowledge and put what their have learned into practice, as well as take the initiative to apply the knowledge to solve practical problems. These are the performance of behavior investment, thereby promoting a better investment in the students learning to cultivate the students "live" capability, "innovation" ability and "contribution" capability.

3 "Causes and Prevention of Myopia," Teaching Case

3.1 Course Objectives

Cognitive goal: to understand the formation of internal and external causes of myopia. Emotional goals: to recognize the confusion caused by myopia, to make students aware of protecting their eyes.

Motor skills objectives: the ability to collect information, make cooperation and exchange, and use multimedia software to display their achievements.

3.2 Contents

Analysis Phase. Divide students into different groups to investigate the number of students suffering from myopia and understand the troubles myopia brought them. Then collect information and pictures on the causes of myopia through the network or the library.

Digital Storytelling Production Phase [5]. Complete program of activities. In order to facilitate activities, students from different groups, and name their own team. Each member is responsible for specific tasks. Then team members discuss the plan activities and arrangements. Here is the plan of a group's activities:

Group Name: Peas Team
General Director: Liu Yang
Dubbing: Zheng Jiangwei
Script writers: Liu Xin
Information workers: Zheng Tong, Zheng Jiangwei
Computer Operator: Wang Shuo
Specific Arrangements of Activities

First step is that members of the group discuss the story theme .The second step is that script writers write the story and computer operator prints it and save it in the computer. The third step is that information workers collect and collate information. The fourth step is that the director guides the production of digital works and dubbing it. The fifth step is to playback digital storytelling, and then proposes amendments suggestions. The sixth step is to upload the digital storytelling to the internet.

Manuscript Writing. Script was written after the students internalize the knowledge output of the process. Comprehensive Practice Courses focus on the various curriculums integration, and pay attention to students' practice, so the Comprehensive Practice Course is more creative than other subjects.

Manuscript Content:
Myopia has become a prominent issue in my school. According to our survey, an average of 4-5 people per class is myopia. What causes myopia? One reason is that the genetic parents, especially the most vulnerable to high myopia genetic parents of the child. Wc can not control this factor. But the day after tomorrow development of our vision is also important, the expert survey, a large children's myopia is due to lack of physical exercise, eating, picky eaters, nutritional insufficiency, coupled with the lack of scientific knowledge with the eyes, and no eye protection, long-term adverse eye habits. It is more dangerous to spend much time watching TV and playing video games. In addition, the computer is one of the most important reasons causing myopia. Students' myopia is closely related to environmental conditions. In terms of students, poor lighting during homework, together with lack of reading and writing habits, forces the students to get close to the book, causing a series of near reflex. Near reflecting lets the biochemistry structural change in the eye wall and increases intraocular pressure to form a longer axial length that is true axial myopia. Therefore, we should seriously do eye exercises every day, for an instant 20 times, making the eyes flexible, comfortable and bright.

Data Collection. After the students write a good script, they will collect information and music in comprehensive practical activities. In this process, as primary students have relatively poor ability to identify information, teachers should provide timely help to guide them to collect online information and save them to different folders according to different types.

Storyboard Series. To alleviate the workload of students, teachers assist students to design a good storyboard series, students according to this template, fill in your own story board with pictures and sound blend together properly.

Edit Digital Storytelling. Students input pictures and sounds collected into the computer, and adjust the picture based on the story boards and the location of the sound, making sound and picture harmonious, in line with the requirements of the theme. After adding some of these images transition effects, as well as rendering the atmosphere add background music to the story, the final output of the story comes out. In the process, it is found that students are more focused on the transition effects and add special effects to avoid distracting the teacher works with the transition effects limit up to a maximum of 3.

Show the Work. Participate in activities with students from fourth and fifth grades, grouping mixed group, stage show in the works, let each group make the work of students participating in activities to demonstrate to the standard evaluation form for group peer review, suggest modifications, and then the revised works were located in their respective class player, form a learning community, promoting the development of students.

After interviewing the lecturer Liu Haitao, it is known that the comprehensive practical activity plays a significant role in students' development, which is summarized as follows:

Digital Storytelling Provides Students with a Platform to Show Themselves. Digital storytelling for students, is a relatively new concept, they like to try and experience. Especially in the completion of the process of digital storytelling, it is to be fully mobilized that language expression, photography skills, aesthetics, music appreciation and computer operations and other areas. So students can easily find their strengths and weaknesses and develop a self-display space.

Improve Students' Comprehensive Knowledge, Performing Ability and Social Practice Ability. In digital storytelling-based activities, students use Powerpoint to create digital storytelling about myopia as their own learning goals, to promote comprehensive application of knowledge learnt in and out of school, explore the theme-related content and activities to complete the activity tasks. Thus the students' hands-on and practical ability are enhanced.

To Ensure the Smooth Implementation of Activities, Teachers Should Provide Necessary Support Services. In the activities, teachers' support services include emotional support and technical support. Emotional support is that in the activity implementation course, the group members may generate intense controversy for a certain issue. When the internal conflict is difficult to resolve, the teacher need to properly analyze problems, make reference, guide students to successfully carry out activities, and encourage students to persevere. Technical support refers to the fact that teachers need to give students the necessary support and assistance in the choice of theme, the flow of digital storytelling, and the application of video software and so on.

References

1. Lamber, J.: Digital Storytelling Cookbook. February 9-20 (2007)
2. The Educational Uses of Digital Storytelling, http://www.coe.uh.edu/digital-storytelling/examples.htm
3. Chen, J.: Application of Digital Storytelling in Education, Shanghai (2006)
4. Peng, X.: On the basic characteristics of "Comprehensive Practice Courses". Social Science Front, 292–294 (2007)
5. Gubrium, A.: Digital Storytelling: An Emergent Method for Health Promotion Research and Practice. Health Promotion Practice,186–191 (2009)

GCM Data Analysis Using Dimensionality Reduction

Zuoling Li and Guirong Weng*

School of Mechanic & Electronic Engineering
Soochow University
Suzhou, China
{20094229048,wgr}@suda.edu.cn

Abstract. Extracting useful information from gene microarray becomes a research in great demand, but because the microarray sample size is small, high-dimensional, nonlinear, traditional statistical learning methods face a challenge of "dimensionality disaster" and "problem of small sample size", therefore, dimensionality reduction becomes a key to pattern recognition. This article uses Principal Component Analysis (PCA) and Local Tangent Space Alignment (LTSA) to reduce the dimensionality of the Global Cancer Map data, and then utilizes Support Vector Machine to classify the data, PCA getting the better result.

Keywords: PCA, LTSA, SVM, GCM, Classification.

1 Introduction

With an increasing number of bioinformatics data sets can be obtained conveniently from the internet data base or some user-generated text documents, finding out information hidden in the gene microarray has gained vast research efforts recently. But due to bioinformatics data have extremely high dimensionality in the feature space and lack labelled samples, it raises new challenges of "dimensionality curse" and "problem of small sample size" to data classification. If the researchers classify the data directly, often fail to achieve the very satisfactory results.

GCM (Global Cancer Map) data set is one of the most notable bioinformatics data first reported by Ramaswamy et al [1]. This gene expression data set represents a collection of 280 various samples including 190 tumors and 90 normal samples. The classification task is to distinguish tumors from normal based on 16,063 gene features. In order to settle the two problems of "dimensionality curse" and "problem of small sample size", both dimensionality reduction to the data sets and a machine learning method fit for small subgroups are essential. In this article, two methods PCA and LTSA are utilized to reduce the dimensionality of the GCM data. And then SVM is used to classify the reduced data, for the decision function of it is only decided by part of the support vectors, this feature can partly overcome the "problem of small sample size" to a certain extent. After comparing the result, PCA performs better.

* Corresponding author, Address: No.178.GanJiang East Road 178, 31#, Suzhou, JiangSu, 215021, P.R. of China.

A. Xie & X. Huang (Eds.): Advances in Computer Science and Education, AISC 140, pp. 217–222.
springerlink.com

2 Dimensionality Reduction Method

Data dimensionality reduction is increasingly conspicuous in research recent years. Map high-dimensional data to low-dimensional space, and low-dimensional data can reflect the information in the original high-dimensional data, this is data dimensionality reduction [2]. At present the data dimensionality reduction methods can be divided into linear and nonlinear dimensionality reduction, PCA used in this paper is a typically linear method, and LTSA is nonlinear.

2.1 Principal Component Analysis (PCA)

The central idea of Principal Component Analysis (PCA) is to reduce the dimensionality of a data set consisting of a large number of interrelated variables, while retaining as much as possible of the variation present in the data set. This is achieved by transforming to a new set of variables, the principal components (PCs), which are uncorrelated, and which are ordered so that the first few retain most of the variation present in all of the original variables [3]. The specific algorithm is as follows [4]:

1) standardize the data,

$$x_{ij} = \frac{x_{ij} - \overline{x}_j}{S_j} \quad (i = 1,2,...,n; j = 1,2,..., p)$$

S_j is the standard deviation of the sample x_j;

2) figure out the convariance matrix Z between variables;
3) analyze the character $Z = U\Lambda U^t$ (Λ is a diagonal matrix made up of eigenvalues of Z);
4) calculate the number of principal components, according to $\frac{\sum_1^n \lambda_i}{\sum_1^p \lambda_j}$ to select principal components.

2.2 LTSA

LTSA (Local tangent space alignment) was firstly put forward by ZHANG Zhen-yue, etc [5]. The main point of this algorithm is the neighborhood of each sample point can be approximately expressed by local tangent space coordinates. The brief description of the algorithm is explained as follows:

1) Constructing k neighbors $X_i = [x_{ij}, \dots, x_{ik}]$, including x_i itself in terms of Euclidian distance in input space.
2) Extracting local information by calculating the d largest eigenvectors $g_1, \dots, g_d$ of the correlation matrix $(X_i - \overline{x_i}e^T)^T (X_i - \overline{x_i}e^T)$, and

setting $G_i = [\frac{e}{k}, g_1, \ldots, g_d]$. Here $\overline{x_i} = \frac{1}{k}\sum_j x_{ij}$, e is a k-dimensional column vector of all ones.

3) Constructing the alignment matrix B by locally summing with initial $B = 0$ as follows : $B(I_i, I_i) \leftarrow B(I_i, I_i) + I - G_i G^T$, $i = 1, \ldots, N$; The neighborhood set I_i represents the set of indices for the k nearest-neighbors of x_i. I is an $N \times N$ identity matrix.
4) Computing the $d+1$ smallest eigenvectors of B and setting the global coordinates $Y = [u_2, \ldots, u_{d+1}]^T$ corresponding to the 2nd to $d+1st$ smallest eigenvalues.

3 Support Vector Machine

Support Vector Machine (SVM) was original introduced by Vapnik [6] and his co-workers, and successfully applied to many fields now. SVM is improved from the best classification surface when linear classified.

In linear classification problem, the classifying function is $x \cdot w + b = 0$, here w is a weight vector and b is a scalar bias. Normalize,

$$(x_i, y_i),\ i = 1, \ldots, n,\ x \in R^d,\ y_i \in \{+1, -1\},$$

Subject to $$y_i[(w \cdot x_i) + b] - 1 \geq 0,\ i = 1, \ldots, n, \tag{1}$$

Through Lagrange optimize method, translate the problem to its duality:

$$\max L(\alpha) = \sum_{i=1}^{m} \alpha_i - \frac{1}{2}\sum_{i,j=1}^{m} \alpha_i \alpha_j y_i y_j K(x_i, x_j), \tag{2}$$

Subject to $$\sum_{i=1}^{m} \alpha_i y_i = 0,\ \ \alpha_i \geq 0,\ i = 1, \ldots, n, \tag{3}$$

here α_i is Lagrange multiplier. Get the best classifying function:

$$f(x) = \operatorname{sgn}\{(w \cdot x) + b\} = \operatorname{sgn}\{\sum_{i=1}^{n} \alpha_i^* y_i K(x_i \cdot x) + b^*\}. \tag{4}$$

In non-linear classification, introduce a slack variable, function (1) is subject to

$$y_i[(w \cdot x_i) + b] - 1 + \xi_i \geq 0,\ i = 1, \ldots, n,$$

And now compute the max of function (2),

subject to $\sum_{i=1}^{m} \alpha_i y_i = 0, \quad 0 \le \alpha_i \le c, \quad i = 1, \ldots, n.$

Compared to linear classification problem, non-linear classification maps input data from original space to a high-dimensional space through $\varphi(x)$, here

$$\varphi(x_i) \cdot \varphi(x_j) = K(x_i, x_j),$$

And $K(x_i, x_j)$ is called kernel function. At present frequently used kernel functions are linear kernel function, polynomial kernel function and RBF kernel function. In this article we use the RBF kernel function:

$$K(x_i, x_j) = e^{\frac{|x_i - x_j|^2}{\sigma^2}}.$$

4 Analysis of Experimental Results

In this experiment, 180 groups of the GCM data are selected as training data, and the other 100 groups of the data as testing data, then use dimensionality reduction methods to reduce the 16063 genes, and then use SVM (RBF kernel) to classify the reduced data, the original data and reduced data are shown below, and the results are shown in table 1:

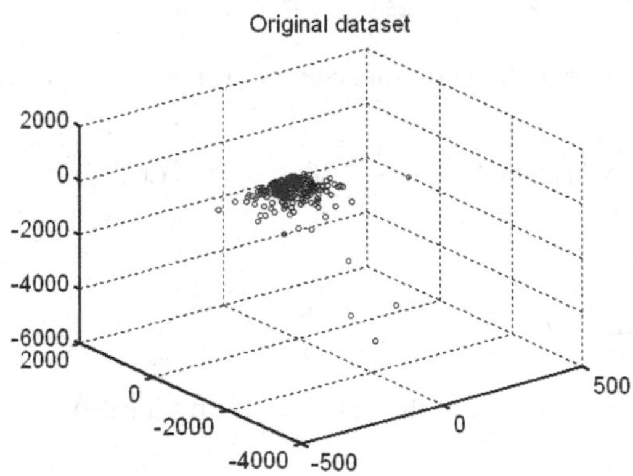

Fig. 1. This figure includes three parts. The first part shows the original dataset in higher dimension, the data are divided into two classes with different colors respectively. And the second is low dimensionality projection of the data reduced to 25-D by PCA. Similarly, the last one presents the 28-D data after dimensionality reduction using LTSA.

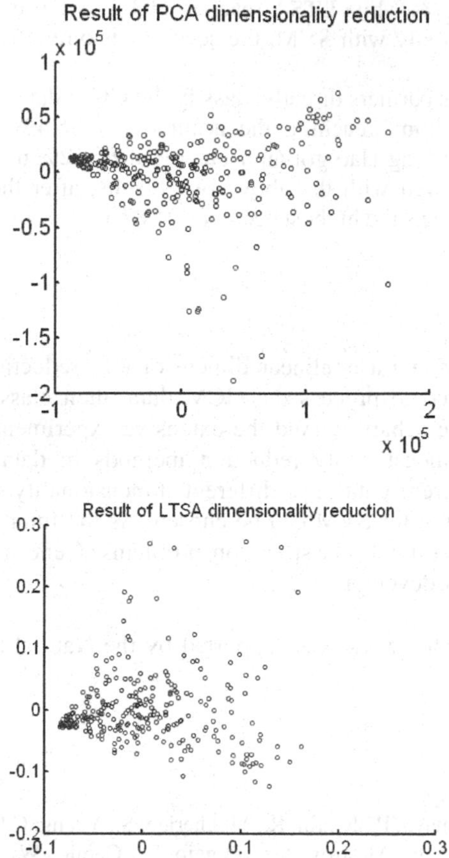

Fig. 1. *(continued)*

Table 1. The accuracy rate of using a linear kernel function

Reduced dimensionality	PCA	LTSA
16	84%	70%
19	83%	75%
22	86%	80%
25	87%	80%
28	78%	84%
31	78%	79%
34	78%	79%

From the table, we can see that different methods and different reduced dimensionality both affect the result of classification, and easily to find that PCA performs better than LTSA, and when it is reduced to 25-dimensionality with PCA and 28-dimensionality with LTSA, two dimensionality reductions both achieve the highest accuracy, PCA gets 87%, while LTSA reaches 84%. If we use the original dataset not reduced directly to classify with SVM, the accuracy is only 70%, and extremely time-consuming.

Lee K Jones and his partners directly classify the GCM data with SVM using a local minimax kernel algorithm, reaching the accuracy of 86.4% [7]. And Ruichu Cai, Zhenjie Zhang and Zhifeng Hao got an accuracy of 86.12% utilizing the algorithm of SVM-RFE [8]. Compared with the above two results, after the PCA dimensionality decreasing, this paper gets the highest accuracy of 87%.

5 Conclusion

In this paper, a linear and a nonlinear dimensionality reduction methods, PCA and LTSA are both applied to process the GCM data, then classify the data sets with SVM, good results are achieved. And the extensive experimental results have shown the superiority of dimensionality reduction methods in data classifying. When it comes to other different data sets, different dimensionality reduction method and different kernel function for SVM can be chosen. As the future research, we will use this method to identify multi-classification problems. Gene microarray experiments have a broad space for development.

Acknowledgments. This work was supported by the Natural Science Foundation of Suzhou (SYJG0934).

References

1. Ramaswamy, S., Tamayo, P., Rifkin, R., Mukherjee, S., Yeang, C.H., Angelo, M., Ladd, C., Reich, M., Latulippe, E., Mesirov, J.P., Poggio, T., Gerald, W., Loda, M., Lander, E.S., Golub, T.R.: Multiclass cancer diagnosis using tumor gene expression signatures. Proc. Natl. Acad. Sci. USA 98, 15149–15154 (2001)
2. Tenenbaum, J.B., de Silvav, Langford, J.C.: A global geometric framework for nonlinear dimensionality reduction. Science 290(5500), 2319–2323 (2000)
3. Jolliffe: Principal Component Analysis, 2nd edn. (2002)
4. Lijing, Weng, G.: PCA for leukemia classification. Applied Mechanics and Materials 43, 744–747 (2011)
5. Zhang, Z., Zha, H.: Principal manifolds and nonlinear dimensionality reduction via local tangent space alignment. SIAM Journal of Scientific Computing 26(1), 313–338 (2004)
6. Vapnik, V.N.: The nature of statistical learning theory. Springer, New York (1995)
7. Lee, J., Fei, Z., Alexander, K., Konstantin, R., Damon, B., Aik, T.: Confident Predictability: Identifying reliable gene expression patterns for individualized tumor classification using a local minimax kernel algorithm. BMC Medical Genomics 1(4), 10–18 (2011)
8. Cai, R., Zhang, Z., Hao, Z.: BASSUM: A Bayesian Semi-Supervised Method for Classification Feature Selection. Pattern Recognition 44, 811–820 (2011)

Application of the BP Neural Network PID Algorithm in Heat Transfer Station Control

Yang Yu and Dongmei Yin

School of Information Science & Engineering, Shenyang Ligong University,
Shenyang, China, 110159
kongkuchen@126.com, dongmei327617@163.com

Abstract. Heat transfer station temperature control system has characteristics of being nonlinear, time-varying and has hysteretic complicated large inertial, and the effect of control is closely related to the algorithms adopted, and the traditional PID controlling can't get good controlling performance for the heat transfer station system. To this problem, design a PID controller based on BP neural network and use increasing PID control algorithm in the Heat transfer station temperature control the system. The MATLAB experimental results prove that the effect of the designed algorithm is better than the traditional PID algorithm, and has a strong reference in the engineering application.

Keywords: BP neural network, PID, Heat transfer station, temperature control system.

1 Introduction

Along with the nonlinear complex degree enhancement of modern industry system and the increasing of controlled object uncertain factors, linear PID controller are often difficult to achieve the satisfactory control effect. However nonlinear PID controller can truly reflect the nonlinear relationship between the between the control volume and the deviation signal, to a certain extent, overcome the shortcomings of linear PID controller. This paper made the initial presentation about composition and principle of the heat-exchanging station control system, and established the BP neural network PID control model of the temperature system of supply water of heat exchange station. Using BP neural network improved the PID parameters, so as to realize the neural network PID control in the temperature system of supply water of heat exchange station.

2 Heat Exchange Station Control System

2.1 Composition of Heat Exchange Station

Hot water pipe network is divided into the first network and the second network, the first network is the network connection between the heat source plant and heat transfer station, the second network is the network connection between the heat exchange station and the heat users. Heat exchange station is connected between the one network

A. Xie & X. Huang (Eds.): Advances in Computer Science and Education, AISC 140, pp. 223–228.
springerlink.com

and the second network. The effect of heat exchange station is adjusting and transforming the heating pipeline heat medium, according to the needs of users assigned to each user.

2.2 The Operation Principle of Heat Exchange Station

The working principle of heat exchange station: the heat source provides high temperature water that is sent all heat exchange stations by the first network. In heat exchange station, hot water of the first heating network through the heat exchanger to conduct heat exchange with circulating water of the second network, transfers heat energy to cycle water of the second network, then from the second network by the heating pipes to the user, cooling cycle water of the second network returns to heat exchange station. Process map of heat exchange station is shown in Figure1.

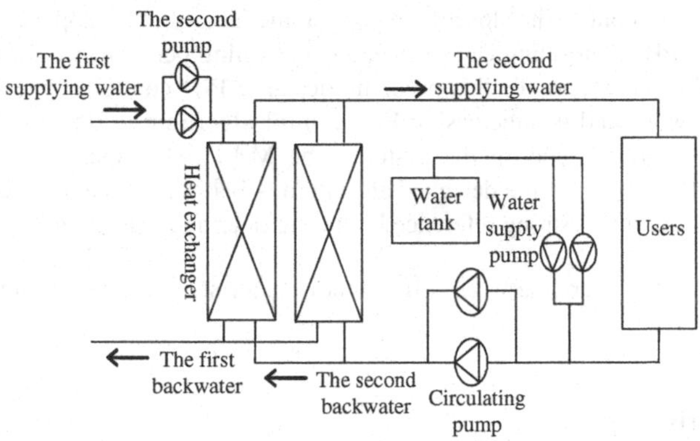

Fig. 1. Process of the Heat Exchange Station

3 Application of BP Neural Network PID Controller

3.1 Composition of BP Neural Network PID Controller

PID controller based on BP neural network composed of two parts, which are PID controller and BP neural network. In which, the PID controller conduct directly closed-loop control to controlled object, and the three parameters K_p, K_i and K_d are online adjustment method, BP neural network according to the operational status of the system to adjust the PID controller parameters, in order to achieve the optimization of operation index. Using BP neural network, can establish the PID controller with K_p, K_i and K_d parameters' self learning, the network structure is shown in Figure2.

In Figure2, O1, O2 and O3 are three input variables of the controller, *i*, *j* and *l* are the network input layer, implicit layer and output layer. w_{ij} is weight of the input layer to the implicit layer, w_{jl} is weight of the implicit layer to the output layer.

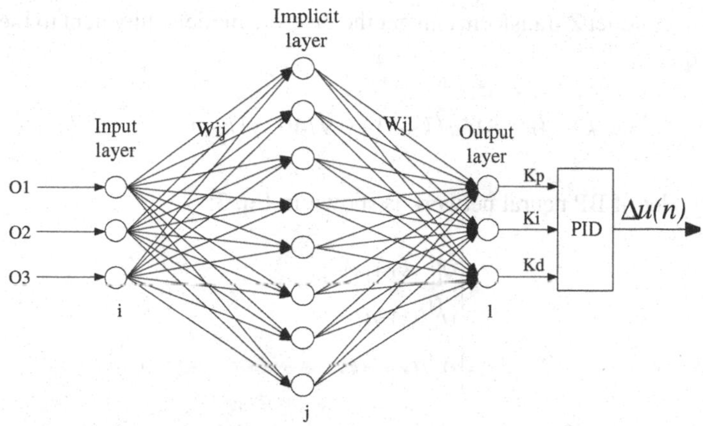

Fig. 2. The Structure of BP Neural Network

3.2 Establishment of the Model of BP Neural Network in Heat Exchange Station

Structure diagram of system is shown in Figure 3.

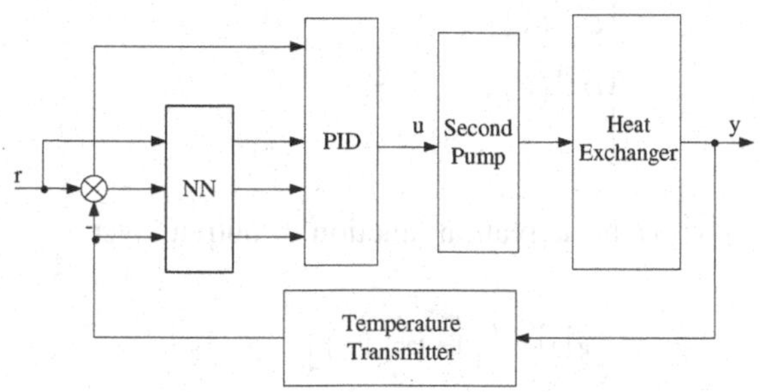

Fig. 3. Structure of the Heat Exchange Station System

This paper takes the PID controller based on BP neural network to use in the heat exchange station control system.

The input signals of the input layer are the given value of controlled variable, practical output value as well as the two deviation value respectively; in the implicit layer, it has a total of eight neurons; the output signals of the output layer are proportional, integral and differential parameter respectively. The dynamic performance of heat exchange station is approximately a delay first order and non periodic segment, the form of transfer function is shown in Eq. 1:

$$G(s) = \frac{0.31}{38s+1} e^{-1.5s} \tag{1}$$

Take Eq. 1 to conduct Z transform can get the discrete model equivalent to the $G(s)$, as shown in Eq.2:

$$y(k)=0.00815u(k-1)+0.97402y(k-1) \tag{2}$$

The input model of BP neural network is shown in Eq. 3:

$$\begin{cases} O_1^{(1)}(k)=e(k) \\ O_2^{(1)}(k)=r(k) \\ O_3^{(1)}(k)=y(k) \end{cases} \tag{3}$$

The input and output of the output layer of the network is shown in Eq. 4:

$$\begin{cases} net_l^{(3)}(k)=\sum_{j=1}^{8} w_{jl}O_j^{(2)}(k) \\ O_k^{(3)}(k)=g\left[net_j^{(2)}(k)\right] \\ O_1^{(3)}(k)=K_p \\ O_2^{(3)}(k)=K_i \\ O_3^{(3)}(k)=K_d \end{cases} \tag{4}$$

In Eq. 4, *l*=1,2,3, $g[\bullet]$ is the activation function of output layer.

$$g[\bullet]=\frac{1}{2}\left[1+\tanh(x)\right] \tag{5}$$

Because of the temperature control system of heat exchange station has certain amount of hysteresis, the using of conventional PID control plan is difficult to control the processes which have longer delay. And because computer is a device for handling of digital signal, so the control system uses the digital PID controller. The digital PID control algorithm is usually divided into position PID control algorithm and incremental PID control algorithm. This paper takes the incremental PID control algorithm for studying, control Equation is shown in Eq.6.

$$\Delta u(k)=u(k)-u(k-1)=K_p[e(k)-e(k-1)]+K_ie(k)+K_d[e(k)-e(k-1)+2e(k-2)] \tag{6}$$

In Eq. 6, $K_i=K_p\cdot T/T_i$, which is integral coefficient; $K_d=K_p\cdot T_d/T$, which is differential coefficient.

4 The Analysis of Results of Simulation

When take the unit step signal as the system input, we get the three important parameters of traditional PID control are $K_p = 23, K_i = 0.4, K_d = 2$, respectively. Simulation curve is shown in the Figure 4, figure 5, Figure 6 and Figure 7.

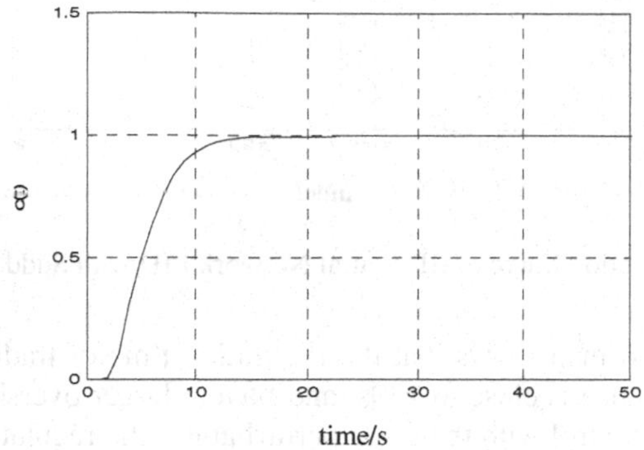

Fig. 4. The Simulation Curve of Traditional PID

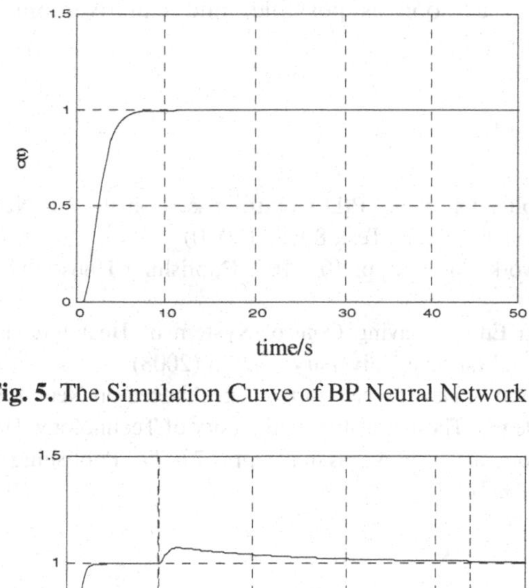

Fig. 5. The Simulation Curve of BP Neural Network PID

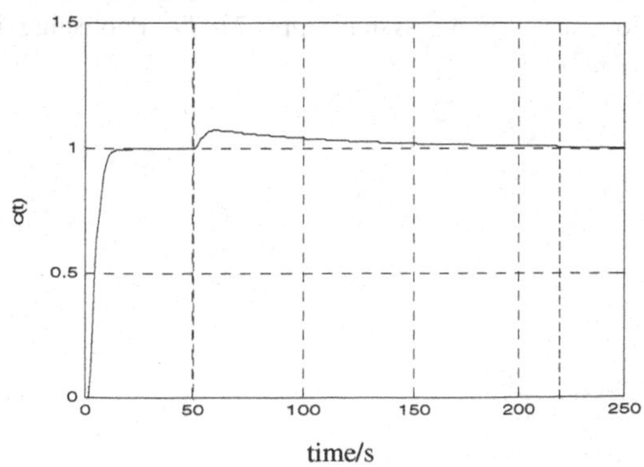

Fig. 6. The Simulation Curve of Traditional PID with Sudden Disturbance

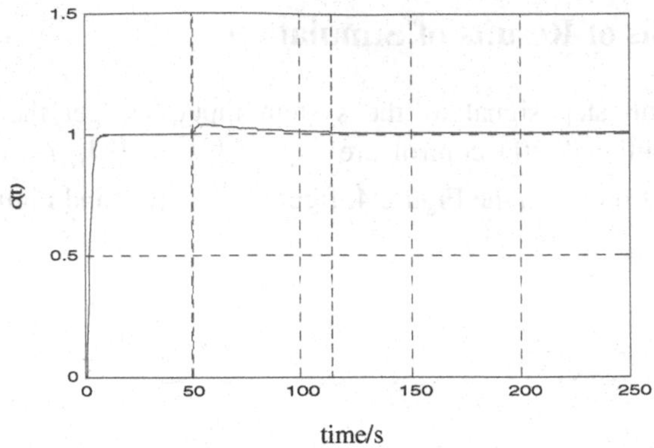

Fig. 7. The Simulation Curve of BP Neural Network PID with Sudden Disturbance

Comparing the two map shows that the regulation time of traditional PID control with external disturbance is close to 170s, and have a larger overshoot; However, BP neural network PID control with the same perturbation, the regulation time only need 65s, system overshoot is within the allowable range. So, comparing of the two simulation results shows that the BP neural network PID control can make the system reach a better stability as soon as possible, and can overcome the influence of disturbance on system.

References

1. Chang, Y.: The Application of the PID Control based on the BP Neural Network in the Thermal Power Station. Electronic Test, 80–82 (2010)
2. Xu, L.: Neural Network Control, pp. 101–103. Publishing House of Electronics Industry, Beijing (2009)
3. Zhang, Y.: Study on Energy Saving Control System of Heat Exchange Station (Master Degree Thesis). Dalian Maritime University, Dalian (2008)
4. Song, X.: The Research of Heating Network Control based on Neural Network Decoupling Controller (Master Degree Thesis). Dalian University of Technology, Dalian (2007)
5. Hua, S., Li, R.: Computer control system, pp. 77–79. Publishing House of Qinghua University, Beijing (2007)

The Influence of Human Resource Management Information System on Human Resource Management

Yan-Hong Yu

School of Business, Shandong Polytechnic University, Jinan 250353, P.R. China
yuyh13@163.com

Abstract. Using the methods of case analysis, the influence of Human Resource Management Information System (HRMIS) on Human Resource Management (HRM) was reviewed and studied. It was shown that HRMIS not only introduced some important theories from other sphere of learning, but also promoted the globalization of HRM, which accelerated the role shift and changed HRM in ways of thinking and behaving. Moreover, HRMIS greatly affect various personnel of HRM, such as chief executive officer, functional manager and common staffs. Furthermore, the development of HRMIS tremendously impact IT industry, including human resources website and HRM Software Company. The results in this paper was expected to propel forward the theories and practices of HRM in China.

Keywords: human resource management, management information system, Internet, influence, measure and evaluation.

1 Introduction

E-HR is the implementation on the project that is about human resource management information system (HRMIS), which is based on the Internet [1]. E-HR is to apply information technology into human resource management (HRM), which can improve the efficient and realize the integration between HRM and other management departments, such as finance, goods flow and customer relationship. From that, it improves the management level and degrades the cost of enterprise.

In China, the development of e-HR is late and its level is comparatively low. However, the increase of e-HR is very fast. In local academy, there are many scholars who study e-HR, but the systematic research on the influence of the e-HR is rare. This paper summarizes the realistic and practical significance and developing process of HRMIS, which expect to propel forward the theories and practices of HRM in China.

2 The Review of HRMIS

2.1 The Developing Process of HRMIS

With the development of information technology and human resource management theory, HRMIS is continuously maturing. Its entire developing process can be divided into four stages [2,3].

A. Xie & X. Huang (Eds.): Advances in Computer Science and Education, AISC 140, pp. 229–234.
springerlink.com

In the late 1960s, computer technology had entered the practical stage, and the size of the business is also growing. It became more and more difficult to calculate and pay the salaries by hands. The first generation of human resource management system - payroll management system came into being. Its main role is to automatically calculate the salary and its user is very rare.

In the late 1970s, the system began to record other basic information of staffs than salary history data. In addition, it generated reports and payroll data analysis functions greatly improved. The system had the functions both handling payroll and personnel information management.

In the early 1990s, the third generation of human resources management system came into being. Its main character was the unified management of all human resources-related data with centralized database. HRMIS had friendly user interface, powerful reporting tools, analytical tools and information sharing.

In the late 1990s and early 2000s, the Internet and Intranet technology became popular and the theories of human resource management developed greatly. In order to respond quickly to internal and external demands and changes, HRMIS became automatic, named E-HR. E-HR can achieve zero distance interaction between the staffs and the human resource managers, except for all functions of HRMIS.

2.2 Web-Based Human Resources Management Information System

Many scholars have defined HRMIS, but there is no consensus. Several of the more mainstream views are as follows.

HRMIS is one module of ERP (Enterprise Resource Planning), which completes the human resources management related business. HRMIS uses information technology to implement human resources management, which makes human resources management computerized and networked in order to enhance human resource management performance. HRMIS is a human resources management system based on Internet / Intranet. It uses a centralized database to process information, make staff self-service, outsourcing and share services. Its aims are to reduce costs, improve efficiency and service delivery staff. HRMIS refers to the establishment of an information platform, which make information technology and human resources management practices meet in the management activities. It is designed to apply for the specific needs of enterprises and sectors and becomes structural tools.

In this article, the definition of HRMIS is that it applies information technology into human resource management (HRM), which can improve the efficient and realize the integration between HRM and other management departments, such as finance, goods flow and customer relationship. It also improves the management level and degrades the cost of enterprise. In one word, this paper mainly studies the influences of the last stage of HRMIS on the human resource management theory and practice in China.

3 The Influence of E-HR on Human Resources Management

3.1 Its Influence on the Theory of Human Resources Management

The Role of HRM Department. In order to meet the challenges of HRMIS, HRM department must improve the qualities of human resource practitioners. Wayne

Anderson, the Amoco's human resources director, pointed out that the ideal human resource managers should start with 25% of the time on strategic human resource planning, 50% of the time to consider the human resources consulting and technology development and 25% of the time to be related to administrative work. Therefore, human resource manager should become the good partner of business operators. In other words, human resource manager in information age is not only a strategic partner, but also a leader of staffs and an agent of change.

The Thinking of Human Resources Management. The greatest influence of Internet is its speed. Human resource managers should speed up their thought, transmission of information and action. In that way, they can quickly respond to the changes of external environment.

HRMIS can greatly improve the efficiency of HRM, thus improve enterprise competitiveness. According to the report of SAP, the use of HRMIS can not only reduce 30% of HRM costs and 30% of the recruitment process, but also improve 25% of staff productivity. Human resource managers will have time and energy to think and practice how to improve the overall level of human resource management and shorten the time of its occupation.

The Innovation of HRM Perspective. Internet breaks through the restriction of time and space, which provide convenience to optimize allocation of human resources in worldwide. On the one hand, Enterprises can train the global sense and ability of their staffs to grasp the most advanced knowledge and philosophy, from which staffs can make greater contributions for enterprises. On the other hand, Multinational Corporation can conveniently use e-HRMIS for remote control of all branches of human resource management. Furthermore, it is more rational for the personnel changes of the various branches to reduce the difficulty of cross-cultural management.

New Theories Have Been Introduced into HRM. E-HRMIS has brought new theories into HRM from other subjects. For example, item response theory and human-computer interaction theory have been used in n-line evaluation. Item response theory and modern measurement techniques are combined together, which greatly improves the efficiency and scientific evaluation.

3.2 The Influence on the Practice of Human Resources Management

HRMIS affects many aspects of business management. For example, E-HRMIS make the teamwork of staffs corporate at any time and any place. The home page of enterprise supports e-HRM in the operational level, which can be used to communicate with customers, suppliers and partners at anytime or anywhere [4,5].

The Influence on Various Personnel. The first personnel are chief executive officer (CEO). E-HRMIS helps the CEOs to change their understanding of human resources from "cost center" to "profit center". They can rapidly grasp evaluation methods of HRM effectiveness through the self-learning modules of company's internal network or specialized training. The direct consequence is to make them become sponsors of the demand for HRM. This in turn will facilitate the development of HRM and make HRM departments become strategic partners of companies. Furthermore, E-HRMIS can supply the CEOs a platform of human resource information query and decision support. They can obtain more accurate information through online queries to make a more scientific decision for business.

The second personnel are functional managers. HRM network is the working platform of the functional managers who involved in HRM activities. Through that platform, functional managers can view the personnel and attendance information, submit recruitment and training plans and approve the regularization, training, leave, retire and other processes. They can also use the system to manage the performance of staffs. From the perspective of functional managers, the main role is reflected to the increase of efficiency.

The third personnel are staffs. Internal LAN makes staffs the opportunity to share information. They can view rules and regulations, organizational structure, key personnel information, recruitment information, the salary, options and performance information, personal well-being total cases, personal attendance and leave, training courses, communication with HR online and so on. In this case, the personnel information and policies of enterprise becomes more transparent and the dialogue between staffs and supervisors becomes more convenient, which will help to enhance the sense of fairness and satisfaction of staffs. It has become an important way to enhance the trust between staffs and HRM department.

The last personnel are HRM department and practitioners. The intelligentification and automation of HRM brings tremendous impact to human resources management theory and practice, which need HRM department and practitioners change their skills and ideas. They should adapt to the learning organization, virtual organization and other modern forms of organization structures. For example, human resource management practitioners should be familiar with the use of computer systems and routine maintenance.

The Influence on IT Industry. The first one is the generation of professional Web sites for human resources [6]. There are many special network companies in human resources services. They have great talent database, powerful database management capabilities, extensive human resources management software and user-friendly interface for all job seekers, workers and company managers. Human resources website can achieve the following items.

Human resources websites make every effort to perfect consultation system of human resources, in order that it can supply human resources strategies and solutions for business users. Online platform can achieve the organic combination of online and offline, ensure the effectiveness and efficiency in the process of cooperation with the clients. Professional human resources websites gather considerable resources to lay a solid foundation for professional ASP services. Researching and developing capability of HRM knowledge can be strengthened through the effective knowledge-sharing resources, which ensure the share of knowledge and resources and serve for customers better. Human resources websites take great advantages of online media to strengthen the academic exchange of HRM. They can also supply evaluation and executive search services for enterprises, which is the “human resource management outsourcing". HRM outsourcing plays an importance role in the development of HRM.

The second one is the emergence and development of HRM Software Company [7]. Human resources management software market is very potential in China. Because of the intensive market competition, companies recognize the importance of human resource management. The needs of related software increase quickly and more and more companies will use HRMIS to manage Human resources.

According to the consultation from CCID Company, human resources management software industry in China grows rapidly. CCID made an investigation on human resources management software market sales in August 2009. The result showed that HRMIS software becomes one of hottest software in the markets. Compared with the last year, there was an increase of more than 40%. So far, it still maintain the good momentum of development.

4 The Prospect of HRMIS Application

The informatization level of human resources management is very low and need to be developed in China. National policies need to be changed vigorously for the promotion of the network application. Relevant policies and regulations of HRMIS industry should be introduced. The national policies must give strong support to promote the development of HRMIS, such as perfecting information network, improving network capacity and transmission speed, and strengthen network and information security system. In short, HRMIS will have the driving force of development only if the supporting facilities improve. Or it will be unable to develop.

The development of HRMIS should be based on the actual situation of the business. A workable stage development plan should be established, through which the development of HRMIS was gradually introduced and improved. In addition, most enterprises have no complete understanding to HRMIS, so relevant training of personnel was needed. During the implementation process of HRMIS project, enterprises should recognize not only the advantages of HRMIS, but also its negative influence, especially the psychological impact. Enterprises should prepare to fight a tough battle during the implementation of HRMIS project.

Advanced management concepts should be used to strengthen the application and guidance of HRMIS. For example, some module should be added, such as staff competency character module, temporary transaction process module and human resource management outsourcing data interface. Advanced experience from foreign countries should be learned to gear to international standards.

HRMIS suppliers should cooperate with professional human resource management staffs. For example, HRMIS system should enter the college classroom and the laboratory of human resources, from which a large number of professionals are trained to promote the development of HRMIS industry.

References

1. Yu, Y.H.: A Study on Web-based Human Resources Management Information System. Shandong University, Jinan (2005) (in Chinese)
2. Zhang, L.: Human Resource Information System, pp. 122–124. Dongbei University Press, Dalian (2002) (in Chinese)

3. Taylor, S., Beechler, S.: Human Resource Management System Integration and Adaptation in Multinational Firms. Comp. Manage., 115–174 (1993) (in Chinese)
4. Liu, G.Y., Xu, K.C.: Development of an Enterprise Human Resource Management Information System. Chin. Manuf. Inf. Eng., 52–57 (2010) (in Chinese)
5. Hao, Y., Lei, S.Z.: The Study of the Application of Human Resource Management Information System. Shandong Electron, 83–85 (2005) (in Chinese)
6. Chen, G.Q.: Studies on Learning Organization: Organizational Learning Capability System, Learning-oriented Human Resource Management System and their Relationships. Chin. J. Manage., 719–750 (2007)
7. Lytras, M.D.: The Role of a "Make" or Internal Human Resource Management System in Spanish Manufacturing Companies: Empirical Evidence. Hum. Factor. Ergon. Man, 464–479 (2008)

The Empirical Study of Investment Opportunities in the Value-Relevance of Accounting Earnings

Yuanyang Fu, He Guo, Danping Wang, and Wanlin Zhang

Zhejiang University, School of Management
Hangzhou, Zhejiang, China

Abstract. The author focused on the investment opportunities of the company, using the Earnings response coefficient to test whether the investment opportunities have impacted in the value relevance of accounting earnings. Our sample includes all unregulated firms that only issue A shares on Shanghai stock market for at least three consecutive years between 2006 and 2008. The analysis is motivated by the growth information hypothesis and the noisy measure hypothesis, both of which are about how and why the value-relevance of earnings varies with investment opportunities. Through the establishment of return - income model, we analyze the investment opportunities of the accounting value-relevance effects. Test results found that the company's investment opportunities on the value relevance of accounting earnings produced a negative effect, indicating that higher the investment opportunities are, lower the use degree of accounting earnings when investors evaluating the investment value are.

Keywords: Investment opportunities, Earnings, Value-relevance.

1 Introduction

Empirical study abroad for investment opportunities and the value relevance of earnings (earnings response coefficient) is not consistent with the conclusions of the relationship. These contradictory findings may be due to different study designs, such as sample selection, unexpected accounting earnings (unexpected earnings), cash flow from operations and accruals basis, the selection of indicators to measure investment opportunities, etc.

In China's market environment, whether the investment opportunities also impact on the value relevance of accounting earnings and whether investors would adjust the trust and dependence degree of earnings information because of the company's investment opportunities will be discussed in this article. The scholars in China have not carried out related researches. Based on this and previous oversea research, when the level of investment opportunities is not the same, in order to learn the way investors using accounting earnings information and the impact on stock prices. The author mainly focused on the impact of listed China's companies' investment opportunities on the value relevance of accounting earnings.

A. Xie & X. Huang (Eds.): Advances in Computer Science and Education, AISC 140, pp. 235–241.
springerlink.com

2 Paper Preparation

Through empirical research, some foreign scholars estimate the earnings response coefficient is between 1-3 (Kormendi and Lipe, 1987;Easton and Zmijewski,1989), the price-earnings ratio as a reasonable estimate of earnings response coefficients, should be 8 to 20. So the estimates derived from empirical research are far less than the actual reasonable value. There are mainly four kinds of explanations: inefficient capital markets invalid, the prices lead earnings surplus phenomenon (Beaver, Lambert and Morse (1980)), noise in earnings(Lev1989), transitory earnings.

This article refers to Brown (1987), Ali and Zarowin (1992), Lev and Zarowin (1999) in his study using the "change and levels " model, to improve the accuracy of the regression model and reduce temporary surplus impact, Pfeifee and Elgers (1999) has used the "current and lagged" model:

$$CAR=\beta_0+\beta_1 IOS+\beta_2 E+\beta_3 E\times IOS+\beta_4 LagE+\beta_5 LagE\times IOS+\sum\beta_0 Contralc$$
$$+\sum\beta_Y D_Y+\sum\beta_I D_I+\varepsilon \quad (1)$$

CAR is the cumulative excess return of year t, E is the accounting earnings of year t, LagE is the earnings of accounting year t-1, IOS is the investment opportunities for years t, Contralc is the control variables, D_Y, D_I are year, industry dummy variables, β_0, β_1, β_2, β_3, ..., βI are parameters.

Collin and Kothari (1989) concluded that when the companies' investment opportunities are larger, the relevance of surplus value is increasing. Ahmed (1994) found a negative correlation between the relationship of investment opportunities and relevance of surplus value. Kumar and Krishnan (2008) pointed out that there are certain problems in the design of these studies.

3 The Research Hypothesis and Research Model

3.1 Research Hypothesis

According to internal resource hypothesis, companies require financial support to explore investment opportunities. Because information asymmetry and the presence of agency cost, the cost of external capital increased. Cash flow operation is the primary resource for internal financing, unexpected operating cash flows allows investors to achieve business investment opportunities for re-evaluation of internal resources, and thus the stock price was revaluated. Un-expected value relevance of the operating cash flow should be rises with the increase of the investment opportunities. Accruals allow investors to understand the company's future operating cash flows to determine the possibility of future investment opportunities. Therefore, the value of accruals associated with investment opportunities should also be positively correlated. Surplus is the total number of operating cash flows and accounting accruals, the value associated with investment opportunities should also be a positive change.

According to noisy measure hypothesis, there is a problem in generally accepted accounting principles, such as, for some capital expenditures (trademarks, patents, R

& D spending) the amortization period is arbitrary. Dechow (1994) pointed out that through proposed timing and matching problem, the accruals improve the value relevance of accounting earnings, but when the amortization of intangible assets is modified randomly, the effectiveness of accruals is greatly reduced. The noise factors of accounting measurement methods result in the operating cash flows and accruals do not reflect the real economic surplus. If the investment opportunities are mainly used for investment in intangible assets, measurement error will be magnified with increasing of the investment opportunities. Rational investors will find more reliable indicators of the information instead of accounting earnings and cash flow, value relevance of accruals decrease when the value of investment opportunities increases.

Thus, according to the two different hypotheses, the value relevance of accounting earnings would increase or decrease on condition of increasing investment opportunities. This article made assumptions from the perspective of economic substance. If the assumption is confirmed, the internal resource hypothesis can be established as well; otherwise it explains that the noise factors influenced the decisions of investors.

Hypothesis: The value relevance of accounting earnings increase with higher investment opportunities.

3.2 Research Model

According to the hypothesis of internal resources or noise hypothesis, with increasing investment opportunities, the value relevance of accounting earnings may be increased or decreased. Therefore, this candidate model was corrected to test the existence of correlation between the value relevance of earnings and investment opportunities. Model is as follows:

$$CAR = \chi_0+\chi_1 IOS+\chi_2 IOS^2+\chi_3 E+\chi_4 E\times IOS+\chi_5 E\times IOS^2+\chi_6 LagE+\chi_7 LagE\times IOS$$
$$+\chi_8 LagE\times IOS^2+\sum\chi_0 \mathrm{Contral\ c}+\sum\chi_Y D_Y+\sum\chi_I D_I+\varepsilon \quad (2)$$

$\chi_3 + \chi_4 IOS + \chi_5 IOS^2$ is earnings response coefficient ERC, which symbolizes the value relevance of accounting earnings. It also stands for the relationship between value relevance and investment opportunities.χ_3 reflects the surplus value-relevance of overall average company (IOS=0) of, χ_4 represents reflects the linear trend that value relevance varies with IOS, χ_5 reflects the changing rate of the trend, which indicates the relationship between value relevance and IOS.

Opportunities for growth itself cannot be directly measured; alternative variables are used in many empirical studies to represent them. This selection of indicators as follows: MKTBKAS, MKTBKEQ, MVAGR, FATMA, E/P.

In this paper, capital asset pricing model is used. And we adopted one-year window period to calculate the consistency of section and the Cumulative Abnormal Return)(CAR) from May of year t to April of t +1,through the establishment of model which is between the company's actual return and rate of return serial distributing character of SHI to calculate the β value, as the risk variable. Continuity of the surplus is set to be dummy variables, when the absolute value of company's earnings changes (the difference between the first t years of surplus and the previous year's surplus) is more than 1, or 0 otherwise. Each rate of issued bonds changes with the expected inflation rate, which is approximately equal to pure interest rate and expected inflation rate. Expected inflation rate can be approximated represented by the capital purchasing price besides risks compensation in financial market. So interest rates should be represented by the average interest rate of 10-year issue long-term bonds. Since the annual-level and industry-level variable is the nominal class variables, the method will be addressed in the regression analysis., we set "1,0"as 2 dummies for three fiscal years, and set "Di1 ... Di11" 11 as virtual variables for 12 industry.

4 Data Sources and Sample Selection

The objects of research are all listed companies which have issued A shares in Shanghai stock market and have continuingly operated more than three years. Since this study is only meaningful for the profitable company under normal operations, so we get rid of the loss company and the ST company from 2006 to 2008.Besides, in this article the cumulative abnormal return (CAR) is calculated in the one-year time window. In order to ensure the validity of the calculation, companies which have cumulative suspended more than 50 days have been excluded. Finally, we get the number of sample companies for the498 for year 2006, 601 for year 2007 and 577 for year 2008, totally 1676 panel data in 12 different industries. Source: wind database.

5 Test Results

5.1 Factor Analysis

At first, we analyze the investment opportunities indicators, except the significance between per share and price earnings ratio (E / P) and the enterprise value of the geometric growth rate (MVAGR) was not significant, other indicators have reached between the level of significance of 0.01. It shows that there are large public components in five indicators. Therefore, in order to simplify the index system, we considered the principal component analysis, so we extract five indicators of investment opportunities as alternative variables among common components.

Obtained two factors by orthogonal rotation: the first factor mainly related to the market value of assets / book value, market value of equity / book value and market value of the geometric growth rate of the three indicators, two indicators are more than 78% load the second factor mainly related to E / P, whereas the market value of fixed assets exert considerable loads on two factors. After the factor analysis, two factors explained the 73% of the four original indicators. This article will use the sum of two factors as the investment opportunities measure, the investment opportunities in the following passage are common factor after the principal component analysis.

5.2 Regression Analysis

In the model, E's coefficient $\chi 3(0.229, p <0.01)$ was obviously positive, indicating that accounting earnings have value relevance, which is equal to the two selected models' results. But $\chi 4$ did not pass the significance tests, the relationship between value relevance of accounting earnings and investment opportunities for the relationship is not obvious. This cross-sectional data may be due to differences in the variance equation of the regression effect affected, so it needs further heteroscedasticity testing.

Table 1. Result of preliminary regression analysis model

	χ_0	χ_1	χ_2	χ_3	χ_4	χ_5	χ_6	χ_7	χ_8	R^2
coefficient	1.49	-0.110	0.012	0.229	-0.065	-0.018	-0.221	0.064	0.019	0.148
T test	6.118***	-5.270***	1.689**	2.424***	-0.870	-1.635*	-3.557***	1.427*	2.144**	

We adopted correlation analysis to the absolute value of the model error term and explanatory variables, except LagE, $|\varepsilon|$ and the explanatory variables are significantly related, p is greater than 0.05, confirming that the results of the regression model exists heteroscedasticity. The author has further use of WLS method to improve the model to eliminate heteroscedasticity, the results are as follows:

Table 2. Result of improved regression analysis model

	χ_0	χ_1	χ_2	χ_3	χ_4	χ_5	χ_6	χ_7	χ_8
coefficient	0.364	-0.107	0.009	0.293	-0.050	-0.013	-0.221	0.056	0.011
T test	4.086***	-12.750***	2.332**	5.814***	-1.530*	-2.209**	-3.557***	3.668***	2.321**

The model can be improved by eliminating the heteroscedasticity, χ_3 (0.293, $p<0.01$), χ_4 (-0.050, $p<0.1$) and χ_5 (-0.013, $p<0.05$) were all significant to some extent, indicating that the surplus value of the related accounting resistance may change with the non-linear investment opportunities fluctuations. As $\chi_3 + \chi_4$IOS + χ_5IOS2 is non-linear formula, the value relevance of accounting earnings levels and rates of change under different levels of investment opportunities cannot be directly derived from the regression equation. In order to calculate the value of the relevant level and rate of change under the levels of other investment opportunities, this article will divide the IOS in the sample into seven parts, each spaced 1.0, from -2.0 to 4.0. In which, the companies whose IOS <-2.0 Accounting for 0.32% of total number of samples, the companies whose IOS <4.0 Accounting for98.5% of total number of samples. With χ_3 + χ_4IOS + χ_5 IOS2 computing earnings response coefficients, representing the value relevance of accounting earnings, using χ_4 +2 χ_5 IOS to calculate the rate of change. We adopt the variance-covariance matrix calculation to calculate the value-relevance level and its rate of change of significance level under seven different levels of investment opportunities.

Table 3. Model regression coefficients of variance - covariance matrix

	E/χ_3	$E{\times}IOS/\chi_4$	$E{\times}IOS^2/\chi_5$
E/χ_3	0.00254		
$E{\times}IOS/\chi_4$	0	0.001068	
$E{\times}IOS^2/\chi_5$	0.00002777	0	0.0000346334

Table 4. Surplus value and its associated rate of change under different levels of investment opportunities

IOS	earnings response coefficient (ERC) ($\chi_3E+\chi_4E{\times}IOS+\chi_5E{\times}IOS^{2)}$)	T test	rate of Change of Value-relevance ($\chi_4+2\chi_5IOS$)	T test
-2.0	0.341	3.915***	0.002	0.050
-1.0	0.330	5.427***	-0.024	-0.691
0.0	0.293	5.814***	-0.050	-1.530*
1.0	0.230	3.782***	-0.076	-2.188**
2.0	0.141	1.619*	-0.102	-2.532***
3.0	0.026	0.209	-0.128	-2.660***
4.0	-0.115	-0.671	-0.154	-2.687***

Ps: ***, **, *Respectively in 0.01,0.05,0.1 (two-tailed) was significantly

When the investment opportunities are at a low level, IOS =- 2.0, the value relevance of accounting earnings changes in the level was greater than zero (0.002), but not significant. From 0.0 to 4.0's 5 levels of investment opportunities, the rate of change is negative, and passed a significant test. This suggests that different levels of investment opportunities have different impact on the value relevance of accounting earnings. When the investment opportunities are at low level, the value relevance of accounting earnings' changes in the trend with investment opportunities are not obvious, but when you raise the level of investment opportunities (greater than 0.0 after), the surplus value increased when the investment opportunities are getting lower, and the higher investment opportunities, the downward trend of the surplus value increased is more obvious. The apparent one-way movements showed that the guess which surplus value-relevance level presents curve changes with the investment opportunities in the previous article can't be established. This is the same with Ahmed's conclusion (1994), whose is opposite with Collins and Kothari's conclusion (1989), but also different with Kumar and Krishnan's conclusion (2008) that Surplus value-relevance will first increase and then decrease with the changing of investment opportunities.

6 Conclusion

In this article, our sample includes all unregulated firms that only issue A shares and have sufficient data on Shanghai stock market for at least three consecutive years from 2006 to 2008.Which provide a basis for the following two questions: whether the investment opportunities also impact on the value relevance of accounting earnings and whether investors would adjust the trust and dependence degree of earnings information due to the company's investment opportunities.

The results showed that, in China's capital market and under existing accounting standards, on the whole, earnings information reported by the listed companies are used by investors. As an important basis for evaluation of company value, earnings information lead to the corresponding stock price volatility, indicating accounting earnings information is of value relevance. If the company's investment opportunities increase, the value relevance of accounting earnings decreases. And higher the levels of investment opportunities are, the faster the reduction of the value relevance are. That is, for companies with high growth, investors use less of the earning information in evaluating the value of the company. The reason may be due to accounting measurement methods' inherent flaws. The earnings information calculated according to GAAP may contain noise. If the increase in investment opportunities makes the noise increases, the value relevance of accounting earnings is reduced. Hypothesis that noise in our environment is sufficiently supported, but does not explains whether the internal resources of the hypothesis are established. The reason may be due to a significant noise impact, which covered up the internal resources factors. In addition, the global economic crisis started from 2007 caused a huge impact on China's stock market, but also undermined investor confidence, increase in short-term speculation adversely affected on the stock market.

References

1. Ahmed, A.: Accounting earnings and future economic rents: an empirical analysis. Journal of Accounting and Economics 17(5), 377–400 (1994)
2. Beaver, W.H., Lambert, R., Morse, D.: The information content of security prices. Journal of Accounting and Economics 2(1), 3–28 (1980)
3. Collins, D.W., Kothari, S.P.: An analysis of inter-temporal and cross-sectional determinants of earnings response coefficients. Journal of Accounting and Economics 11(2/3), 143–181 (1989)
4. Easton, P.D., Zmijewski, M.E.: Cross-sectional variation in the stock market response to accounting earnings announcements. Journal of Accounting and Economics 11(2/3), 117–141 (1989)
5. Fama, E.F., French, K.R.: The cross-section of expected returns. The Journal of Finance 47(5), 427–465 (1992)
6. Kormendi, R., Lipe, R.: Earnings innovations, earnings persistence, and stock returns. The Journal of Business 60(3), 323–345 (1987)
7. Kumar, K.R., Krishnan, G.V.: The value-relevance of cash flows and accruals: the role of investment opportunities. The Accounting Review 83(4), 997–1040 (2008)
8. Lev, B., Zarowin, P.: Boundaries of financial reporting. Journal of Accounting Research 37(2), 353–358 (1999)
9. Pfeiffer, R.J., Elgers, P.T., Lo, M.H., Rees, L.L.: Additions evidence on the incremental information content of cash flows and accruals: the impact of errors in measuring market expectations. The Accounting Review 73(3), 373–385 (1998)

On Software Professional Studio Construction in Higher Vocational Schools and Its Meaning

ShuYan Yu*

Department of Management and Information, Zhejiang Posts and Telecommunication College, 312016, Shaoxing, Zhejiang, China
shuyanyu1231@yahoo.cn

Abstract. Higher vocational education advocates the study mode of the combination of productive labor and social practice. A studio is an effective platform for such a learning mode. Through this professional studio both teachers' and students' ability are improved. In addition achievements in science research can be promoted. In this paper the author based on the software professional studio construction proposes a construction mode for software professional studio in higher vocational schools. And the effect as well as the importance are illustrated in this paper.

Keywords: studio, software technology, second class, project.

1 Introduction

In the author's visits of some software professional teaching in some higher vocational schools, the author found that most students there had less interest in their major study, lacked the initiative to learn their major and feared that they did not have enough ability to learn their major well. Besides that, students had different foundations, because some students became software majors not because of their own choice. In addition, the software major is not an easy thing to master, due to which some students lost their heart. Thus, a dilemma occurs: on the one hand some students feel that they can learn more than what teachers' teach in class and other students think that what teachers teach is beyond their capacity.

How to promote those students' professional ability who are eager to learn and sustain their passion to this major and take up this major as their future career and help other students who feel less devoted to this major becomes a big challenge.

In some software companies, they are in great need of software talents. During the author's visits to those companies, it was found that they did not attach much importance to the degree of one's education. What they take heed of is the student's practical ability and future development [1].

How to increase student's ability in some practical projects and make students grasp some excellent ways of learning and in the end demonstrate enough practical ability to the potential employers is what has been considered.

* Corresponding author: ShuYan Yu, NO.474 Shanyin Road Shaoxing City Zhejiang Province, 312016.

A. Xie & X. Huang (Eds.): Advances in Computer Science and Education, AISC 140, pp. 243–247.
springerlink.com

After one year of construction, it is discovered that besides some special measures in daily teaching, the construction of professional studio is one of the most efficient means to resolve those problems mentioned above.

2 Research Background

The earliest studio recorded with the name Bauhaus can be traced back to the beginning of the 20th century founded by Walter Gropius. In which theory and practice were both paid attention to. They opened thirteen studios covering different fields to cultivate students' practical ability [2]. Such a teaching mode was deemed weird at that time by traditional colleges, which, however, turned out to be the most general mode in modern arts and design teaching throughout the whole world.

Currently, the mode of studio teaching is employed by some higher vocational schools of arts in Zhejiang Province, which combines the project teaching with studio. Such a teaching is based on studio and employ projects through the whole process. Now the Higher Vocational school of Jinhua and the Higher Vocational school of Ningbo City have founded studios in succession. Under teachers' guidance students can take projects from the society and finish the project so as to improve their professional ability.

The studio proposed by us is different from those mentioned above. First, our studio is not available to all the students but just accessible to those who have strong passion for this major, teachers included. Second, the work stated in this paper is not included in daily teaching, but is after daily teaching, which falls into second class. Third, the studio in this paper aims to promote the ability of those outstanding students and exert influence on all students in this major and even all students from the whole school.

3 Running of the Studio

The studio contains professional teachers and outstanding students which aims to promote both teachers' and students' professional ability throughout the projects and competitions. The studio is run in two modes-daily running and project running, which is illustrated in detail as the following.

3.1 Recruitment

Twice of recruitment are carried out each year. The autumn recruitment is held together with the recruitment of students' associations. Fruits will be exhibited in public to catch more eyes and in the meanwhile notices of recruitment are attached in the department's website and in the major's website. On spot recruitment only accept the way of contact of those students who are willing to join the studio. After that, time of interview will be noticed through bulletin board and notice in the websites. The interview is usually held one or two days after the recruitment exhibition. During the interview students' motive, strong points and professional foundation will be paid attention to. Interviewers are composes of key member students in the studio and teachers. Interviewees are graded during the interview. After the interview all the interviewees will be given a comprehensive evaluation to select the best ones. The result will be released in the department's and the major's websites.

3.2 Software and Hardware Construction of the Studio

There are two requirements for the site of the studio. One is it is better to be located near the professional teachers' office and the other is that it should be away from the noisy streets. The requirement for hardware in the studio is relatively low. The tables can be arranged in a circle so that discussion is easy. The proper amount of computers and a projector will be sufficient for discussion and lectures.

Software environment is more important, which includes name of the studio, slogan of the studio and environment and titles of the members so that it is easier to have a common cognition and values.

With the efforts of all studio members, the studio is named Ace studio, the logo and PPT model and titles have been handed in for use. Student members designed the backdrop for the studio and the whole room demonstrates a warm and progressive atmosphere for study and work. Students feel it more like a home.

3.3 Daily Operation

To run the studio smoothly, a teacher is designated to be in charge of the studio. A studio leader is selected out of student members, who sees to the daily running of the studio. Based on the professional field and service projects, three teams are formed: mobile phone software development, web development and frontier design and development. Each team has a teacher for guidance. When a certain project is being developed, a project team will come into being.

Our studio has detailed regulations of achievement measurements, fund management and report, aiming to regulate member's behavior and measure their achievements.

Each week a discussion of new technology and original ideas is held. The discussion can be anything related to the gains and loss in the recent projects, the new technology and one's original ideas. Anything related to the studio service project and technology is encouraged. Through such sharing, members can learn from each other, help resolve problems they encounter in the projects and have deeper understanding of their projects.

One technology lecture is delivered each month. The lecturer will be either teacher from our school or expert in a company and such a lecture can extend students' knowledge scales, enlarge their knowledge and broaden their horizon.

After the recruitment each time, the new members are divided into relative teams according to their own will. Old members will assign some exercises, model projects to help them improve their capacity and in the end enable them to participate in the real projects.

3.4 Project Management

When there is a project, relevant teams will be formed. The detailed project will be in the charge of the project manager, and team members just obey regulations in this project. Projects are categorized into three aspects: competition projects, non-production projects and production projects.

Competition projects refer to those competitions of all levels, such as college, municipal level and provincial multimedia design competition. Also municipal, provincial, and national software design competitions are included. Usually, professional teachers will be designated to guide the competition and in charge of the training. Multimedia competition will see a self-formed team by students and guided by teachers.

Non-production projects refer to those that are used to train new members or assignments by the professional teaching and researching office and the studio, such as the posters for lectures, a temporary system, studio logo and so on. Such projects tend to be completed by students, and checked by the studio.

Production projects refer to those projects that need investment from the school or outside school. The key points include the outstanding courses websites, research projects of teachers and other projects from outside school. Usually these projects have strict requirements such as quality requirement, time requirement. But these projects can also benefit students who participate in them. Such projects usually go with teachers' participation.

4 Achievements and Meaning of the Studio

Studio, as part of the practice basis in our majors is a necessary complement to our teaching. Through the practice in the projects, students experience the whole process and gain much beneficial experience. Through "learning in doing and doing in learning" students not only consolidate their current knowledge but also cultivate their practical ability and develop their cooperative spirit. Additionally, through the projects, students are initiated to learn, to resolve the problems they have and develop their own ways of study. In joining the studio, each student has at least one work completed by themselves, which is a best product presented to their future employer. Besides these, the fulfillment makes students more confident and more willing in their study, which is of much necessity in higher vocational school students. The members in our studio this year have achieved outstanding feats, and all the third year students are willing to have practice or find a job in software companies. Most companies speak highly of the interns from our studio.

Studio has exerted some radiative effects among majors. After a period of time, studio members have relatively better academic performance and practical ability than other students. Their autonomous learning attitude attracts other students to focus on their study too, which indirectly promote the learning atmosphere in class. When asked the reasons they chose to be members of the studio many interviewees expressed the opinion that studio members have a different learning attitude and learning mode which is more positive and the learning atmosphere is nice too.

Studio is not only a place for teachers to accumulate work experience and improve teaching ability but also a place demonstrating our scientific research power. Through all kinds of projects from the school or outside, teachers can improve themselves as well as direct students to better their ability. Practical projects enrich teaching too, which in the meanwhile promotes the creation of new professional research fruits.

5 Conclusion

Studio on the one hand, improves students' ability and makes them have the initiative to study, to resolve their problems and enjoy their study. One student once said in a QQ group conversation: "study is happiness", which I believe is the return of our studio. On the other hand, studio also improves teachers' ability. Only through a mutual benefit can the quality of professional talents be improved.

References

1. Yu, S., Xu, Z., Zhang, M.: Research of Demand for Professionals in Computer Applications of Vocational Colleges and Thinking of its Train Model. Journal of Shangqiu Vocational and Technical College 2, 52–54 (2010)
2. Bauhaus, http://zh.wikipedia.org
3. Shi, B.: Building the Teaching Workshop for the Implementation of Work and Study. Education and Vocation 20, 166–167 (2011)
4. Jiang, X.: Studio of Software Technology in Vocational Schools. Journal of Binzhou Vocational College 4, 54–57 (2010)
5. Zhao, Y.: Research of "Studio Room System" Model of Ideological and Political Work. Data of Culture and Education 6(229), 57 (2011)
6. Ren, J.: Discussion of the System of Higher Interior Design Studio Teaching Model. Journal of Changsha Social Work College 1, 99–102 (2011)
7. Chen, G.: Specialty Construction Research on Higher Vocational Education Based on the "Professionalization" Strategy. Exploration and Research of Vocational Education 1, 19–21 (2009)
8. How to Train Vocational Art Professionals, Experts Pushing Teaching Studio System, http://www.zjjybkzs.com

The Research on Teaching Ideas of "Data Structure and Algorithm" in Non-computer Major

Zhao Wang[1,2]

[1] Institute of Software, School of Electronics Engineering and Computer Science, Peking University, Beijing, 100871, China
[2] Key Laboratory of High Confidence Software Technologies, Ministry of Education, Beijing, 100871, China
wangzhao@pku.edu.cn

Abstract. For the main problems and difficulties the non-computer professional students may face in the learning of data structures and algorithms course, this paper, based on years of teaching experience, discusses some ideas in the teaching of data structures and algorithms, such as using algorithm design methods as clues to introduce various types of algorithms, using data logical structure as modules to organize course contents, through real life examples to improve students interest, emphasis and application ability.

Keywords: non-computer major, Data Structure, algorithm, teaching ideas.

1 Introduction

American Association for Computer Machinery(ACM) and the American IEEE Computer Society(IEEE-CS), in order to meet the educational trend of globalization, issued a Computing Curricula. Computing Curricula 2001 (CC2001) and the latest CC2005 classified "data structures and algorithms" as an undergraduate required courses in the computer and information technology-related disciplines [1] [2]. In China, "data structures and algorithms" has become an undergraduate required information technology foundation course in non-computer professional science and engineering majors [3]. Data structures and algorithms are two complementary and inseparable aspects to the design of the program."Data structure" does research on the problem of information presentation, while "algorithm" does research on the problem of information processing. This course is the only way to understand computer principle and to master programming skill. It is of great significance to train and improve the students' computational thinking ability. However, There are many differences between the teaching of "data structures and algorithms" in non-computer major and computer major. The author of this paper combines the problems she encountered when teaching the course at Peking University, gives some ideas and experience, hope to promote the development of this discipline.

A. Xie & X. Huang (Eds.): Advances in Computer Science and Education, AISC 140, pp. 249–254.
springerlink.com

2 Data Structures and Algorithms Teaching Status and Problems in Non-computer Major

2.1 Student Interest in Learning is Not High

In the teaching process, the first problem encountered is that students' interest is not high. This is mainly caused by the following aspects:

(1) The course content is highly logical and abstract, students may have some difficulties at the beginning of learning, so that they feel the course is difficult, loss of interest in earning.
(2) This course requires relatively high both in theories and in practices, even if you understand the theory, you may encounter other difficulties in practice. Some students reflected that they understood on the class, but they still can't write programs. Computer programming itself is not a simple and easy thing, even computer professionals, regarded who write programs as "IT workers".Non-computer science students encounter difficulties in programming feel more difficult to understand, dull and boring.
(3) For non-computer science students, their instinctive feeling is the course is not their major's courses, they will not think highly of it. What's more, they do not know the relation between this course and their major, may have a feeling that even learn now, it's no use in the future, so their motivation to study is not high. And the courses are generally set up as a basic course in the first or second grade, students don't understand their professional knowledge well, how to establish the link between the course and their majors is also a teaching difficulty.

2.2 Students' Programming Language Bases Are Relatively Weak, Very Different

The course "Data structures and algorithms" introduces the basic theory of data structures and algorithms , students learn to rationally organize the data, effectively process data, grasp the methods of algorithm design and analysis, train and improve the abilities of the students to analyze and solve problems using computer, improve their programming level. Current "data structures and algorithms" textbooks are often described using a kind of programming language, such as C, C + + , JAVA and other computer languages. Program design is the Advanced Placement course of this course, due to the limited class hours, non-computer science students usually have relatively shorter period of time to learn programming languages, have fewer opportunities to practice. In contrast, many students don't grasp C programming language solidly, are only familiar with simple C language syntax, can't using struct, pointer and other data types well, even they don't understand them. The data structure course involves pointer, struct and other data types at the very beginning, which make students generates fear to the program at the start. As this course is the basic course for all science and engineering students, teachers are limited, a class usually has more than 100 students, their basis varies widely.

3 Teaching Strategies and Ideas

To the status and problems of the above, I take the following teaching strategies and teaching ideas.

3.1 Use of Primary and Auxiliary Two Clues

This course takes the textbook "algorithms and data structures - C language description" which is edited by Professor Zhang Naixiao in our university. Consistent with most of the textbooks, this textbook also takes data logical structure as its primary clues, gives an introduction to linear structure, tree structure, set structure and graph structure sequentially. When giving the introduction of each data structure, their storage structure and associated algorithms are discussed. Some of the more important algorithms are discussed in separate sections, such as sorting and algorithm design methods. The process of teaching is nearly the same as sequence of the textbook, when explaining the correlation algorithm, the auxiliary clue is included. The another clue is the algorithm design methods: greedy method, divide and conquer, backtracking, dynamic programming and branch and bound method. For example, the Huffman algorithm, the map coloring algorithm, the minimum spanning tree construction algorithm (Prim algorithm and Kruskal's algorithm), the shortest path of weighted graph algorithm (Dijkstra algorithm) are specific algorithms designed based on greedy method. Binary search, quick sort and merge sort algorithms are specific algorithms designed based on divide and conquer method. When I explain the algorithm, not only describe specific process of each algorithm, but also focus on the analysis of common design ideas and methods of all algorithms. It will not only avoid the students to learn so many complex algorithms by rote, have the feeling of contents is many and chaos, be boring and afraid of difficulty, but also stand on a certain height, teach the students problem-solving ideas and methods. Teach others to fish and they will fish for a lifetime.

3.2 Using Real-Life Examples to Guide Everyone's Interest in Learning

Although data structure and algorithm do researches on the computer's operating objects and their relationships and operations in non-numerical computation programming, data is sign of objective things, the relationship between objective things naturally exist, so data structure is naturally occurring. For example, the paths between the cities and the cities are naturally occurring, we can using vertices of the map --a type of data structure to represent cities, the edges of the map to represent the paths between the cities. Algorithm, broadly speaking, is a specific problem-solving methods and a description of the steps, and we do everything in accordance with certain methods and procedures, therefore, there is algorithm in doing anything. According to the popular song "see or not see," I give such conclusion about the data structures and algorithms, "holidays, miss or not miss, it will come on schedule; friends, see or not see, I feel around; friendship, far or nearly ,always in my heart; blessing, say or not say, always surround with you; data structure, whether you use or do not, he is naturally occurring; algorithm, whether you realize or not realize, you do use them. "In the problem solving using the computer, select the appropriate data structures and algorithms, we can reduce the algorithm's time complexity and improve the effectiveness of the procedure, while in real life, when we consciously use the data structures and algorithms, it will also raise our efficiency in doing things. Brute-force method, all possible solutions of the problem may be listed from the solution space, use the given test conditions to determine. It is the most direct solution to the problem and the most basic method. It is a simple algorithm, but computationally intensive, In order to improve efficiency, we need to adopt other more effective algorithm strategies, such as the greedy method, divide and conquer, and backtracking and so on. Greedy method is an algorithm design method which achieves the desired optimum through local optimum at each step. For the study of a course, is the same truth, listen every class well , do every homework well, master the knowledge which should learn at each period, do not leave any knowledge you don't understand at each period, so you can finally grasp all knowledge of the curriculum best through the best grasp of knowledge at each stage. We graduated with honors from elementary school to high school, graduated from the high school in the best state, then will graduate from the college with best results, in every stage of life if we can play the best of ourselves, we can ultimately obtain a satisfactory life. Through guidance, the students suddenly realized that they have unconsciously used greedy method for a long time. Finally, encourage students consciously use of greedy method in the study of this course. Greedy method, we not only have to meet several times in this course, he will accompany with me a long way in life. The basic idea of divide and conquer method is "divide and rule", it divide a large problem into two or more smaller sub-problems similar to the original problem, solve the sub-problem firstly, and then combine the various sub-problems, that is, "divide" "rule", "integration" [4], and in daily life, the wisdom "reduce major issues to minor ones, and then minor issues to naught" [5] is similar to and consistent with the idea of divide and conquer method. The basic idea of backtracking is to "go forward if you can, if dead go back, find another way to go," sounds have the same meaning with an old saying "fall back a step then brighter". When you are at a life crossroad, perhaps this strategy will help you. It is not used just to solve the maze problem in the course. These examples not only help students to understand the algorithm, but also make students feel the algorithm are closely related to them, the importance of the course , inspire the students' interests to learn the course.

3.3 Programming Language Test and Review

To prevent C programming language from being an obstacle in the study of this course, in first lesson, I arranged a simple C language programming problem, I have written a basic framework of this program, but set a number of knowledge evaluation point in it, such as: can you compile and run the results correctly, the header file, function declarations or pre-allocated, memory allocation, test memory allocation is successful, memory release, end of the string handling, transfer function parameters and function return, etc., that is intentionally designed a number of errors, let the students correct errors as much as possible and submit before the next lesson. Although it's a hard work to check everyone's homework, but the teacher can get a clear impression of the students' weakness in C programming language, and can explain them in the second lesson, make up the C language weaknesses of the students, lay a certain foundation for follow-up study.

3.4 Analogy Method

Although there are many knowledge points in this course, these knowledge points are not alone. If you do not grasp the relation and difference between them, it is not easy to understand and remember. Using analogy teaching method, teachers can deepen the students understanding of knowledge by comparing the similarities and differences between different points of knowledge. For example, comparing the link list with a head node or without a head node, the students will have a deep understanding of the meaning and role of the head node. In addition, finding different data structures and algorithms to solve the same problem, is benefit to help them build the spirits of seeking increasing perfection, be positive and innovative. For example, this course introduces five types and ten kinds of sorting algorithms, the design ideas, implementation process, time complexity and application of various algorithms are comprehensively compared. I often encourage them, if you want to become smarter, you must try to think from a different perspective. This is a course make you more intelligent, once understanding, lifelong benefit.

3.5 Full Use of Multimedia Teaching Tools

Adopt appropriate display formats depending on the teaching contents. For example, the teaching courseware adopts PowerPoint which can be a vivid display of the course content; Source code examples adopt two display forms, the first is webpage which can demonstrate the source code completely, the other form is TXT format which is suitable to compile; Homework comment sums up the solving ideas, methods and common errors for each question, it use WORD document which can display Chinese words, programs and images well; In order to quickly check the learning effect of the students, we use network tests, after the tests are finished, the students will be able to quickly know their achievements and answers of the question; For some algorithms which are difficult to understand, using visual FLASH animation. Vivid animation is adopted to show abstract and difficult course content. Animation are designed specifically for some problems which are difficult to understand, for example, the use of pointer variables, a program of creation of the link list, algorithm of pattern matching without backtracking, the minimum spanning tree construction （Prim） algorithm, so that the corresponding difficulties become easy to be accepted and understood.

3.6 Establish Course Website

This course has been using a course website for several years. With the support of Peking University “teaching network”, a new version of the course website is construction in 2011. The course website provides "course notification, lesson plans, teaching teachers, teaching content, course homework, my grades, study guides, course resources, teacher-student interaction "nine modules. Teaching content and course information are organized by tree structure, a clear structure, self-contained, and the "tree structure " happens to be a kind of data logical structure. "Teaching content tree" is not only an access guide to course contents, but also a vivid example of the practical application of the tree structure, a very good reflection of the old saying “learn to meet practical needs”. Teaching contents is stored according to "chapter", there are teaching courseware, animation, source code examples, instructional videos, homework submission and comment folders in each chapter. Web site style is lively, fresh and lovely, create a happy learning environment, promote to learn happily. Fully use of various communication forms relying on PKU "teaching network", such as course notification, discussion boards, e-mail ets. Students reflected the course website is very convenient, comprehensive and detailed, clear, is a very good platform.

4 Conclusion

These teaching strategies and ideas, achieved good results through our practice. Students said that the biggest gain was that their rigorous logic thinking had been established; Programming requires care, patience, willpower and perseverance; Algorithm is really a feast of thinking and wisdom, only when you pay sweat, and continue to make unremitting efforts ,challenge yourself ,you can get the greatest benefit. Algorithm is not only related with the computer, procedures, it is a reflection of learning, a life of learning.

References

1. The Joint Task Force on Computing Curricula IEEE Computer Society Association for Computing Machinery. Computing Curricula 2001 Computer Science (EB/OL) (2001), http://www.acm.org/education/education/education/curric_vols/cc2001.pdf
2. CC 2005. The Overview Report of Computing Curricula 2005(EB/OL) (2005), http://www.acm.org/education/curric_vols/CC2005-March06Final.pdf
3. Zhang, M., Xu, Z., Tang, S.: The knowledge architecture and teaching practice of data structure course. Computer Education (3), 89–91 (2004)
4. Zhang, M., Zhao, H., Wang, T., Song, G.: Experimentation on data structure and algorithm. Higher Education Press, Beijing (2011)
5. Huang, Q., Tang, S.: Non-computer Professional Data Structure Teaching Practice and Innovation. Computer Education (3), 38–42 (2011)

Application-Oriented Intercultural Engineering Education in Automotive Engineering and Service

Yanfen Mao[1], Ming Chen[1,*], and Hans Wiedmann[2]

[1] Automotive Engineering & Service, Sino-German College of Applied Sciences, Tongji University, No.4800 Cao'an Rd, 201804 Shanghai, China
{maoyanfen,chen.ming}@tongji.edu.cn

[2] Mechatronics & Electrical Engineering, Esslingen University of Applied Sciences, Esslingen D-73728, Germany
hans.wiedmann@hs-esslingen.de

Abstract. The paper describes how the theory-oriented education principles of China's universities are combined with the practical-oriented education methods of Germany's universities of applied sciences in CDHAW (Chinesisch-Deutsche Hochschule für Angewandte Wissenschaften), Tongji University. As one of four majors in CDHAW, Automotive Engineering and Service (AES) translates those theoretical and practical requirements into an integrated new education framework. After seven years implementation, AES, as an emerging inter-discipline gained a good education evaluation from AQAS (Agentur für Qualitätssicherung durch Akkreditierung von Studiengängen) due to its own practical-oriented education model and good educational results.

Keywords: application-oriented education, intercultural engineering education, learning by doing, automotive engineering and service.

1 Introduction

Today engineers must possess both, greater breadth of capability and greater specialized technical and managerial competence [1]. Globalization is changing the way in which engineering work is organized and in which companies acquire innovation. The development and execution of IT-based service projects is usually accomplished by dividing the functions into a dozen or so components, each of which is carried out by a different group of engineers and managers. These groups are often located in several different places around the world [2]. And that increases convergence between nations in technology, education and economic growth [3]. As an emerging inter-discipline, automotive service engineering has attracted significant attention due to the result of globalization of development in automotive industry and the serious shortage of automotive service engineers.

Automotive Engineering and Service (AES) is one of four majors in Sino-German College of Applied Sciences (Chinesisch-Deutsche Hochschule für Angewandte

* Corresponding author.

A. Xie & X. Huang (Eds.): Advances in Computer Science and Education, AISC 140, pp. 255–260.
springerlink.com

Wissenschaften, CDHAW). CDHAW is an educational project of the Chinese Ministry of Education (MoE) and the German Federal Ministry of Education and Research (BMBF). Participants are Tongji University and from German side a consortium of Universities of Applied Sciences. The main goals are to foster engineers which are capable of doing qualified engineering work immediately after finishing their studies. Beside of gaining the required technical competence the students are guided to develop intercultural and soft skills which prepare them best to work inside Chinese-German companies.

2 Requirements and Features of AES in CDHAW

2.1 Requirements on AES

Nowadays and future requirements of automotive engineering & service are:

Application-Oriented Education: According to the forecasts of 2010 to 2015 of China's automotive service industry, the qualified personnel who has strong knowledge of field, problem-solving ability, basic engineering skills, engineering design, professional attitude, and teamwork is seriously short [4]. Nowadays young engineers coming out of Chinese universities require one or even two years of practical training inside the companies until they can start doing a high qualified job autonomously. This point especially requires integration of practical application-oriented education methods into our universities.

Inter-disciplinary Education: The last few years automotive service engineering has grown rapidly and evolved to utilize new materials and components like plastics, aluminum, biological materials, semiconductors, software, and new production methods like gluing, welding of aluminum, and development of high reliant software and hence it comprises different disciplines like mechanics, electronics, informatics, etc. This demands for engineers with more broad education and gain profound practical engineering background in multi-disciplinary fields. Therefore, multi-disciplinary education is an essential requirement for the whole automotive industry chain: from auto spare parts or subsystem suppliers like Bosch to after-sales shops.

Intercultural Education: The globalization of automotive markets demands for intercultural educated engineers especially in China, where lots of other nations' automobile manufacturers are searching for employees which are capable of understanding the philosophy and culture of other nations, especially German.

2.2 Unique Features of AES

Automotive Engineering & Service has several distinctive features comparing with other automotive service engineering in China.

Firstly, it built with CDHAW the dual education system project. In engineering education, German universities of applied sciences (Fachhochshule: FH and Hochschule: HS) have distinctive features from another higher education system named technical universities (Technische Universität: TU). The universities of applied sciences comprise study programs which tightly connect state-of-the-art technology and scientific theory to a practical, "hands on" education model. The effectiveness and success of this kind of engineer-oriented education have been proved with the popularity of their graduates. Chinese high education has advantages in theory-oriented principles. Therefore, AES in CDHAW combines the advantages of Chinese education system with Germans university of applied sciences system.

Secondly, AES is not focused on the after-sales topic as other similar majors. Learning from the successful experiences of German partner universities, AES defined its orientation as service-oriented automotive technology, and determined the four specialized subjects automotive service-oriented products, recycling, diagnose, and special purpose vehicles (Fig. 1).

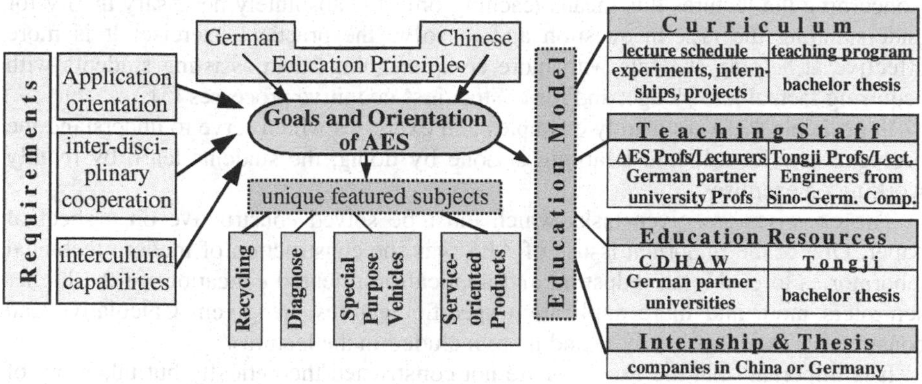

Fig. 1. Operation Concept of AES in CDHAW

Thirdly, AES is not only educating Chinese students, but also German students from 10 partner universities in the 7th and 8th semester. That requires AES to offer all 7th semester lectures and Bachelor thesis supervision in English or German. It enables German students to pass their last academic year inside this program in China to reach a double bachelor certification same as Chinese students.

3 Application-Oriented Education Methods and Implementation

Engineering education has emphasized technical contents and has all but ignored professional obligations to the public. Engineering education formerly has largely been developed by educators rather than practitioners, but collaboration between the two segments is essential if engineering curricula are going to impart a thorough understanding of what is required for an engineer to best serve society today [1].

During the first 3 semesters there are not many differences to any other Chinese higher education institutions, except that our students have to learn German language. From 4th to 6th semester however there are an increasing number of lectures accompanied by laboratory experiments and given by German Professors. The gradually change over to the new education methods in the higher semesters comprise the introduction of more practical topics and integrate practical skills into the studies. From 7th semester, Chinese students who meet the requirements can finish their studies with two final semesters in Germany and German students will come to China as well.

The curricula of AES study courses are derived from the German partner universities. Curricula needs to reflect what we understand about the practice of engineering now and into the future, and also needs to connect to what we know about our students, how they learn, what motivates and excites them, tapping into their idealism and concerns.

3.1 Practical-Oriented Lectures and Exercises

Concerning the lectures this means teaching only the absolutely necessary theory for understanding the task in question and to solve the practical exercise. It is more effective at helping students with more complex things than assisting students with acquiring factual knowledge, and foster students' cognitive processes [5].

In principle there are mainly examples and exercises which serve to understand the problems and its solution. Learning is done by doing, the students learn by mainly working on exercises.

The exercises are given tasks which must be solved constructive on a sheet of paper. One of the important issues of AES was the construction of modern technical laboratories to enable an industrial and application oriented education. In the higher semesters more and more application specific lectures are given. Calculative and constructive exercises are more and more included in the lectures.

It is important that the exercises are not constructed theoretically but taken out of the praxis instead. The exercises should deal with the problems from the engineers' practical task. From the teachers this teaching method requires a deep understanding of the subject in concern. The teachers also must think about what kind of projects should engage the best of engineering talent and knowledge in the years ahead.

3.2 Application-Oriented Experiments and Projects

Except basic laboratories in other colleges of Tongji University, AES has five feature laboratories, named automotive electric and electronics, automotive sensors and actuators, automotive service-oriented products, special purposed vehicle, and automotive CAN-bus systems laboratory.

Take CAN-bus laboratory as an example for explaining the laboratory planning and contraction (Fig.2). In CDHAW, the major of AES has two lectures involving CAN bus technology. One is automotive electric and electronics, and the other is automotive service products. In another major named Mechatronics, CAN bus will be added to one of the lectures that already exists. Therefore, CAN bus laboratory is a common experiment platform for AES and Mechatronics.

In nearly every field of application engineers usually first need to make a design, to construct a model before starting the realization phase, e.g. design of software, electric/electronic hardware, construct a simulation model, design a functioning principle, etc. Concerning the practical education in laboratories that means the students have to perform practical exercises and experiments, perform projects guided by teachers and engineers, and then stepping over to self-guided projects. Students have to make reports and written documentations on all their laboratory-work.

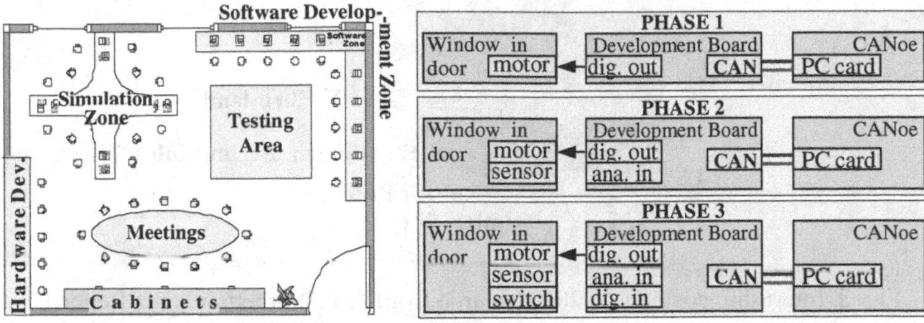

Fig. 2. Layout of CAN-bus laboratory and project schedule for window opener

In the 7th semester, all students should take a course named *Project II* which includes project management (kick-off meeting, weekly report, presentation and all related documents). One of the project topics is about Robotino which is a mobile robot from FESTO. As one kind of Automated Guided Vehicles (AGVs), Robotino is equipped with a camera and is used to do projects on automotive assistance systems like recognition of patterns to read traffic signs like speed limits and give warnings to the driver, dynamic obstacle collision avoidance and moving targets tracking e.g. for Lane Departure Warning (LDW) Systems or Automatic Parking Systems, and so on.

3.3 Engineering-Oriented Internship and Thesis Work

CDHAW tightly cooperates with lots of German industrial companies like FESTO. With the help of those partners the students are prepared to achieve their bachelor level with a high attitude to practical research and development. One part of this education process is to perform internship under industrial conditions in the field of their major. In their last 8th semester the students have to make an industrial internship, preferably in Sino-German companies, such as Volkswagen, Bosch, Daimler, etc. (Fig.3). AES students perform during their thesis engineers tasks, and every student has supervisors one from company and from AES as well.

The students get contact and learn soon during their studies the company-culture and can exercise to work on practical development tasks. This is actually not state of the art in China. So in the beginning actually it is difficult to find a company which offers an internship because the companies do not know and have no experience with that form of education. However after the last four years of implementation, more and more companies realized that they can get benefits from that kind of cooperation with AES. They know the students well from their internships and can decide which job is best for them or which students are best for the company. They can take advantage as the new Bachelors can start to work efficiently under industrial requirements immediately after graduation. They can outsource simple development tasks to AES projects for students. Thus the development-cost is reduced.

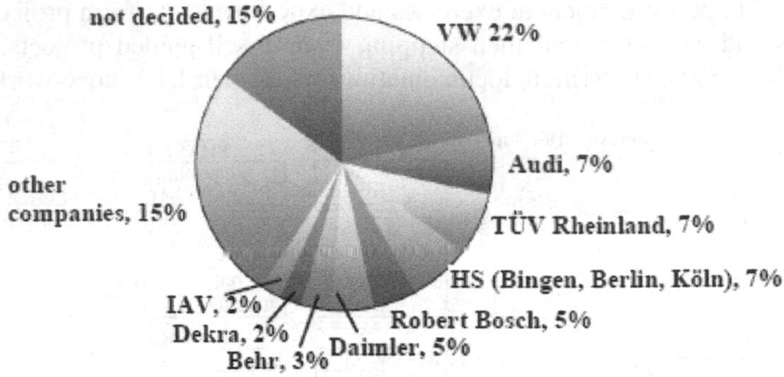

Fig. 3. Internship positions in 2010 (Source from Prof. Norbert Schreier Report)

4 Results

CDHAW started its operation in 2004 under the roof of Tongji University in Shanghai. In 2008 the first graduates got their double bachelor degree, a Chinese bachelor of science at CDHAW and a German bachelor of engineering at one of the German partner universities of applied sciences. The companies which employed those graduates are much contended and up to now we can already see that companies develop more and more attention to CDHAW.

The MoE evaluated CDHAW in 2007 with a favorable rating. In 2008, all study courses have been accredited by AQAS (Agentur für Qualitätssicherung durch Akkreditierung von Studiengängen) with a satisfying result.

References

1. Galloway, P.D.: The 21st-Century Engineer: A Proposal for Engineering Reform, pp. 25–32. American Society of Civil Engineers, Renton, VA (2008)
2. Vest, C.M.: Context and Challenge for Twenty-First century Engineering Education. Journal of Engineering Education, 235–236 (2008)
3. Ratchford, J.T.: Globalizing Engineering Education: Public Policy Impacts. In: Proceedings of China-US Bilateral Seminar on Engineering Education, pp. 142–166 (2002)
4. Li, S.M.: Investment and Forecast Report on China Automotive Industry, 2010-2015. China Investment Consulting (2010) (in Chinese)
5. Zydney, J.M.: The effect of multiple scaffolding tools on students' understanding, consideration of different perspectives, and misconceptions of a complex problem. Computers & Education 54, 360–370 (2010)

Research of Private Courier Enterprise Based on E-Commerce

Bo Zhang

Huazhong University of Science and Technology, Industrial Engineering 21,
Hubei Wuhan, China
zhangbo901227@163.com

Abstract. The prosperity of e-commerce has been driving the rapid development of private courier enterprise. However, local companies are facing the pressure both from multi-national enterprises and the disorderly competition of local rivals, which leads to the continuous decrease of price, while the operating cost keeps growing. This industry is confronting a bottleneck of development. This paper proposes possible countermeasures from four aspects. Firstly, broaden the channels for finance with the support of the state. Second, speed the integration of the industry in order to complement one another. Third, implement service standard to regulate market order. Fourth, improve the profitability with more innovation and creation.

Keywords: Electronic Commerce, Private Express Enterprise, B2B.

Electronic commerce, known as e-commerce, indicates carrying out commercial activities over the Internet and other computer networks, including online shopping, electronic trading, electronic banking, as well as electronic service provided by government, such as online tax and online custom declaration, etc. With the development of B2C, C2C, private express enterprise, of which the main business is e-commerce, has grown dramatically as well. On the other hand, e-commerce has realized prominent development based on courier, playing an increasingly significant role in consumption and circulation field. The platform and scope of cooperation between e-commerce and express enterprise keep the continually expanding while these two important industries are now forming a relationship of supporting, coordinate and prosperity on the Synergist.

Since the year of 2007, driving by the abnormal prospering domestic e-commerce, the future and prospects of express industry have maintained bright and encouraging. However, in order to grab market share, low price competition and mutual suppression of prices keep pushing down the market price, resulting an increasingly lower profit. Furthermore, as operating cost, including labor and energy price, goes up every day, numerous express companies have zero or even negative profit, making the industrial profit distribution improper day by day. Therefore, domestic express business is exposed to a “cold winter” rather than a favorable situation. The increasing speed of foreign express enterprise’s access to Chinese market has brought huge change to industry structure. Domestic companies not only have to burden the adverse impact from foreign competitors, but also need to face a disorder local competition.

A. Xie & X. Huang (Eds.): Advances in Computer Science and Education, AISC 140, pp. 261–264.
springerlink.com

This industry is confronting a bottleneck of development and struggling to survive under the duel pressure. This paper proposes some countermeasures from several prospects: implement of express service standard, modification of the integration of the industry and the transition to modern service business.

1 Broaden Financing Channel

The primary financing problem for private express enterprise lies in the fact that it does not meet the standard of the mortgage for commercial loan. With the reference and experience of foreign financial support to small/medium enterprises, it is proposed that the government should establish proprietary institutions to provide policy loans, or guarantee instead of direct investments for express enterprises under good development condition. As a matter of fact, express industry is not an exception in China, financing difficulty has already become a common as well as prominent problem with wide concern for all small/medium enterprise. Although the guarantee company can play the role of the bridge connecting banks and enterprises, its high premium is beyond express companies' ability due to their low profit margin. With the coming of post financial crisis era, how to broaden financing channel for express enterprises deserves the government's attention of all levels, with practical solutions. Only with the great support of the government, can both express industry and e-commerce expand and develop simultaneously.

2 Speeding Up the Integration of the Industry

The booming development in e-commerce requires a faster, safer and higher standard for express service, which will inevitably wipe out the small, inferior companies. In terms to the winners in the market, cooperation and reinforcing each other's strengths should be the mainstream in the integration of the industry. For instance, in spite of the fact that EMS has no competitiveness with private express companies in aspects of flexibility and price, it has an incomparable advantage over the rests due to its wide agents which can cover almost all rural area. In addition, the privilege of letter delivery under the new China Mail Bill helps to reserve its special and specific market space. Therefore, it is possible for private express enterprises and EMS to complement each other's advantages. What's more, private express companies should abandon the stereotype of pure competition and form strategy alliance as megamerger to augment their own strength by opening logistic resources, constructing agents corporately and "pooling car".

3 Implementing Service Regulation and Standard

Without regulations and standards, "disorder" has been taken as a normal state in the express industry for e-commerce. How to expand service agent should be the primary issue, however, under the pressure of capital and operation costs, franchise mode is more favored by private express enterprises in order to roll out the market rapidly. Only the delivery in main lines and First Tier Cities is operated by big express companies, which are taken place by local express companies in Second and Third Tier Cities, this leads to problems of independent accounting and various service standards in different region, in which the price and service quality is determined by local market. Worse still, for the sake of self-profit maximizing, some branches even make injurious to other branches, sometimes even the whole enterprise, resulting market mayhem in this industry. So, the standardization of logistic service maintains an important link in the transition to modernized service industry. In terms to major express companies, it is also proposed to explore ceaselessly in the aspect of service standardization, which not only help to unify the executed standards in their own branches, but also provide template for the whole industry. Recently, Xingchen Express Company tries to provide a standardized and unified service among its 1524 branches and agents so as to deal with a scientific, systematic and professional competition in express industry's future competition.

4 Exploring Service Innovation and Creation

Modified China Mail Bill has been put in force since Oct 1ST 2009, which will greatly enhance the threshold for market entry and survival costs of the enterprise. The new Bill takes the Practice for Access System, regulating the qualification in aspects such as capital scope, service capacity and safety, which indicates a new challenge and opportunity for express service operators. System of Letters Monopolization further cut down the profit of private companies, which make ends meet by downsizing and reducing employee training costs. This severe competition environment makes service innovation and creation an even more necessary strategy. There is no fixed pattern in it, as long as it is economical and efficient. Innovation and creation involves, but not only, exploiting to the full one's favorable conditions and avoiding unfavorable ones, providing customized service, replace container with container's cage to minimize product damages while maximize volume , extending business via healthy customer relationship is also in the domain of innovation and creation. Express delivery enterprise with innovation and creation thinking is able to get rid of the old profit mode (which is realized via freight premium), by making profit in other channel rather than traditional package delivery, including commission revenue from insurance agent and receiving agent, as well as sales commission, initial fee and information fee, etc.

In summary, private express logistic industry is confronting bottleneck in all aspects of capital, management and service, breakthrough in broaden financing channel, speeding up the integration of the industry, implementation of standardized service regulation and innovative exploration is paramount in order to get rid of malignant cycle of low price competition, and head towards a sustainable developing mode emphasizing the optimization in customer relationship and improvement in core competetiveness.

References

1. Chaudhury, Abijit, Kuilboer, J.P.: E-business and E-commerce Infrastructure. McGrw-Hill (2002)
2. Gao, X.: Research on improvement of private express delivery enterprises competitiveness. Market Modernization 2, 15–17 (2009)

3. Zhao, X.: Evaluation research on express delivery service based on grey target theory. Corporation Economics 5, 159–161 (2009)
4. Anonymity, Construction of marketing system with closed circle management. Modern Post 10, 6–9 (2010)
5. Xie, Z., Han, F.: The impact of new China Mail Bill on Chinese express industry. China Commerce 11, 18–23 (2009)

The Considering about Chinese Enterprises to Establish E-Training Platform

Jianlin Qiu

Guangxi University of Finance and Economics, Nanning, P.R. China, 530001
qiujianlin@163.com

Abstract. In order to maintain the competitiveness of talent, many Chinese enterprises have increased their investment in corporate human resources training. With the extensive application of computer network technology, the e-Training relying on a computer network technology has also been rapid development; Meanwhile, e-Training in the development of the various constraints encountered in the process, but also hindered the e-Training in the development of Chinese enterprises. This study explores these limitations, and propose solution ideas and methods to promote e-Training in the development of Chinese enterprises.

Keywords: Human Resources Training, Computer network, e-Training, limiation.

In order to maintain the competitiveness of talent, many companies regard the great focus on the intellectual investment, invest heavily in staff training.

As the growing usage of network tools in education and learning ,the mean using computers and networks to conduct staff training to improve overall organizational performance in enterprises has been more and more common. Entrepreneurs have to look into the training management information—e-Training, thinking to use the network means to achieve effective management of training, thus bringing the company to build a knowledge-based enterprises

At the same time, to meet the needs of corporate e-Training development, many of eHR software vendors have introduced e-Training software module or to enhance the proportion of e-Training module in the ERP platform in order to help companies promote the use of e –Training better.

1 The Introduction of E-Training

Typical Web-based e-Training to the exchange between the user and the computer network instead of face to face communication with teachers. The curriculum that seem to have no teachers is just actually a teacher-centered in fact. The only difference is that the computer network and a series of pre-designed routines instead of the teachers to guide the participants to assess, strengthen and encourage other.

Watching from the uses purpose, E-Training is a training management system for corporate and HR staff to manage the trainings' resources, processes and performance.

A. Xie & X. Huang (Eds.): Advances in Computer Science and Education, AISC 140, pp. 265–268.
springerlink.com

Although the e-Training in which the employees, officers or other departments and individuals have rights to use interface, but this use of information's is order to manage the training process more effectively.

Watching from the use object, The object of e-Training is mainly to executives, and employees in the system are in a passive state. There are strict distinction on categories of personnel permission to use the resources and processes in the e-Training system.

Watching from functional areas, E-Training is a training management tool, the overall management of corporate training , including OJT (on job training), employee self-learning and e-Learning training and all forms of training. Enterprises through the training system to ensure proper implementation of the training process, reflecting the resources of scientific training and performance is reasonable, supported by corporate decision-making and human resources planning.

The following is the flow chart of e-Training:

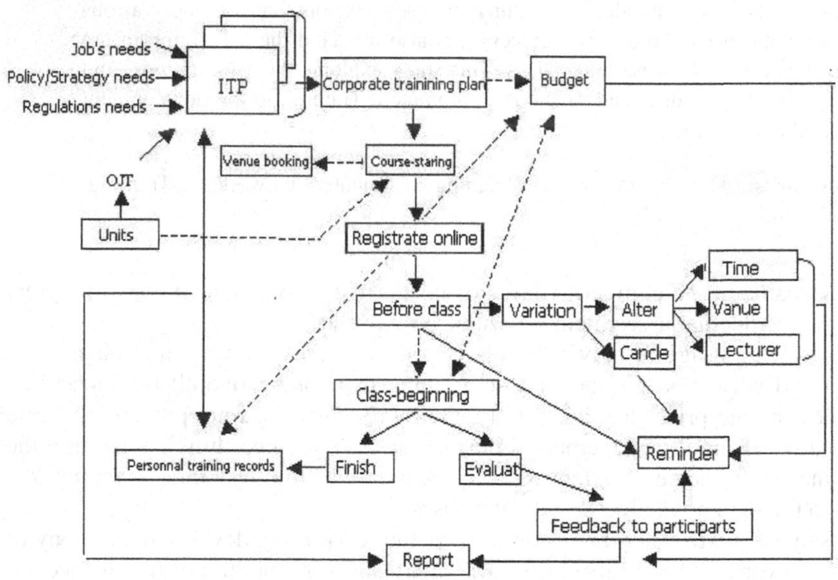

2 The Advantages of E-Training

The advantages of e-Training in the training of management and improve human resources management are in the following three aspects:

First, full of powerful decision support capabilities. After using the e-Training, a company's training resources and forms have been unified into a single platform to manage and effectively with other enterprise information modules for data sharing and docking. Managers can have access to real-time dynamic control the utilization of enterprise resources for training; the implementation of corporate, departmental and individual training plan; the feedback of staff about the training curriculum and the entire system.

Second, the open and efficient self-management. After using the e-Training, the management of corporate training have achieved standardized processes and automation. Corporate training process without any additional equipment and training investment, rapid integration into enterprise workflow management to achieve the registration, examination, classes, graduation, etc. According to the online registration approval , the HR staff can adjust the training implementation plan on time. E-Training is a more effective way to organizing in-firm which can make in-firm training to achieve "any time, any place" can be carried out, thereby reducing the direct costs(such as instructor salaries ,printing and distribution of materials, purchase of training equipment and other expenses) and indirect costs(transportation costs, travel expenses, lost income, fees,etc.). This is undoubtedly very attractive for business. For example, PNC Bank in Pittsburgh of the United States installed a set of online training system to adapt to the individuals' speed, for each trainee is expected to save about 40% of the cost.

Third, the scientific and effective training evaluation system. the training effectiveness assessment Module in e-Training can be timely and effective assessment of the situation of each training course times. E-Training can quickly grasp the rate for participating students participating in the training, assessment and examination results, etc., assess each training courses and also assess the individuals, departments and the business case. Participants are also able to assess every individual course and instructor's teaching quality.

3 E-Training in the Chinese Enterprise Application Status and Their Limitations

In Chinese enterprises, the traditional mode of training and education still occupy the mainstream. Many Chinese companies have tried to establish their own e-Training platform, produced the e-Training Knowledge Warehouse including the training, electronic courseware, varieties of materials, procedures templates. Some companies even organized into a virtual campus. But most of the Chinese companies just focus on how to maximize the number of e-Training, and the measures adopted to enable the participants to interact with the training materials and other issues.

E-Training method can lean something distributed, but they are not very suitable for general working time and environmental requirements. For example, e-Training courses should be designed on the basis of a fixed order to create attractive interaction. However, as for the sound and multimedia etc. to stimulate students' attention are limited by the networks, multimedia, and other aspects of environmental, the ones who are in the open, the smaller the staff office space at participating e-Training to learn may be hindered the others' work. In addition, using e-Training method, the requirements for the users' discipline, willpower and other are very high. Lack of face to face communication with teachers, users are often not very clear for some feedback confused. Many people say that the e-Training requires learners to be better communication skills, self-training and computer knowledge. For the above reasons are that e-Training is difficult to integrate with the real working environment, thereby limiting the e-Training widely used in Chinese enterprises.

4 The Principle of Chinese Enterprises to Establish E-Training Platform

To realize the performance of e-Training platform , the Chinese companies to build e-Training platform should follow the following four principles:

First, e-Training platform software should be combined with their own characteristics. Although a relative mature E Training software have been produced, but because of the he differences of company's products or services, business e-Training system requirements are also unique. Enterprises should combine their characteristics, development or joint development of their e-Training platform software

Second, openning is the basic features of the system. That openning, refers to employees not only easy to use e-Training platform of training resources, using the platforms to achieve their required knowledge and skills, and training consulting and evaluation;,but also the platform and eHRM's other modules can achieve communication and interaction, to form an organic whole or skills, especially with corporate cash flow, logistics management system to share data interactively.

Third, Self-training is the concep to build the platform. In the course of e-Training training , because the trainers and trainees do not communicate face to face, lacking of direct supervision and control, the trainees are in the state of self-management all the time. This requires trainees to lift the "learning, training, promoting" up for his strong vision, and then Train with high intensity of the subjective motives voluntarily. To make the trainees from passive to active, and enhance the enthusiasm of the trainees

Forth, promoting continuous-learning is the goal of the platform. Enterprises to establish e-Training platform is mainly to improve the skills of employees. Companies hope that the employees can gain new skills, new knowledge through the platform and apply them to work, while sharing with other employees. If the employee use these skills to improve the quality of their works, and this process do not end with the end of training, This means that the companies have achieved a continuous learning, and fully tap and zoom in e-Training platform performance.

5 Conclusion

Construction of e-Training platform is a way for Chinese enterprises to strengthen the effectiveness of corporate human resources training and save training costs. It will have a bright future. Due to various constraints, e-Training in the development of Chinese enterprises face many limitations, which requires corporate managers constantly strive to explore, to break through these limitations, thus promoting the development of e-Training.

References

1. Illinois, R.A., et al.: In: Liu, X. (trans.) Human resource management to gain competitive advantage. Renmin University of China Press (2005)
2. Zhao, Y., Wang, J.: The construction of the regional education consortium based on Sharing of educational resources, vol (5). Higher Education Press (2006)

Empirical Study on Undergraduate Employability and Training Mechanism in Chinese Independent Institutes

Yong Chen

School of Business, Zhejiang University City College
NO.51 Huzhou Street, Hangzhou, China, 310015
camychen5@hotmail.com

Abstract. Based on literature retrospective related to undergraduate employability, the paper had designed the questionnaire. Through the survey of independent institute graduate, application of exploratory factor analysis and means comparison, the paper conducted a quantitative analysis on the components and enhancement mechanism of undergraduate employability. The components of undergraduate employability are subject capacity, personal attributes, communication skill, teamwork ability, innovation and entrepreneurship and career planning capacity. To set appropriate goals and strategies, curriculum design, reasonable arrangement of the system of training, guidance to the students in career choice and development planning and establishment of industry-university cooperation mechanisms are the key factor of enhancement mechanism.

Keywords: undergraduate employability, enhancement mechanism, quantitative analysis.

1 Introduction

With the reform of employment system for college students in China after higher education reform of the late-1990s, undergraduate labor market and employment patterns have undergone a fundamental and radical change, especially after expansion of China's higher education enrollment. In the increasingly competitive situation of employment, employers are unwilling to massive new university graduates, but tend to recruit job seekers who have practical experience. One of the reasons for such a phenomenon is that college students employability capacity is insufficient. The topic is important because the job-market is still in transition.

From the papers review about undergraduate employability, there is no uniform definition on employability. Harvey (2001) noted that explicit concepts of employability is uncommon. The researcher may summarize the different connotation or the definition from the different angle. But whatever definition, and scholars agreed that students ' employability refers to graduates to gain and sustain jobs, as well as in have access to career sustainable development. (Yorke&Knight,2004; Rothwell etal.,2009; Boden&Nedeva,2010;Zhu,2009). According to related research literature at

A. Xie & X. Huang (Eds.): Advances in Computer Science and Education, AISC 140, pp. 269–274.
springerlink.com

home and abroad, this paper found that there is still a lack of the empirical analysis about the composition and affecting factors about the employability of university students in China , especially for the Independent Institute.

The paper touches upon several dimensions of the undergraduates' employability, but will focus upon the way academic achievement during college affects undergraduates' employability. To preview main results, the paper found that subject knowledge and skill, generic skill, personal attributes, career management sill are the four dimension of employability. Formulation goal and strategy about employability, establishment of mechanisms for interaction with the employer, the ability-oriented ways of training are the key factor to improve undergraduates' employability. The paper is composed of five sections. Section 2 introduces the relevant literature. Section 3 describes our data. Section 4 presents our main results. Section 5 concludes this study.

2 Literature Review

Employability was defined as a set of ability to enable individuals to better employment, and competent to his career, including achievements, understanding and personal qualities(Boden&Nedeva,2010). Rothwell etal. (2009) et al. also take the university student as the objects, proposed that undergraduates' employability refers to individual access to sustainable employment opportunities commensurate with their level of qualification. The undergraduates' employability is set of capacity, refers to the graduate to be able to obtain the employment opportunity as well as to obtain the success in the operating post. This views of employability remain dominant in Chinese higher education(Zhu,2009).

Yorke & Knight's (2004) proposed that USEM model embodied four broad, inter-related components: Understanding, Skills, Efficacy beliefs, Metacognition. Fugate, Kinicki, and Ashforth (2004) considered employability as a psycho-social construct which'subsumes a host of person-centred constructs' as a 'synergistic combination' of career identity, personal adaptability, and social and human capital. Dacre Pool&Sewell (2007) pointed out that key element of employability includes the following areas: career development learning and experience, degree subject knowledge, understanding & skills,generic skills (including enterprise skills),emotional intelligence. Xie Zhiyuan (2005) argued that employability includes basic ability, professional competence and individual special abilities. Xiao Yun et al (2007) argued that the employability include basic practical ability, knowledge expansion capability, capacity for innovation.

Scholars generally believe that universities play an important role to enhance students ' employability, but also professional talent training capabilities for the quality of higher education and reflect the important ways. The study on improvement of undergraduates' employability have focused primarily on curriculum design, practice strengthening education, strengthen the various skills training and so on, how to achieve enhanced employability (Yorke & Knight, 2004). Sun (2007) thought the university should strengthen the relation with the enterprise, signed a contract of supply and demand and protect the employment of school graduates. Otherwise, previous studies were lack of quantitative empirical analysis on the impact factors of Chinese college students employability and the improvement mechanism, just qualitative analysis.

3 Methods

To address the research aim of the paper, a survey was developed that incorporated many of the elements from the university bachelor graduates. The undergraduate employability belong to the individual level, the required data can not be obtained from publicly available information, so this data collection using the questionnaire.

This questionnaire is designed to mainly focus on the employability of students and universities to improve the undergraduates' employability. The questionnaire includes the following three aspects: the components of employability, the key factor of employability enhancement and the basic conditions of respondents. The survey conducted primarily through two approaches. One is to draw on visiting alumni activities, out in letter form indirectly; the other is through direct visits to enterprises graduate. A total of 541 questionnaires were received, of which 8 questionnaires to fill out incomplete, another 9 questionnaire with little difference to be rejected, 524 valid samples were actually obtained, and the effective response rate was 78%.This article utilizes the SPSS16.0 software, carries on the descriptive statistical analysis to the investigation object basic situation, and exploratory factor analysis.

4 Results

4.1 The Components of Undergraduate Employability

First, the survey objects are already participating in the work of graduates. Second this survey is the ratio of men and women graduates, different graduating year's number is quite, and the conclusion to be reached high reliability. A series of principal components analyses (PCA) were conducted to explore the dimensions of employability and the related measures. The ratio of items to respondents was 13:1. Factor analysis of the Bartlett test of sphericity statistic is 1264.326, $P = 0.000$, KMO index of 0.837, indicating that factor analysis is feasible. The Kaiser–Meyer–Olkin measure of sampling adequacy was .837, above the recommended value of .6, and Bartlett's test of sphericity was significant at $p < .001$. Kaiser's criterion (eigenvalues of one or more) extracted six components. Having specified a six-factor solution, and using varimax rotation, the six components accounted, respectively, for 12.988, 12.697, 11.223, 10.788, 10.361 and 8.907 percent of the variance. The solution is presented as Table 1.

From the factor analysis results, the first component includes three items: profound subject knowledge, interdisciplinary knowledge and variation specialized skill, name after subject capacity. The second component has four items: the sense of responsibility, emotional control, positive attitude and the self-understanding capacity, named personal attributes. The third component is consisting of three items: a good listener, verbal ability and ability of empathy, named communication skill. The fourth component has three items: defining its role as a member of a team, the ability to share information and the ability to use teamwork to deal with planning and crisis management, named teamwork ability. The fifth component has three items:

Table 1. Varimax rotation, combined scales for employability

	Component						
	1	2	3	4	5	6	Communalities
Profound subject knowledge	.599						0.643
Interdisciplinary knowledge	.589						0.635
Variation specialized skill	.523						0.546
The sense of responsibility		.788					0.687
Emotional control		.552					0.676
Positive attitude		.595					0.647
The self-understanding capacity		.560					0.546
A good listener			.649				0.652
Verbal ability			.567				0.565
Ability of empathy			.608				0.515
Defining its role as a member of a team				.801			0.675
The ability to share information				.626			0.578
The ability to use teamwork to deal with planning and crisis management				.666			0.742
Consciousness of Entrepreneurship and innovation					.823		0.568
Ability to translate ideas into action					.584		0.691
Ability to independently identify issues					.850		0.845
Career-related capacity display						.866	0.614
Career choices and planning knowledge						.765	0.725
The ability to understand organizational culture						.713	0.757
Eigenvalues	2.993	2.035	2.019	1.771	1.446	1.399	

consciousness of Entrepreneurship and innovation, ability to translate ideas into action and ability to independently identify issues, named innovation and entrepreneurship. The last component has three items: career-related capacity display, career choices and planning knowledge and the ability to understand organizational culture, named career planning capacity.

From exploratory factor analysis, six dimensions are reasonable distribution, each indicator in the corresponding have higher load. It follows that the employability of the structure of college students is generally acceptable.

4.2 The Key Factor of Employability Enhancement Mechanism

The employability of university undergraduate is through access to learning in university. Using SPSS16.0 software on influence factors of college student's employability, this paper obtain the conclusion: training on employment ability of college students in universities, was very important to their ability to access to employment.

According to the comparing means analysis, teaches experience and deep expertise, ,social practice, cultivation mode, curriculum design , mechanism of interaction with the employer, the means of above factors are 6 or above 6, and apart from teaches experience and deep expertise of standard difference is greater than the 1, other factors of variance is less than 1. The results indicates that above factors on college students employability training is important, the recognition of graduates about the importance of these factors is high, more consistently believe that these factors are very important.

5 Conclusion

Enhancement of undergraduates employability, whose aim is to enhance the quality of personnel training, and improve the quality of higher education. According to the positive analysis, the undergraduate employability is composed of six components, namely subject capacity, personal attributes, communication skill, teamwork ability, innovation and entrepreneurship and career planning capacity.

After China entering the popularization of higher education, diverse trends within the higher education system is inevitable. The goal of Independent Institute is to train application-oriented talents, to enhance the employability of students should become its important objectives and strategies. To set appropriate goals and strategies, curriculum design, reasonable arrangement of the system of training, guidance to the students in career choice and development planning, the establishment of industry-university cooperation mechanisms, these are the key factor of undergraduates employability enhancement mechanism in Chinese Independent Institute.

Acknowledgement. The paper obtains the subsidization from Hangzhou Philosophy and Social Sciences Fund, project NO. B11JY06. It is also funded by the Hangzhou electronic commerce laboratory.

References

1. Harvey, L.: Defining and measuring employability. Quality in Higher Education 7(2), 97–109 (2001)
2. Yorke, M., Knight, P.: Self-theories: some implications for teaching and learning in higher education. Studies in Higher Education 29(1), 25–37 (2004)
3. Rothwell, A., et al.: Self-perceived employability: Investigating the response of post-graduate students. Journal of Vocational Behavior 75, 152–161 (2009)
4. Boden, R., Nedeva, M.: Employing discourse: universities and graduate's employability. Journal of Education Policy 25(1), 37–45 (2010)
5. Zhu, X.-S.: On the employability of university students to develop. Higher Education Exploration 4, 34–37 (2009)
6. Fugate, M., et al.: Employability: A psycho-social construct, its dimensions, and applications. Journal of Vocational Behavior 65, 14–38 (2004)
7. Dacre Pool, L., Sewell, P.: The Key to Employability: Developing a practical model of graduate employability. Education & Training 49(4), 277–289 (2007)

8. Xie, Z.: Study on cultivating undergraduate employability. Education Development Research I, 90–92 (2005)
9. Xiao, Y.: Study on students ’ employability and social demand differences. Higher Education Exploration 6, 130–134 (2007)
10. Sun, C.-Y.: Study on enhancement of the employability. China Education Research 11, 87–88 (2007)

Research and Practice on the Open Teaching for the Authentication of AutoCAD Senior Application Engineers

Qibing Wang

Engineering Training Center, Huaihai Institute of Technology, Lianyungang, P.R. China, 222005
wqb123926@163.com

Abstract. In order to make the students in different levels and different professional backgrounds quickly and efficiently achieve the authentication level of AutoCAD senior application engineers in a relatively short period of time, we have carried out the teaching research from the following four aspects: the AutoCAD drawing foundation, the AutoCAD precise drawing, the AutoCAD three-dimensional drawing and the AutoCAD secondary development. The practice has shown that the teaching activity that is implemented by the well-designed and typical examples have achieved good teaching results, which has realized the desired purpose.

Keywords: AutoCAD, Examples, Research and Practice.

1 Introduction

Along with the rapid development of computer technology, the computer aided design has gradually replaced the traditional manual drawing, and AutoCAD drawing course has become compulsory course in engineering and polytechnic universities [1]. AutoCAD software authentication program examination is an application and professional technical proficiency examination implemented to increase the digital design capacity of students in engineering and polytechnic universities. The guiding ideology of this examination suggests that the examination shall make for the demand of relevant fields on professional engineering design personnel, and shall be helpful to promote the increase of teaching quality of various courses in engineering and polytechnic universities [2]. The requirements of AutoCAD software authentication program examination include: the examinees shall understand the function of AutoCAD and its application technology in the design; master its basic orders and methods; have capacity to combine the professional design demand with functions of software organically; be able to apply their knowledge and methods comprehensively to increase the design application and development capacity of the major. The requirements of senior application engineers include: the examinees shall be able to draw X-Y scheme of product proficiently and produce three-dimensional diagram of product; have the capacity to perform secondary development on original system to extend its functions by means of secondary development tool; increase the automation design degree of product; and master certain skills. Currently, many universities and colleges have opened independent course of AutoCAD in the teaching

A. Xie & X. Huang (Eds.): Advances in Computer Science and Education, AISC 140, pp. 275–280.
springerlink.com

according to the requirements of majors and development goals of students. As the software has strong functions, so it is unpractical to master the whole software in limited class hours. It has been a problem that needs careful consideration in the course of AutoCAD to complete better teaching of AutoCAD.

The training course is open to the public, and the students participating in the training have different levels and different education background: there are students in grade 1, 2, 3 and 4; some students have learned the software with certain basis, and some have learnt nothing about the software; there are students majoring in construction, machinery, electrics, and geomatics engineering. Then how shall we give lessons? How can we promote students to achieve the authentication level of AutoCAD senior application engineers in short time? Based on R&D experience of various years and research on teaching of recent years, the author has developed a teaching method focusing on the example teaching, and obtained good teaching effect by giving lessons including AutoCAD drawing basis, AutoCAD precise drawing, AutoCAD three-dimensional drawing and AutoCAD secondary development through classic examples. The research and practice on the open teaching for the authentication of AutoCAD senior application engineers has been introduced from the above four aspects as the reference for colleagues.

1.1 The Research and Practice on the Teaching of AutoCAD Drawing Basis

There are many orders on drawing and editing in AutoCAD. How can students master the orders rapidly and efficiently? In traditional teaching method of AutoCAD, various orders may be introduced according to functions of software, which costs numerous class hours. Meanwhile, students can not fully perform their subjective initiative for teachers speak a lot in the class. As a result, students often know what teachers teach at that time, but they can not apply orders they have leant according to practical condition when they draw example graph. Furthermore, teachers fail to integrate the examples with the class, so students have low enthusiasm on class, resulting in undesirable teaching effect. Instead of giving lessons according to the orders, the example teaching has integrated the drawing technique with the drawing process of examples, and introduced the orders in the application process. The teaching method can attract attention of students and stimulate their appetite for further knowledge due to direct and practical features. It shall be noted that example for one order shall be selected appropriately, and the orders that shall be taught through the examples shall be arranged elaborately.

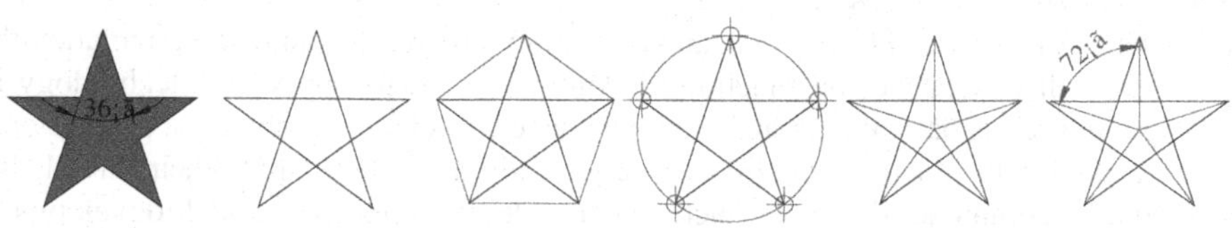

a **b** **c** **d** **e** **f**

Fig. 1. The drawing style of the five-pointed star

The figure 1 (a) is the example of five-pointed star that the author has used in the first class. The five-pointed star is consisted of 10 lines, with each included angle of 36°. Then how can we draw the line quickly? How can we guarantee that the included angle is 36°. How many drawing techniques there are? Then the author introduces the contents of the class: introducing various drawing techniques of straight line; opening polar axis tracking; and setting the increment angle as 36. Then the figure 1 (b) is completed quickly. How can we delete the lines in figure (b) quickly? Then the cutting in modification is introduced. It shall be noted that the selection object shall be chosen and the median shall be deleted. Then how can we solve the problem of colors? How can we finish the figure for application? How can we apply the figure to documents? How can we set the passwords for confidentiality? The teaching process follows the rule of proposing problem, analyzing problem and solving problem.

The above five-pointed star is completed by opening polar axis tracking with increment angle of 36. Then are there any other methods? The five-pointed star in figure 1 (c) is completed by drawing pentagon, which introduces the drawing technique of pentagon: the five-pointed star figure 1 (d) is completed by dividing the circle into five parts after introducing the format setting of points, drawing technique of point, and drawing technique of circle; the five-pointed star in figure 1 (e) is completed through annular array by opening orthogonal (realize by shifting F8) and drawing a perpendicular line, which introduces the editing method of array; the five-pointed star in figure 1 (f) is completed by drawing two repeated perpendicular lines (notice: open the object catching and start the endpoint), spinning 72°based on the bottom endpoint, and then imaging constantly based on any side of the angle to achieve other three sides, which introduces the editing method of spinning and imaging. Of cause, there are other methods. Due to the space limitation, the paper will not introduce those methods.

1.2 The Research and Practice on AutoCAD Precise Drawing

The above paper has introduced AutoCAD drawing orders and editing orders through various drawing techniques of examples with simple structure to promote students

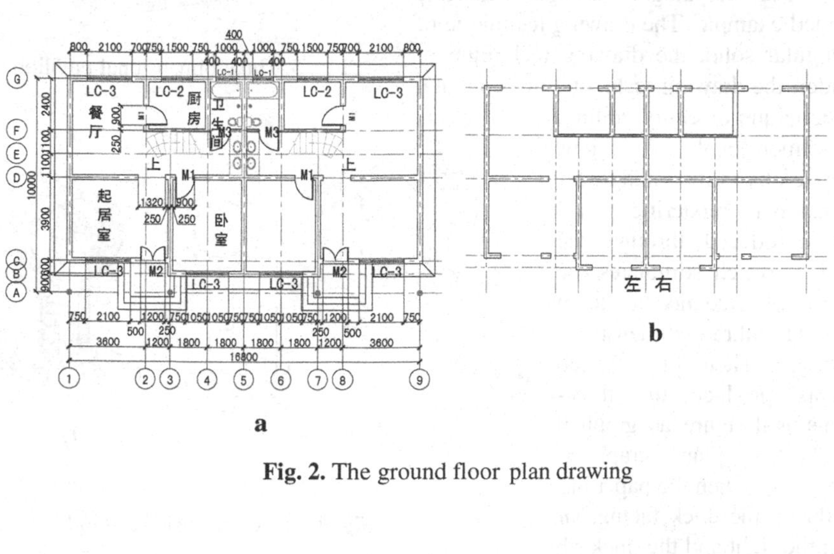

a

b

Fig. 2. The ground floor plan drawing

master the skills quickly. However, the figure 2 (a) seems complex, which is difficult to read. Then how can students read the figure easily? How can perform precise drawing according to the size of figure? Then the layer setting is introduced. Various settings including axis, wall line, door and window, mark and word can be set as a layer. According to the size in the figure, the left axis can be drawn firstly. Then the right axis can be drawn through mirroring process. The wall line can be drawn on the basis of axis. Then the right wall line can be drawn through mirroring process, as shown in figure 2 (b). How can we draw sections including door and window, bath, stinkpot, and axis sign quickly in figure 2(a)? Then the contents including segment and attribute, design center, tool pane, and segment editor are introduced. The drawing efficiency can be increased by mastering the above methods. Then how can we perform the size marking and word marking in the figure 2 (a). Then the setting of marking format and setting of word format have been introduced. Reasonable marking format and word format can not only promote the drawing efficiency, but also make the figure simple and clear. Due to the space limitation, the research and practice on AutoCAD precise drawing is introduced through the above example. Then the paper will introduce the research and practice on AutoCAD three-dimensional drawing.

1.3 The Research and Practice on AutoCAD Three-Dimensional Drawing

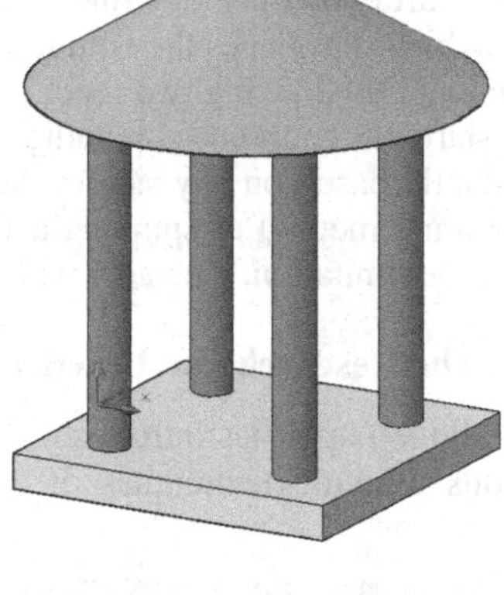

Fig. 3. Three-dimensional pavilion

The new version of AutoCAD comes out every year since 2003, with increasing capacity of three-dimensional drawing. How can students master the skills quickly and efficiently? Similar to the above, the required orders can be taught through elaborate designed example. The drawing technique of rectangular solid, the drawing technique of cylinder, the drawing technique of cone, and the technique of entity editing, three-dimensional view, dynamic observation and three-dimensional rendering have been introduced through the figure 3, which stimulates the interest of students to learn AutoCAD three dimensional drawing. Then how can students produce the three-dimensional figure as graph b according to plane graph as figure 4 (a)? Then the paper has introduced the thick facing bar 14 on the right and the thick rib

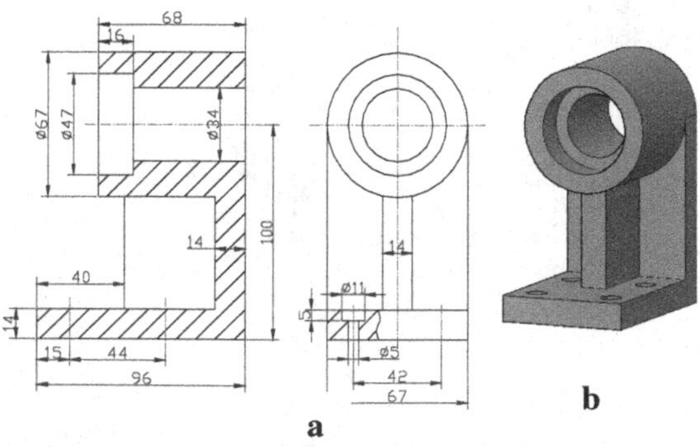

Fig. 4. Three-dimensional stand

plate 14 in the middle by defining region and stretching. The cylinder 67 with external diameter of φ is completed by drawing three concentric cylinders and subtracting the three cylinders. Then the cylinder is united with the completed facing bar, rib plate and baseboard in the bottom. The four countersinks are deleted through subtracting the bottom baseboard. As a result, the three-dimensional stereogram in figure (b) is completed. Then the paper has introduced contents including entity editing, three-dimensional view, dynamic observation, three-dimensional rendering, three-dimensional operations and quality features. Other three-dimensional drawing orders and editing orders will be introduced through other examples instead of introducing here.

1.4 Research and Practice on Teaching of AutoCAD Secondary Development

As the application software of CAD, AutoCAD has been widely applied in various fields including machinery, traffic, spinning, aviation, and civil construction. With open architecture, AutoCAD allows users to perform more convenient, professional and standard drawing and design by developing and customizing according to their demand. During the drawing process, the users may implement the same drawing tasks repeatedly. For example, in the underground pipeline survey, there are pipelines

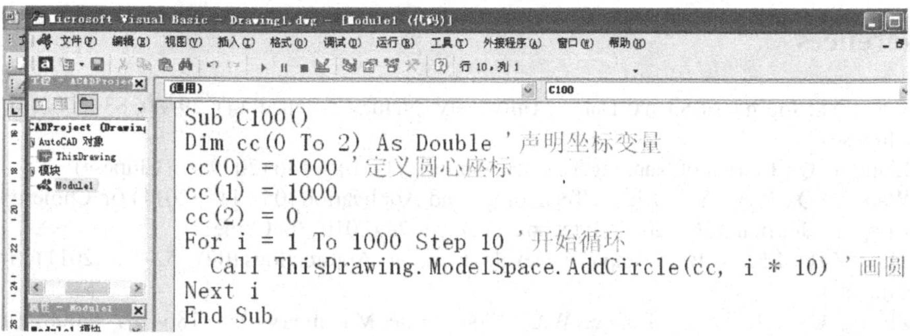
Microsoft Visual Basic - Drawing1.dwg - [Module1 (代码)]
文件(F) 编辑(E) 视图(V) 插入(I) 格式(O) 调试(D) 运行(R) 工具(T) 外接程序(A) 窗口(W) 帮助(H)
行 10, 列 1
(通用)
C100
ThisDrawing
模块
Module1
Sub C100()
Dim cc(0 To 2) As Double '声明坐标变量
cc(0) = 1000 '定义圆心座标
cc(1) = 1000
cc(2) = 0
For i = 1 To 1000 Step 10 '开始循环
Call ThisDrawing.ModelSpace.AddCircle(cc, i * 10) '画圆
Next i
End Sub

Fig. 5. VBA Program

with different attributes in a street (communication pipeline, military pipeline, watering pipeline and sewage pipeline). The survey crews shall measure the feature point coordinates for thousands of various pipelines by means of total station, hen produce comprehensive pipeline graph by connecting pipelines with the same attributes in AutoCAD. It is a very stuffy, repeated and low-efficient work to connect these feature points as pipeline graph in AutoCAD manually. If we can perform secondary development on the platform of AutoCAD and customize application program according to the drawing task of pipelines, we can complete the drawing task of pipelines automatically with high efficiency [3]. Currently, common development languages include Auto LISP, C＃, SCR, Object ARX, VB, and VBA [4-9]. As various students have the programming basis of VB, so we can introduce VBA development language gradually. Firstly, we can teach the simple example: draw 100 concentric circles. How can we write program to draw 100 concentric circles quickly? Firstly, we open the macro window through the macro submenu in tool menu of AutoCAD model space, or through shortcut key Alt+F8 to fill in C100 as the name of macro. Then we write the program as figure 5 after clicking "New" and "Confirm" buttons. Then we press the button of "Operate" in the macro window of macro submenu in tool menu. Then in the drawing space of AutoCAD model space, the 100 concentric circles with increasing radius of 100 have been realized. No other examples are introduced due to space limitation.

2 Conclusion

In order to promote students to achieve the authentication level of AutoCAD senior application engineers in short time, the above paper has introduced the teaching method from the four aspects including AutoCAD precise drawing, AutoCAD three-dimensional drawing and AutoCAD secondary development through classic examples, which has achieved good teaching effect and expected goals.

Acknowledgements. Fund project: educational reform fund project of the university (director, No. 5509002); project task for the 2011 provincial modern educational technology research (director, No. of task 2011-R-18824).

References

1. Wang, H.: Journal of Shanxi Datong University (Natural Science) (4), 79–80, 83 (2008) (in Chinese)
2. Liang, D.Q.: Journal of Nanning Teachers College (1), 144–146 (2009) (in Chinese)
3. Wang, H.Q., Jiang, Y.J.: Digital Technology and Application (05), 119 (2011) (in Chinese)
4. Dong, P.: Journal of Ningbo Polytechnic (10), 76–78 (2010) (in Chinese)
5. Liu, W.H., Liu, H.B.: Computer Development & Applications (02), 52–54 (2011) (in Chinese)
6. Zhang, C.C., He, L.L., Kuang, W.C.: Coal Mine Machinery (05), 186–187 (2008) (in Chinese)
7. Yao, J., Jiang, Z.F.: Mechanical Engineer (07), 27–29 (2007) (in Chinese)
8. Mo, Y.M., Gan, W.D.: Popular Science & Technology (03), 31–32 (2011) (in Chinese)
9. He, D.Z., Cao, L., Zhao, S.B.: Journal of Architectural Education in Institutions of Higher Learning (05, 147–150 (2010) (in Chinese)

The Application Research on Determination Method of Poor University Students Based on the Interval Density Cluster

Huang Zhenzhen

Department of Politics and Law,
Zhoukou Normal University, Henan, China
huangzhenzhen1982@163.com

Abstract. Poor university students determination is related to important social issues of education equity. In this paper, a density clustering method based on the interval is brought forward, which can effectively solve the problem that fuzzy c-means method heavily depends on the initial cluster center randomly generated. Through simulation, the piecewise linear membership function of the average daily food consumption is determined, which solves the problem that the poor university students is hard to be defined.

Keywords: poor university students determination, interval density cluster, poor, living in poverty, developing in poverty, not poor.

1 Introduction

Poor university students' financial assistance is related to important social issues of education equity, poor university students determination is the difficulties of poor students' university working. In the U.S., demand for financial assistance for poor students is by the difference between the family cost of supply and demand determine[1], in Canada, determination the economic status of students is based on the student's need, which not only takes into account economic factors, also considered the combination of factors, such as family, school and regional[2]. Poor university students' determination of developed countries is built on a sound tax system, however, in the process of determination of poor students in Chinese universities. As the determined standards are not uniform in different regions, family economic situation fraud evidence, the high cost of determined, and other reasons has been caused poor university students determination not objective.

According to the provisions of the Ministry of Education, at first, the majority of domestic colleges and universities verifiable evidence of poverty, then combined with the performance of students, counselors' judge, field research and other methods determined poor university students. However, due to the randomness of student performance, Counselor insufficient information, too much cost about field research and other reasons lead to poor university students' financial assistance be simplification, which will appear funded in turn, average distribution and so on. In this

A. Xie & X. Huang (Eds.): Advances in Computer Science and Education, AISC 140, pp. 281–285.
springerlink.com

paper, an interval density clustering method based on interval density is put forward. According to the dynamic consumption of the university students, the cluster center is quickly obtained by using this method, and the piecewise linear membership function is determined too, in a word, determination method of poor university students is a relatively simple and fast algorithm.

2 Poor University Students

Poor university students are difficult to pay their school during and basic living costs through themselves and their families. According to poverty level can be divided into four groups: poor, living in poverty, developing in poverty and not poor. Poor is that students have access to financial support very little, lack of protection of life, can not afford school fees. Living in poverty is that students can solve basic livelihood problem and the tuition, but the solution level is low. Developing in poverty is that students can solve life problems, but it is limiting their own development because of the economic conditions.

Poor university students is discrimination based on certain theories and methods, do the evaluation of the degree of difficulty, through a certain method to poor students from the total students can be selected out. The average daily consumption of food in school is important indicators of poor university students' determination, which is calculated as (1).

$$x = \frac{\sum_{i=1}^{m_1} a_i}{m_1} + \frac{\sum_{i=1}^{m_2} b_i}{m_2} + \frac{\sum_{i=1}^{m_3} c_i}{m_3} \tag{1}$$

Where, a, b, c represent breakfast, lunch, dinner consumption, m_1, m_2, m_3 representing the times of school breakfast, lunch and dinner.

Average daily consumption of food reflects the extent of the problem of student life, according to their circumstances can be divided into four groups: poor, living in poverty, developing in poverty and not poor.

3 Density Clustering Method Based on the Interval

In this paper, a density clustering method based on the interval is brought forward, first, according to the interval density, the data is classified, and then the clustering center is obtained by the density[3-5], which effectively avoids the cluster results falling into a local minimum.

For the data $X = \{x_1, x_2, ..., x_n\} \subset R$, the density clustering method is expressed as follows:

Step1: Set threshold τ of the interval density, and number of cluster center q;

Step2: Set width d of the interval, and divide the data set X into interval S_i with width d;

Step3: Scan the data, and get the density of the interval, that is the number of the data in each interval;

Step4: Select the interval S_k with the largest density, and search S_{k-1} from the left, if the density threshold is larger than τ , search S_{k-1} until the density threshold is smaller than τ; search S_{k+1} from the left, if the density threshold is larger than τ, search S_{k+2} until the density threshold is smaller than τ; at last the continuous intervals l with density threshold larger than τ is obtained.

Step5: Calculate the density index of each data x_i:

$$D_i = \sum_{j=1}^{n} \exp\left[\frac{-\left\|x_i - x_j\right\|^2}{(0.5r_a)^2}\right] \tag{2}$$

Select the data x_{c_1} with the largest density as a cluster center, and delete the continuous density interval l;

The field radius r_a in equation (2) is:

$$r_a = \frac{l}{2} \times d \tag{3}$$

Step6: Execute step 4 and 5 continuously until the cluster center q is obtained, and define the cluster center as w_m, $m = 1, 2, \ldots, q$.

In the above method, r_a represents radius of the continuous interval l, and is always decided by equation (3). In practical analysis, the cluster center q has no effect on the cluster results, as a result, radius r_a is determined by equation (3), and the computational complexity is smaller, the center appears earlier.

4 Simulation

By using the density cluster method based on the interval to get the cluster center, the piecewise linear membership function can be obtained. In this paper, the average daily food consumption data from the department of politics and law in Zhoukou Normal University is selected, some data distribution as shown in Fig. 1, and the piecewise membership function are determined by Matlab.

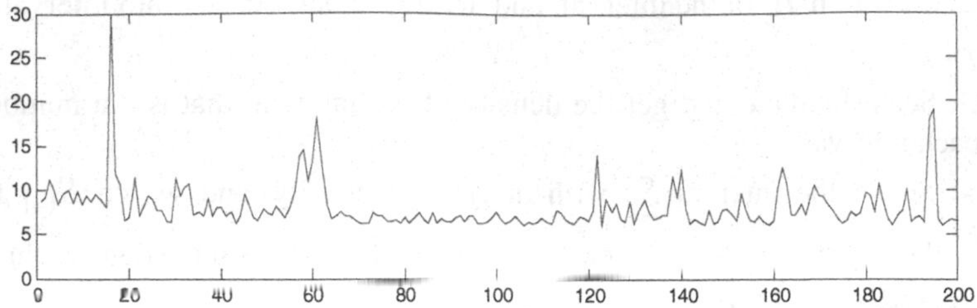

Fig. 1. The average daily food consumption data

On the basis of the initial cluster center q obtained by the density clustering method, the cluster center w_m and some other parameters, the concrete process of piecewise linear membership function is described as follows:

Step1: Unify universe of the data, and determine the number of the linguistic value. In this paper, membership function of the average daily food consumption from the department of politics and law in Zhoukou Normal University is searched, and data is mapped to a unified universe space, there are four linguistic values: poor, living in poverty, developing in poverty and not poor, that is to say, $q = 4, 1 \le k \le q$.

Step2: Determine the type of the membership function, that is to say, living in poverty and developing in poverty mean the middle linguistic value, which are expressed by triangle membership function, poor and not poor mean the bipolar linguistic value, which are expressed by monotone membership function.

Step3: Calculate the clustering center w_m by the piecewise density cluster, and the obtained centers are $w_m = 6.7281$, w_m=9.0342, $w_m = 12.3780$, $w_m = 19.6059$.

Table 1. Parameters of piecewise linear membership function

	Linguistic value	Type of the function	Parameter
1	poor	$f(x,a,b) = \max\left\{\left[\min\left(1, \frac{e-x}{e-d}\right)\right], 0\right\}$	$d = 6.7281$ $e = 9.0342$
2	living in poverty	$f(x,a,b,c) = \max\left\{\left[\min\left(\frac{x-a}{b-a}, \frac{c-x}{c-b}\right)\right], 0\right\}$	$a = 6.7281$ $b = 9.0342$ $c = 12.3780$
3	developing in poverty	$f(x,a,b,c) = \max\left\{\left[\min\left(\frac{x-a}{b-a}, \frac{c-x}{c-b}\right)\right], 0\right\}$	$a = 9.0342$ $b = 12.3780$ $c = 19.6059$
4	not poor	$f(x,a,b) = \max\left\{\left[\min\left(\frac{x-d}{e-d}, 1\right)\right], 0\right\}$	$d = 12.3780$ $e = 19.6059$

Step4: According to the vicinity principle, the parameters of the triangle and monotone membership function are shown in Table 1.

Step5: According to the parameters in Table 1, the membership function is shown in Fig. 2.

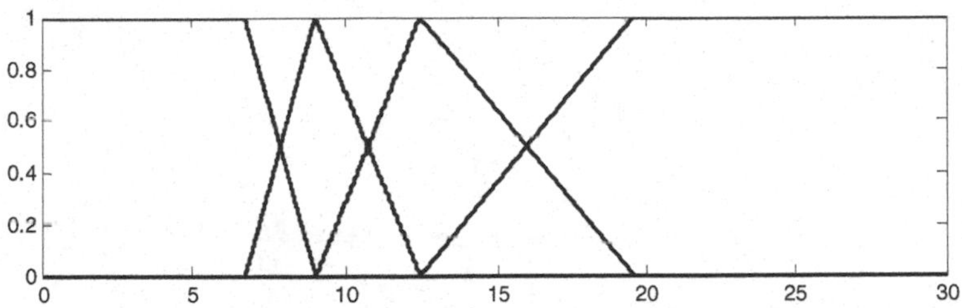

Fig. 2. Piecewise linear membership function of the average daily food consumption

5 Conclusion

According to the case study, the density cluster method brought forward in this paper has no iteration, in this method, the initial cluster center is obtained by the density of interval, which can effectively solve the problem that fuzzy c-means method heavily depends on the initial cluster center randomly generated, and obtain the center rapidly. By using the density cluster method proposed in this paper, Piecewise linear membership function of the average daily food consumption are obtained, which is a new and simple method to define poor university students.

Acknowledgments. The research of this paper is supported by the natural science foundation of Education Department in Henan province in China. (Grant NO. 2011B510021).

References

1. Creech, J., Davis, J.: Merit Based versus Need Based Aid, p. 56. The American Council for Education and the Oryx Press (1999)
2. Usher, A.: Are the Poor Needy? Are the Needy poor? The Distribution of Student Loans and Grants by Family Income Quartile in Canada. Educational Policy Institute, p. 11 (2004)
3. Zadeh, L.A.: Fuzzy sets. Information and Control 8, 338–353 (1965)
4. Liu, Q., Zhang, H.-H., Huang, Z.-Z.: Analysis of oil drilling working state based on fuzzy clustering method(FCM). Journal of Zhoukou Normal University 5, 47–49 (2009) (in Chinese)
5. Liu, Q., Du, Y.-D.: The normal cloud model of fuzzy inference prediction method for petroleum drilling accident. In: Qi, L.I.U., Yuan-dong, D.U. (eds.) 2011 International Conference on Electric Information and Control Engineering, vol. 2, pp. 1816–1821 (2011)

Course Design and Practices of Teaching Undergraduate's Database System

Hongji Wang[1] and Qingqin Yan[2]

[1] Software School, Xiamen University,
Xiamen, Fujian, China
whj@xmu.edu.cn

[2] School of Marxism, Nanchang University,
Nanchang, Jiangxi, China
qingqinyan@163.com

Abstract. This paper describes the course design and practices in teaching undergraduate's database system (DBS) to software engineering (SE) and digital media engineering (DME) students at the software school of Xiamen University, China. Although the majority teaching objectives of DBS are the same for SE and DME students, there exist minor emphasis differences between them. How to achieve their respective goals under the unified teaching framework is a meaningful challenge. In our course design, by introducing advanced (graduate-level) materials to SE and multimedia database technology materials to DME, all the students can not only obtain the deep understanding of the database system concepts, but also master the solid practical development skills related to their specific areas.

Keywords: undergraduate teaching, database system, course design.

1 Introduction

The software school of Xiamen University, founded in Feb., 2002, is one of the 35 national exemplary software schools approved by the Ministry of Education and the former National Development and Reform Commission of China. The school offers undergraduate and postgraduate education in software engineering and digital media technology. According to the literature [1], one of the goals of establishing exemplary software school is to foster the high-level practical talents who can meet the industrial community requirements for software technology, such as system engineer, system designer, system architecture and project manager etc..

Nowadays database technology is used pervasively. Database-related courses play an important role in teaching SE and DME. In our school, DBS is taught as an introductory and fundamental database management system (DBMS)-related course every year. This course is arranged for all the junior students at the same semester, and it is compulsory to SE students and elective to DME students. On one hand, the backgrounds of SE and DM students are different. For example, most of SE students have mastered at least a programming language such as C/C++, or Java, which is prerequisite for our course. However, most of DM students show lack of

A. Xie & X. Huang (Eds.): Advances in Computer Science and Education, AISC 140, pp. 287–291.
springerlink.com

programming experiences, this will lead to difficulties for them in learning this course, especially when need to do good job in course project. On the other hand, the training objectives of SE and DME are different. SE emphasizes on the software development and maintenance abilities, while DME concentrates on the digital media technologies. One issue we should consider is how to combine their specific needs into our course effectively. In such cases, how to bridge the gap between SE and DME, and satisfy their specialized subject requirements under the unified teaching framework is an interesting problem.

The unified teaching framework (UTF) we designed is composed of three parts. That is, regular class teaching, experimental teaching, and course project based on the practical topics. The design principle of UTF is to promote students to gain the deep understanding of the principles of DBS, and master the software development skills needed in the future effectively and efficiently. Through the practices of three continuous academic years, *i.e.*, 2008-2009, 2009-2010, 2010-2011 academic years, the result is satisfactory.

The remainder of this paper is organized as follows. Section 2 presents the contents of DBS course, and analyzes some facts in teaching activities. Section 3 gives the course design scheme. Section 4 discusses the experimental results. Section 5 concludes this paper.

2 Course Analysis

What topics should be covered in the course teaching is the issue we need to take into account seriously. According to the literature [2], the typical database management system topics in computer science and SE curriculum are organized into the following seven categories:

(1) **Theory and Concepts:** Introduction, ER model, conceptual design, relational model, relational algebra, relational calculus.
(2) **Applications:** SQL, embedded SQL, database APIs, application development, Internet applications, XML.
(3) **Storage and Organization:** Data storage, tree index, hash table, buffer management.
(4) **Query Processing:** External sorting, nested-loop execution, merge-sort execution, index only execution, optimization of individual operators.
(5) **Transaction Management:** ACID properties, concurrency control, lock-based and timestamp-based, crash recovery, 2-phase commit.
(6) **Design and Tuning:** Scheme refinement and normalization, physical design and performance tuning, authorization and database security.
(7) **Advanced Topics:** Distributed and parallel databases, object-oriented databases, deductive databases, data warehousing, data mining, information retrieval, geographical information systems, multimedia databases, etc.

Briefly speaking, categories (1) and (2) can be viewed as application-oriented topics, and categories (3) -- (6) as system-oriented topics. The main topic of application-oriented is modeling and design, which focuses on creating an optimal database from a given set of user requirements, while the main topic of system-oriented is about internals of database systems, which cover such issues as query optimization, transaction processing and storage structures. From the practical view point, we require our students to implement a practical project that should follow the principle of database application system development. To do this well, students need some essential capacities, such as deep understanding of database system concepts, good programming skills, good communication skills, and excellent teamwork spirits. Usually, application-oriented topics are covered at the undergraduate level, while system-oriented topics are introduced to the graduates served as advanced database technology.

In our school, the cultivation objectives of SE and DME students are different. For SE students, we always emphasize the abilities related to software development life cycle, in which software development ability is the fundamental and key capacity. For DME students we emphasize their abilities aiming to express the artistic effects by using digital media technology. Therefore, after completing our course, SE students should grasp the development method of database application systems; DME students should deal with multimedia-related data such as video and audio. Meanwhile, we noticed that most of SE students always have a hunger for knowing how the DBMS works, and DME students want to know how the DBMS manage the multimedia data.

Considering above differences of SE and DME, we should make corresponding adjustments on the course from both contents and practices under the unified teaching framework in order to achieve SE and DME respective goals.

One thing we need to mention here is about J2EE. J2EE is a course that is taught at the same semester as DBS. It is elective to SE students only, and no DME student is required to learn it. This arrangement directly leads to the different impacts on our teaching. Apparently it is beneficial to SE students. Because J2EE could be used in the database application systems development seamlessly.

3 Course Design and Practices

According to the previous section's analysis, in this section we present our detailed course design and operations. Firstly, the contents of the course are divided into two parts: a practical part and a theoretical part, in which practical part is divided into two parts: a laboratory assignments part and a project part. Secondly, we start the course with the practical part. According to our experience our students are better motivated to tackle the theory behind databases once they have learned about their use. Another advantage is that we can let the students come into contact with a database system earlier in the course through laboratory assignments. In parallel with the theoretical part the students perform a project using the knowledge gained in the practical part. Thirdly, we define a number of modules consisting of lectures and laboratory assignments. Most of the modules are general and used in both SE and DME. Some of the modules are specialized and used in SE or DME separately.

In theoretical part, besides regular teaching, we introduce some system-related materials of a concrete DBMS, especially administrative materials, for SE students (take Oracle 10g for example), and some materials of multimedia database for DM students respectively.

In laboratory assignments part, we choose Oracle 10g as experimental tool. The reason why we make such decision is as follows. (1)Oracle is an advanced database management technology tool, which covers many mainstream database technologies; (2) its market-share is nearly 70%, which means that students have the advantages in finding good jobs if they use Oracle earlier.

Before getting started to laboratory assignments, we first install an Oracle 10g server in a server computer and create an account for every student, then every student can login the database server in two ways: client way and web way. In the former way, a student should install Oracle 10g client in his/her own computer and make the configuration properly. In the latter way, it is very simple for students to have access to the database server. What he/she need is only a browser and connected network. No matter what way the students take, the Oracle server is always available in 7×24 hour, students can do their laboratory assignments in anytime and from anywhere. Besides these, we also provide a number of on-line resources, including general information about the set-up of the laboratory system, an Oracle manual and an FAQ.

4 Experiment Evaluation

The grading in this course is based on the class attendance, the class assignments, the laboratory assignments, the project work, and a final written exam. Here we list the grading of continuous three academic years below, *i.e.*, 2008-2009, 2009-2010, 2010-2011 academic year. The related results are depicted as Table 1, Table 2, and Table 3, in which Table 1 and Table 2 are about the grading of SE students and DME students, respectively, Table 3 is about the feedback from the students.

Table 1. The grading of SE students

	Semester 2008-2009	Semester 2009-2010	Semester 2010-2011
Above 90	20%	18%	22%
80-89	33%	35%	36%
70-79	28%	27%	25%
60-69	18%	20%	15%
Less than 60	1%	0%	2%

Table 2. The grading of DME students

	Semester 2008-2009	Semester 2009-2010	Semester2010-2011
Above 90	22%	20%	24%
80-89	35%	33%	36%
70-79	25%	28%	23%
60-69	16%	18%	15%
Less than 60	2%	1%	2%

Table 3. The satisfaction of students to this course (score)

	Semester 2008-2009	Semester 2009-2010	Semester 2010-2011
Class 1	98.6	92.2	93.3
Class 2	92.3	93.6	92.6
Class 3	90.4	94.7	95.8

From Table 3 we can see that the average satisfaction of students to our course is 93.7, which shows that students' evaluation to our teaching including course design is excellent.

5 Conclusions and Future Work

This paper discusses an experiment in teaching database management basics at the undergraduate level by combining standard course material with advanced topics. Besides working on traditional exercises with database tables, SE students should explore the implementation aspects of the DBMS, which can promote students to gain deep understanding of how the DBMS works; DME students should master the multimedia database management technology, which is very useful in working IT-related jobs.

At the same time, we also have observed that database management technology varies very fast, we will introduce new database technology to the teaching activities to keep state-of-the-art for our students based on their needs and future development.

Acknowledgments. This work is supported by the Teaching Reform and Teaching Quality Assurance Program of Software School, Xiamen University, China.

References

1. Zhang, Y.: The thoughts on establishing exemplary software school. China Higher Education 10, 5–9 (2004) (in Chinese)
2. Yang, L., Sanver, M.: Teaching Database in Two Courses: Reconciling Theoretical Framework with Practical Considerations. In: Procedings of ASEE Conference, Salt Lake City, UT, pp. 13489–13494 (2004)

The Impact of Use Context on Mobile Payment Acceptance: An Empirical Study in China

Luzhuang Wang[1] and Yongzheng Yi[2]

[1] Zhejiang University City College, 310015, Hangzhou, China
zuccjohn@hotmail.com
[2] School of Management, Zhejiang University, 310058, Hangzhou, China
yongzhengyi@gmail.com

Abstract. By adding two related constructs (use context and perceived risk), this paper explored the impacts on the using of mobile payment based upon UTAUT model. With empirical study from 111 questionnaires, the results indicate that performance expectancy and effort expectancy are key determinants for the users to adopt the mobile payment. Use context, via its significant influence on performance expectancy and effort expectancy, has an indirect affect on user behavior. Furthermore, inconsistent with previous researches, there is no significant relationship between perceived risk and behavioral intention, neither between social influence and behavioral intention. Also there is no significant influence of facilitating conditions on user behavior.

Keywords: E-Commerce, Mobile payment, Use context, UTAUT.

1 Introduction

Mobile commerce as a newly raising business model is penetrating into our lives. The greatest advantage of m-commerce is it can achieve business process regardless of time and location. As an important part of m-commerce, mobile payment can be realized anytime and anywhere, breaking the bottleneck of traditional e-commerce. As a result, m-payment is considered to be the key to the m-commerce development.

The statistical data released by Ministry of Industry and Information Technology of China shows that the number of mobile phone users has reached 929.8 million by July, 2011, among which 3G users are 98.2million[1]. This indicates that m-payment has great potential market in China.

Although people generally embrace positive attitude towards the future development of m-payment, the extent of m-payment among real consumers is now still lower than expected.

Some studies have explored factors affecting consumer adoption of m-payment. Niina[2] made a qualitative research on the adoption of m-payments, suggesting that the relative advantages of m-payment depend on certain situational factors such as lack of other payment methods or urgency, but he did not propose a research model and test it empirically. By literature review Tomi et al.[3] proposed a framework of four contingency and five competitive factor, but most of the factors were only addressed by

A. Xie & X. Huang (Eds.): Advances in Computer Science and Education, AISC 140, pp. 293–299.
springerlink.com

exploratory and early phase studies. Based their framework on TAM and IDT, Niina et al.[4] studied impacts on consumer behavior and human-computer interaction, and found that use context is a significant determinant for consumers' intention to use the m-commerce.

However, there are quite few researches about acceptance and use behaviors in China, especially empirical researches. This paper tried to extend UTAUT model by adding use context factor and perceived risk factor in order to find out impacts upon user's intention for m-payment in China.

2 Research Model and Hypotheses

2.1 Basic Model

Information systems (IS) researchers have made significant efforts in building theories to examine and predict the determinants of information technology (IT) acceptance, such as TRA, TPB, TAM, IDT and so on. UTAUT was based on studies of eight prominent models in IS adoption research, which had improved theoretically and its explanation power had achieved 70% proved by Venkatesh empirically[5].

This study tried to modify and test the constructs of a revised UTAUT model and its explanatory power in the context of m-payment in China. The authors proposed:

H1. Performance expectancy has a direct positive effect on user's behavioral intention.
H2. Effort expectancy has a direct positive effect on user's behavioral intention.
H3. Social influence has a direct positive effect on user's behavioral intention.
H4. Behavioral intention has a direct positive effect on user's use behavior.
H5. Facilitating conditions has a direct positive effect on user's use behavior.

2.2 Use Context and Perceived Risk

Because users' concerns and needs vary with the context in which they use m-payment, the m-payment meeting users' needs in a specific context will provide the best value to them[6].Therefore, it is believed that users' perception of the ease of use (EE) and the usefulness(PE) of mobile services may vary in different contexts.

H6. Use context has a direct positive effect on performance expectancy.
H7. Use context has a direct positive effect on effort expectancy.

Beside perceived benefits (i.e., performance expectancy and effort expectancy), innovations usually come with risks[7]. In the context of E-services, perceived risk conceptualized as the likelihood of privacy invasion, has been found to be a particularly critical concern among consumers [8].

H8: Perceived risk has a direct negative effect on user's behavioral intention to use m-payment service.

From above arguments, a revised UTAUT model is proposed in Fig.1.

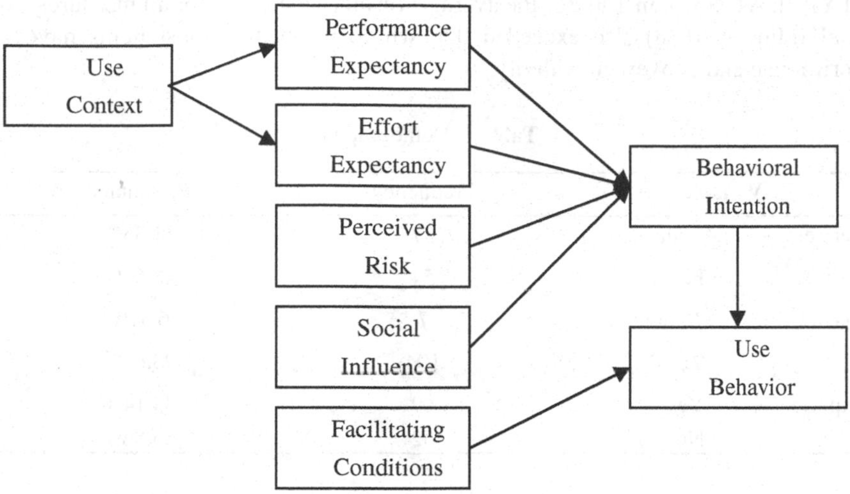

Fig. 1. Proposed research model for m-payment acceptance

3 Research Methodology

To evaluate above model, empirical data were collected using a self-administered questionnaire distributed online. As a pilot test of a whole research, a total of 111 usable questionnaires, mainly from university students, were collected during September 2011, and then assessed by structural equation modeling technology.

Based on the research model proposed, eight key constructs was measured with multiple items. These items were either adopted or adapted from the extant literature, except for the use context which was self-developed. Each items was measured with a seven-point Likert scale, ranging from "strongly disagree"(1) to "strongly agree"(7).

4 Empirical Results

The demographic profile of the respondents is reported in Table 1. It shows that for the respondents all were no more than 30 years old, 51.35% were female, and near half had used the mobile payment service before the time of the survey.

The proposed research model was evaluated by SEM. The data obtained were tested for reliability and validity using confirmatory factor analysis (CFA). The model includes 27 items describing 8 latent constructs: performance expectancy, effort expectancy, use context, perceived risk, social influence, facilitating conditions, behavioral intention, use behavior. The CFA was computed using the AMOS 6.0. The composite reliabilities of the measures included in the model ranged from 0.73 to 0.90 (see Table 2). All were greater than the benchmark of 0.60 recommended by Bagozzi and Yi[9]. As shown in Table 2, the average variance extracted for all measures except social influence(0.38) also exceeded 0.5, which meant that most items have good discriminate and convergent validity.

Table 1. Demographics

Variables		Frequency	Percentage
Gender:	Male	57	51.35%
	Female	54	48.65%
Age:	20 or under	7	6.31%
	21-30	104	93.69%
Experience:	Yes	49	44.14%
	No	62	55.86%

Table 2. Assessment of the construct reliability

Variables	The composite Reliability(>0.6)	Average variance extracted (>0.5)
Performance expectancy	0.87	0.69
Effort expectancy	0.78	0.55
Use context	0.90	0.74
Perceived risk	0.90	0.75
Social influence	0.73	0.38
Facilitating conditions	0.83	0.52
Behavioral intention	0.83	0.62
Use behavior.	0.81	0.67

The final results are in Table 3 and Fig. 2. As shown in Table 3, some of the model-fit indices were lower than their recommended values. It needs further study to further understand the problem.

Table 3. Model evaluation overall fit measurement

Quality-of-fit measure	Recommended value	Structural model
/df	<3.00	1.835
GFI	>0.90	0.747
AGFI	>0.80	0.677
CFI	>0.90	0.872
NFI	>0.90	0.762
TLI	>0.90	0.848
RMSEA	<0.08	0.087

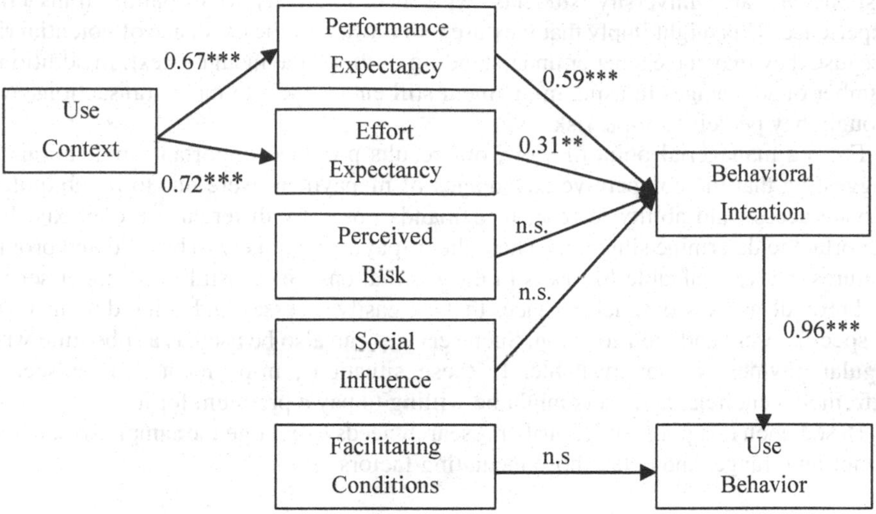

Fig. 2. The estimated structural model

Note : Numbers represent path coefficients, ** p<0.01., *** p<0.001., n.s. not significant relationship

Out of total 8 hypotheses, 5 were supported, as shown in Fig.2. Inconsistent with previous research, there are no significant relationships between PR and BI and neither between SI and BI. Also there are no significant influence from FC on UB. However, PE and EE were found to positively and significantly relate to BI. And the result indicated that BI has a significant effect on the UB. This means that users' behavioral intention is an important determinant of system usage.

In addition, UC is a very important factor in influencing the adoption of mobile payment. It significantly and directly influences both PE and EE.

5 Discussions

Above results suggest that users consider two different types of benefits of mobile payment service: performance-related usefulness and the spatially and temporally improved service access enabled by mobile technology.

The performance expectancy and effort expectancy factors affected behavioral intention directly. It appears that performance expectancy and effort expectancy all indirectly influence the use behavior through behavioral intention to use which is consistent with work of Venkatesh[5].

From a theoretical perspective, it seems reasonable that a higher perceived risk in mobile payment will lead to a lower rate of behavioral intention to use, resulting in lower m-payment use. Furthermore, perceived risk was believed to be a predictor and barrier to online transactions, and expected to negatively influence user' intent. However, above analysis suggests that there is not significant relationship. The real reasons for this finding are not quite clear yet. But with respect to perceived risk, the respondents are university students who had relatively rich online transaction experience. This might imply that they are more aware of the existence of potential risk because they may have a better understanding of the m-payment context. In addition, a number of advantages in using m-payment still entice users to make transactions even though they perceived some risk.

From a managerial point of view, our results provided important implications by suggesting that the competitive advantages of m–payment were due to its ubiquitous service access and ability to react to demands posed by different use contexts. It is important to determine situations where the m-payment is likely to be used and provide features that are valuable for users in those situations. Successful m-payment service will provide users with timely services that are easily accessed and tailored to the needs of specific users and their location. Such services can also be used as a substitute when regular payment is not available. In these situations, m-payment can be seen as extremely beneficial and users might be willing to pay a premium for it.

Based upon this pilot study, further research needs to enlarge the sample size, extend sampling arrange, and detail those mediating factors.

6 Conclusions

From an empirical study of mobile payment service adoption, it finds that use context, a new concept in a revised UTAUT model, is a significant determinant for uses' performance expectancy and effort expectancy in using the mobile payment service, implying that users valued the benefits of the mobile payment service in situations where other payment alternatives were not available, or where there were queues at points of sale.

Furthermore, inconsistent with previous researches, there are no significant relationships between perceived risk and behavioral intention, and neither between social influence and behavioral intention. Also, there are no significant influence from facilitating conditions on use behavior.

References

1. Ministry of Industry and Information Technology of China, http://www.miit.gov.cn
2. Mallat, N.: Exploring Consumer Adoption of Mobile Payments-A Qualitative Study. Journal of Strategic Information Systems 16(4), 413–426 (2007)
3. Dahlberg, T., Mallat, N., Ondrus, J., Zmijewska, A.: Past, Present and Future of Mobile Payments Research: A Literature Review. Electronic Commerce Research and Applications 7, 165–181 (2008)
4. Mallat, N., Rossi, M., KristiinaTuunainen, V., Oorni, A.: The impact of use context on mobile services acceptance: The case of mobile ticketing. Information & Management 46, 190–195 (2009)
5. Venkatesh, V., Morris, M.G., Davis, G.B., Davis, F.D.: User Acceptance of Information Technology: Toward a Unified View. MIS Quarterly 27(3), 425–478 (2003)
6. Figge, S.: Situation-Dependent Services—A Challenge for Mobile Network Operators. Journal of Business Research 57, 1416–1422 (2004)

7. Cho, J.: Likelihood to Abort An Online Transaction: Influences from Cognitive Evaluations, Attitudes, and Behavioral Variables. Information and Management 41(7), 827–838 (2001)
8. Lwin, M., Wirtz, J., Williams, J.D.: Consumer Online Privacy Concerns and Responses: A Power-Responsibility Equilibrium Perspective. Journal of the Academy of Marketing Science 35(4), 572–585 (2007)
9. Bagozzi, R.P., Yi, Y.: On the Evaluation of Structural Equation Models. Journal of Academy of Marketing Science 16(1), 74–94 (1988)

An Empirical Research on Correlation of Communication Companies' Compensation Gap with Enterprise's Performance

Luzhuang Wang and Hui Yu

Zhejiang University City College, Hangzhou, 310015, China
zuccjohn@hotmail.com, 284499453@qq.com

Abstract. This paper discusses mainly the relationship between executives' compensation gaps and enterprise's performances of Communication industry, with data from Shenzhen and Shanghai stock exchanges. The equation of linear regression is built, through correlation analysis and inspection, confirming the correlativity between variables. Main results are that the payment systems of Communication Services and Equipment Manufacturing industry conform to Behavior Theory, but there's a positive correlativity between the compensation gap of general manager to other junior managers and enterprise's performance.

Keywords: Communication Company, Executives' Compensation Gap, Influence Factor, Company Performance, Empirical Research.

1 Introduction

As one of the incentive system, compensation has a deep connection with company's development. So, how to set a scientific compensation system is gaining more and more attention of executives, especially human resource managers.

Contract to quite few Chinese empirical results, there're many empirical researches in this area, discussing the effect to compensation structure with Tournament Theory [1]. Lamber etc. found that compensation gap between executives would increase with position grades, especially the gap between CEO and the second chief [2]. The correlation of compensation gap and performance is positive. With the data from Denmark agent, Eriksson [3] concluded the compensation gap had a positive correlation with selling profits. Beginning by relative positioning of compensation, Cowherd etc. [4] studied the compensation gap and performance between executives and common staff. The result showed positive correlation, meanwhile, the staff with higher compensation has weak sensitivity to compensation gap than the lower ones.

Lin Junqing etc. [5] take "absolute compensation gap" and "relative compensation gap" into research, finding that the compensation gap of executives had obvious positive correlation with company's performance. Aimed at the state-owned enterprise Liu Chun & Sun Liang found that the correlation of performance and compensation gap between executive and staff was obviously positive, and companies' performance from coastland areas had lower sensitivity to compensation gap than these from inland areas [6]. Wang etc. found the listed companies' executives' compensation gap

A. Xie & X. Huang (Eds.): Advances in Computer Science and Education, AISC 140, pp. 301–306.
springerlink.com

from Liaoning province had a positive correlation with performance. Wang Haibo etc. had similar conclusion through studying [7].

Few researches were based on Behavioral Theory, and they found that in industry in which executives favor to collude, smaller compensation gap would bring higher stock return [8], compensation gap having a negative correlation with company performance. This is opposite to what Main etc. (1993) concluded. Zhang (2008) [9] studied the compensation gaps both internal executives and executives with staff, and found the two gaps both had negative infection to performance. Lu (2009) [10] found that when controlling the endogeneity of executives' internal compensation gap, the performance would result in a larger compensation gap, while ignoring endogeneity, there's no obvious positive correlation between the two objects.

Literature analysis shows that most researches are done before 2000. For Chinese listed companies, the correlation research of compensation gap and company performance needs recent empirical results because of the severe change of security environment. And there lack of liner expression in literatures correlating the dependent variable (DV) and controlled variable (CV).

2 Research Methodology

Sample companies are listed in Shanghai and Shenzhen. Usually one company will define the compensation standard of the next year by the last year's performance, so authors choose company performance data from 2008 to 2010 matched with executive compensation data from 2007 to 2009 for analyzing. Only the normal A Stock listed companies were selected.

Communication industry is taken as sample industry, because it has been developed very fast in recent years, and the performance and compensation structure are undertaking great changes. Analyzing the sample has typical value to understand company executives' compensation level and the development state of China. Moreover, based on locations, authors separate sample companies into coastland area and inland area to analyze the area difference.

2.1 Independent Variable

The compensation gap of executives is the independent variable. Compensation refers to generalized economical compensation, including salary, bonus, welfare and other numbers offered by cash. It's pretax income of executives reported by annual reports. Based on the former five positions, we defined four different compensation gap, representing Absolute Gap between executives in different position and the Relative Gap between executive compensation and payroll. Gap 1 = annual compensation of president – annual compensation of vice president, Gap 2 = ACP - annual compensation of CFO, Gap 3 = average annual compensation of executives – average annual compensation of secretary and supervisor, Gap 4 = total compensation of executives/ total of payroll payable.

2.2 Dependent and Control Variables

Return of equity (ROE), rate of net profit and net assets, represents enterprise assets utilization degree. Earnings per share (EPS), rate of net profit and annual total shares of common stock, reflects the common stock gain profit level, measures the company profitability.

Considering the hysteresis quality influence the compensation to performance, combining executive compensation gap influence factors, from the references, set the company size, the company areas, and the share concentration ratio for the control variables, add two factors of asset ratio and the state-owned ratio. Because of the hysteresis quality influence, the year study of controlled variable is the same as compensation, one year early to the performance data.

According to Zhang J. and Zhang H. [12], the performance will bc impacted by location of companies (Area), so divided companies into coastal area, economically more developed provinces and cities like Beijing, Tianjin, Liaoning, Shanghai, Jiangsu, Zhejiang, Fujian, Guangdong, Hainan, and Shandong, and inland areas of other provinces and cities.

Company size (Size) measures the natural logarithm of the end of the book value of the company's total assets. Ownership concentration (Herf) is the sum of the square of the Top 10 shareholders' holding ratio. State-owned stock ratio (Stock) is the shareholding rate of state-owned shareholder in the enterprise. According to the Ultimate Theory of property rights, there's approximately 84% China's listed companies was eventually state-controlled [13]. Asset-liability ratio (Debt) is used to examine the influence of debt.

3 Hypothesis and the Model

This study selects factors which has connection with compensation gap and enterprise performance. This article analyzes from the two aspects of the enterprise internal and external influences, and then raises the research hypotheses.

H1: there's a positive correlation between regional development level and the compensation gap.
H2: There's a positive correlation between industry development level and the compensation gap. Because of the space, this article will not do specific analysis.
H3: There's a positive correlation between "Size" and "Gap".
H4: There's a negative correlation between "Herf" and "Gap".
H5: There's a negative correlation between "Stock" and "Gap".

Using the software SPSS to do descriptive statistics and the correlation analysis. Through the P test to determine whether there is a correlation between variables, through the stepwise regression to establish the linear equation between variables.

Area difference analysis is used to observe whether the correlation between enterprise performance and compensation gap as well as other factors to have gradient changes in different areas.

Using two dependent variables, authors compare the models, validate the conclusions, and make the results comparable to further enhance the application value.

With above hypotheses and corresponding analysis, following model suggested.

$$ROA=\beta0+\beta1Size+\beta2Herf+\beta3Stock+\beta4Debt+\beta5Gap1+\beta6Gap2+\beta7Gap3+\beta8Gap4+\varepsilon$$

The correlation between EPS and compensation gaps, as well as other control variables could be obtained by way of similar to what ROA analyze does.

4 Empirical Results

Extreme value and average of compensation gap of Communication Industry listed in Table 1. It shows the gaps are relatively average, from 11.92, 11.89 to 15.77, and maximum only 5-10 times.

Table 1. Index of Compensation Gap Data of Communication Industry

Gap Index	Sample Size	Minimum	Maximum	Average
Gap1	93	0	130.77	11.92
Gap2	93	0	132	11.89
Gap3	93	0.56	99.7	15.77
Gap4	93	0.05	111.50	21.00

Table 2 reflects the Mean, Standard Deviation and Pearson correlation coefficient of variables. It shows that there is positive correlation between compensation gap and enterprise size, reaching 0.384, 0.264 and 0.398 respectively, while the performance and the gap indexes is weak or irrelevant. In the control variables, the correlation coefficient between the state-owned ratio and enterprise performance is -0.419, a weak negative correlation. There might be a weak positive correlation between Gap2 and the enterprise performance, with index 0.241.

Table 2. ROA-Average, Standard Deviation and Pearson Coefficient of the Variables

	ROA	Gap1	Gap2	Gap3	Gap4	Size	Debt	Stock
Gap1	0.146							
Gap2	0.241	0.837						
Gap3	0.06	0.557	0.475					
Gap4	0.011	0.096	0.033	0.008				
Size	0.043	0.384	0.264	0.398	0.216			
Debt	0.061	0.207	0.002	0.289	0.01	0.579		
Stock	0.419	0.045	0.085	0.218	0.074	0.115	0.227	
Herf	0.205	0.044	0.141	0.02	0.188	0.203	0.102	0.536

The study focuses on all the correlation between variables, so adopt stepwise regression method, and only select variables with big correlation coefficient. Because there are linear relations between ratio of state-owned shares and Gap2 with the performance, so choose state-owned shares (Stock) as a control variables, Table 3 reflects the fitting results. Sig value showed that the enterprise performance with state-owned ratio and Gap2 have correlation, supported by Pearson coefficient.

Table 3. Communication Industry ROA- Variable Model Summary

Model	R	R^2	A. R^2		Q. Sum	df	M. Sq.	F	Sig.
1	0.419a	0.176	0.167		702.718	1	702.718	19.433	.000a
					3290.608	91	36.161		
				Total	3993.327	92			

a. Prediction Variable: (Constant), Stock

b. Dependent Variable: ROA

Table 4. Communication Industry ROA - Model Coefficient

Model		Non-standard Coefficient		Standard Coefficient	T	Sig.	Correlation		
		B	S.Error				Zero-order	Partial-order	Part-order
	(Constant)	10.947	1.021		10.721	0			
2	Stock	0.113	0.026	-0.402	4.298	0	-0.419	-0.413	-0.401
	Gap2	0.079	0.036	0.207	2.209	0.03	0.241	0.227	0.206

a. Dependent Variable: ROA

Table 4 reflects the correlation regression model analysis results. According to the test, the following expression established.

$$ROA= 10.947 -0.113Stock +0.079Gap2 +\varepsilon \quad (1)$$

Using EPS as the dependent variable to build model, the result is as Expression 2.

$$EPS= -1.389 -0.003Stock +0.009Gap2 -0.08Size +\varepsilon \quad (2)$$

From Table 1, it can be imagined that in communication manufacturing and service industry, the compensation level is relatively equity, and the presence of extremes might due to the fast development of this industry which leads to high level rewards for high level management in some companies.

Therefore, among those newly developed industry, it should be focused upon to design a relatively fair compensation structure, while increasing market share and profit level, in order to strengthen the inner harmony. For those bigger companies, it is important to increase the compensation gaps between high level management to some extent, especially the gap of CEO with others, which will both increase and retain such an important management force.

5 Empirical Conclusion

From empirical results, it can been concluded that in Chinese Communication Industry there are a positive correlation between company size and compensation gap, and a correlation between enterprise performance with state-owned share and the compensation gap of CEO to CFO, with the performance negatively relating to the state-owned share and positively relating slightly to the executive compensation gap of CEO to CFO. Such a conclusion deduced from communication manufacturing and service industry in China seems more close to the Behavioral Theory for explaining compensation structure instead of Tournament Theory.

Acknowledgments. This paper is supported by the construct program of the key laboratory in Hangzhou, China. Data is from CSMAR database (Guo Tai An).

References

1. Li, J.: The Relevance Theory of the Uncertainty, Top Executive Pay Gap and Company Performance: Competition Validation. Accounting Review (42), 23–53 (2006)
2. Lambert, R.A., Larcker, D.F., Weigtlt, K.: The Structure of Organizational Incentives. Administrative Science Quarterly 38, 438–461 (1993)
3. Eriksson, T.: Executive Compensation and Tournament Theory: Empirical Tests on Danish Data. Journal of Labor Economics 17, 262–280 (1999)
4. Cowherd, Douglas, M., Levine, D.I.: Product Quality and Pay Equity between Lower-level Employees and Top Management: An Investigation of Distributive Justice Theory. Administrative Science Quarterly 37, 302–320 (1992)
5. Lin, J., Huang, Z., Sun, Y.T.: Pay Gap,Firm Performance and Corporate Governance. Economic Research Journal (4), 33–35 (2003)
6. Liu, C., Sun, L.: A Study on Relation of Salary Difference and Firm Performance: Evidence from State-owned Enterprises. Nankai Business Review 13(2), 30–39 (2010)
7. Wang, Y., Hao, X.: The Influence Factors of Executive Pay Gap between Internal Teams-the Empirical Research based on the Listed Company from Liaoning Province. Economic Forum (7), 157–159 (2010)
8. Hambrick, D.C., Siegel, P.A.: Pay Dispersion within Top-management Groups: Harmful Effects on Performance of High-technology Firms. Academy of Management Proceedings (1), 26–30 (1996)
9. Zhang, Z.: The Empirical Research of the Enterprise Internal Salary Gap's Impact to the Future Performance of Organization. According Reach (9), 81–87 (2008)
10. Lu, H.: A Study of Top Executive's Pay Gap and Firm Performance Based on Endogeneity. Soft Science (12), 22–29 (2009)
11. Wang, H., Han, S., Du, L.: An Empirical Research on the Top-management's Compensation and Enterprises Performance of Listed Companies in Hubei Province. Contemporary Economics (6X), 53–54 (2006)
12. Zhang, J., Zhang, H.: Performance, Market Value and Cooperate Governance of the Private Listed Compa nies in China. World Economy (11), 1–13 (2004)
13. Liu, S., Sun, P., Liu, N.: The Principle of Ultimate Ownership, Ownership Structure and Enterprise Performance. Economic Research (4), 51–62 (2003)

On the Principles of College Chinese Textbooks Compilation

Zhonghua Li

Department of Humanity and Social Sciences
Henan Institute of Engineering
Zhengzhou, China
lindalee2009@yeah.net

Abstract. College Chinese teaching has been recovered for more than thirty years. In the thirty years, many scholars have written many kinds of textbooks, some of high quality and some of poor quality. To compile a good College Chinese Textbook, the compiler should follow the following four basic principles: First, the principle of mother tongue education; second, the classic principle; the third is the transcendent principle; the fourth is the principle of openness.

Keywords: College Chinese, textbooks compiling, principles.

1 Introduction

In the late 1970s Kuang Yaming, Su Buqing initiatived to restore College Chinese teaching in chinese dolleges and universities. From then to now, thirty years have passed. In these thirty years, the course of College Chinese flourished. The majority of colleges and universities have set up the course, and this course has widespread concern in society, the Ministry of Education issued formal documents to encourage, promote colleges and universities to open the course. The quality and quantity of research papers on College Chinese teaching have an unprecedented increase. Although College Chinese teaching encountered some difficulties, but these difficulties are difficulties in the development of this course which will certainly be overcome.

The prosperity of College Chinese teaching can be seen from the compilation of the textbooks of this course. In the National Library's collection catalog of 2006, the College Chinese textbooks have as many as 1402 kinds. In those textbooks, many are of high quality, such as the two by Wang Po-ko, Chen Hong as chief editors have won the national title of quality materials. But as you can imagine, there are many shoddy textbooks whose compilation are for the pursuit of scientific research and the job classification. Some materials are published only for profit. Higher education teachers must have achievements in scientific research, teachers with no scientific research can only serve as little role. Although the university language teachers are often in vulnerable groups, their titles low, scientific research ability low. But also because of this, they often focused scientific research on writing textbooks. Coupled with the press enterprization, the autonomy of choosing textbooks, and other factors,

A. Xie & X. Huang (Eds.): Advances in Computer Science and Education, AISC 140, pp. 307–312.
springerlink.com

compiling teaching materials is becoming easier and easier. More and more college chinese teachers choose compiling textbooks as the breakthrough of their scientific research.

According to the author' many years of teaching practice, a good textbook should be written according to the following principles.

2 The Four Principles of College Chinese Textbooks Compilation

First, the principle of mother tongue education.

The so-called mother tongue, generally refers to a person's native language. As for the mother tongue of the Chinese nation, it generally refers to the Chinese. Mandarin Chinese has almost become the common language of the chinese nation. Chinese speaking ability can be obtained naturally in childhood by a person's parents and relatives, and this ability continues to increase in the process of growth. But the Chinese reading and writing skills is not naturally available. It must be obtained through school, and one must make efforts to gain this ability.

Today's society was summed up as "Image Era", and the social impact of television is everywhere. it also caused a considerable impact on the development of contemporary college students' mother language ability. Many college students' ability to speaking and writiing is far away from the expectations of their parents and the society. In 2009, some scholars made a survey in Tsinghua University, Renmin University of China, Beijing Foreign Studies University and the Central Academy of Drama. From the survey results, we can see that the mother tongue ability of college students is indeed worrying. Students' language foundation is weak, and there is a serious shortage in their writing skills. Although there is college chinese course in all the six universities, only one of them made college chinese as a required course for all the undergraduate students. In other five colleges and universities, " college chinese" was either only an elective public course for undergraduate students themselves to choose whether to attend, or a required course for only a few professional students.It is known that failure to include"college chinese " in the compulsory undergraduate courses is still quite common in colleges and universities in our country at present. Contrary to this, foreign languages are compulsory courses for undergraduate students in all our country's colleges and universities. Generally, the course capacity of "college chinese" in all colleges and universities across the country, whether it is compulsory or elective, is only 2 credits, about 36 hours of classroom teaching hours. The number of "college chinese" teachers are seriously inadequate. In the six colleges and universities, several colleges and universities now have only one or two teachers to engage in college chinese teaching. So few teachers to take the task of teaching college chinese for thousands and even tens of thousands of undergraduate student's, reflects the current marginalization of mother-tongue education in colleges and universities. Compare the resources investment of colleges and universities in foreign language education, the weak position of mother-tongue education becomes more prominent. Currently, foreign language courses in public colleges and universities across the country is not only compulsory, but the course capacity is generally between 8-14 credits, classroom hours is between 140-250 hours, the teachers engaged in foreign languages teaching in a public university are often up to dozens of people. Even taking into account the importance of foreign language education to the country's modernization, taking into account that foreign language education is relatively more difficult than mother tongue education and needs more time, it is not normal that the difference between the curriculums natures, the class allocation and personnel of foreign language education and mother tongue education is so huge.

Different from universities in mainland China, mother-tongue education in universities in the United States, in Hong Kong and Taiwan has played an important role and has taken an important position. The important position of mother tongue education is institutionally protected by including it in the compulsory courses system. This arrangement guarantees the necessary input for mother-tongue education.

Chinese is the world's oldest, richest in cultural heritage spoken system and writing system. In the history of thousand years, Chinese suffered multiple tests, tuning more and more strong and becoming one of the world's most important languages. The United Nations listed her among the five working languages. In spite of all this, we can not sleep without any anxiety. We must see soberly that, in today's world cultural competition, Chinese characters is confronted with hitherto unknown crisis. Chinese is a United Nations official working language, but the actual use is far less than the English. English is the most important reason for the crisis of Chinese language. At home the situation is also this. Junior middle school entrance examination needs to have an English test (although the state policy prohibiting Junior middle school entrance examination, but in fact the Junior middle school entrance examination is not only in existence but also competition is increasingly fiercer). In college entrance examination, English is one of the important subjects. Master's degree entrance examination, PhD entrance examination, English has become the key subjects for competition. To get professional title needs to pass English test, to go abroad needs to pass English test. English grade certificate is needed to find a job, to attend international academic conference needs to submit articles in English and speak in English, to do scientific research needs to read the English materials, to publish papers needs to contribute or write English abstract ... ... Don't you see, English training is like a raging fire, but young people's chinese ability is insufficient seriously. College should be an important stage to strengthen the students' chinese ability, whereas the university Chinese teaching is severely marginalized.

There are divergent views on the nature of College Chinese teaching. Some people say that College Chinese is college literature, some say that it is university culture. This argument will provoke unnecessary controversy. The nature of college Chinese teaching should be positioned as a mother tongue education. "Only positioned as the mother tongue higher education courses can we ensure college Chinese teaching the university foundation course and compulsory core course status." Learning their mother tongue well is not only a basic human right of Chinese college students, but also their basic obligations. People with outstanding achievements are often those to master their mother tongue very well. Su Buqing, Qian Xuesen, Hua Luogeng is Chinese examples, Einstein is an example of foreigners. When Mr. Wang Bugao taught College Chinese courses at Tsinghua University, he was deeply moved by Tsinghua students' high level of chinese language and their tireless learning spirit. He drew the conclusion that Tsinghua students' deep love for their mother tongue is one of the reasons for their overall development. This is indeed the case. The English level of those Chinese college students who had passed the band-4 of band-6 exam and who can talk fluently with foreigners generally does not exceed their Chinese level. Mother tongue is the foundation of thinking, the carrier of Chinese traditional culture, the window of most knowledge. Mother tongue gives a profound impact on memory and understanding ability. Now the college chinese textbooks usually selected classical-language articles, some textbooks attempted to include foreign literature but often these forengn literature issuperficial and can not achieve its proper role. From the perspective of mother tongue education, college chinese textbooks should be all Chinese works. The teaching of foreign literature would be in some elective courses such as the "appreciation of foreign literature" rather than as a compulsory part of college chinese course.

Second, the classic principles. The cultural heritage of the Chinese nation in the long history has strong vitality. The national spirit is the strong momentum for the continuation, development, and growth of the Chinese nation. In modern times, quite a few people in the face of Western culture lost national self-confidence and become national nihilist. In the May Fourth period, the saying that Chinese character will be abolished prevailed. It seemed inevitable that chinese characters will die out and be replaced by phonetic characters. The facts have proved that to abolish Chinese characters does not work. Today, national nihilism has lost market. More and more people recognized the value of the precious national heritage, more and more people thought that the great rejuvenation of Chinese nation must be accompanied by the great revival of national culture. This is the cultural background of the boom of Guoxue, this is also the cultural background of college chinese course to attract the attention of the whole society. The reason why college chinese attrated universal concern is that it carries national heritage and national spirit which other courses can not carry. Therefore, the compilation of college chinese textbooks must pay attention to classic principle. Famous writers' famous works are generally required for the selection. Almost the whole history of China's literature is the classical laguage era. Classics are the essence of traditional Chinese culture. Chinese traditional ideas such as the harmony of Heaven and people, reverence for life, learning from nature, focusing on ethics, self-cultivation and so on are embodied in the classics.

Third, the principle of transcendence. Traditional college chinese textbooks generally stressed that "Less is more." Former college chinese textbooks alse stressed "less is more". Since the release of college chinese with Mr. Wang Bugao as chief editor., a growing number of college chinese teachers recognized the "double super" concept in college chinese textbooks compilation. The so-called "double super" is the principle of much more contents in textbooks than teaching needed, of teaching students more difficult contents than common students can accept. Such materials can be different from high school chinese textbooks, and will not be dubbed as "fourth grade chinese in high school". It can play the role of a summary of primary and secondary chinese, of the beginning to study systematically the classical chinese literature. It can make the students to find the sense of going to college.

Since the May Fourth period, there has been a bad tendency of being increasingly simplistic in the field of language and literature. For example, some people advocate the abolition of Chinese characters and the implementation of a phonetic system, the abolition of traditional characters and the implementation of simplified characters. The good aspect of this is that it made the speaking system going with the writing system, narrowed the distance between books and the working masses. But there are some negative effects. The most obvious negative effect is that young people's feelings of Chinese traditional culture became more and more weak. Language is the result of human brain evolution, the law of which is from simple to complex, because the evolution law of human social life, human thoughts, human feelings, and the human brain is from simple to complex. Language is a tool of human communication. Of course it is subjected to the evolutionary law of human society to become more and more complicated. College students have been learning chinese for many years from primary school to high school. In the university stage, if college chinese is not difficult at all, it must be tasteless like drinking water, and cause the college students have a sense of loss. They will look down on the College Chinese, and they will not find the good feeling of going to college.

Outstanding literary works will not be bland, but be somehow difficult. Even literature experts can not boast to fully understand "chancery", "The Analects of Confucius", "A Dream of the Red Mansions", and other classics. Faced first time to the profound literary classics, the Students probably have a sense of hesitance, looking at the classics as a little mountain which is hard to climb. But after their lessons, they will get a little harvest. Reading for many times, each time there will be new harvest. Enjoying many times, the appreciation will get deeper. Only through their own personal taste the harvest to their own. Only through this way and through the students' repeated appreciation of the literary classics, can the students recognize the greatness of traditional Chinese culture , can they have deep feelings to the traditional Chinese culture.

The difficulty in the principle of descendency is not that the more difficult is the better, but should be moderate. Repeating the contents of high school or even junior high school is of course not good. It will not be welcomed by students and teachers and it will be against the law of education. However, if the material is too difficult, the teacher will feel unable to start when they see the material, the students can not understand after a semester of learning. This will dampen the enthusiasm of teachers and students, will also affect students' interest. Teachers and students will not have a sense of accomplishment. Therefore, textbook writers must compile materials in accordance with students' aptitude, considering most students' acceptance ability and language base.

In general, the language foundation of famousuniversities' undergraduate students is better than that of the common universities' undergraduate students. The language foundation of common universities' undergraduate students is better than that of the vocational college students. For these three kinds of students the compiler should write different language teaching materials. Some materials have different versions, each adapted to different students. This is very good. For example, the college chinese by Mr. Wang Bugao as chief editor has three versions, the whole version, the simplified version, and the Vocational version. The three versions are with different difficulty degrees, so adapted to the three different types of students. In addition, liberal arts students are different from the science and engineering students, students of military academies are different from the ordinary students. The preparation of college chinese materials should also take into account these different situations.

Fourth, the principle of openness. The establishment of a large chinese concept is needed. Pre-Qin writers' prose and historical essays although many articles of which are not in the strict sense of the literature, should be incorporated into the college chinese textbooks. In ancient China, there is no strict division of subjects. Philosophy, literature, history were often put together with no clear natural boundaries. Writers tended to be philosophers, politicians and historians at the same time.

The principle of openness should also require the college chinese textbooks to be inclusive and free of ideas. In fact, this is the very difference between college chinese textbooks and middle school chinese textbooks. Middle school students are not adults. They lack the ability to distinguish right and wrong. But college students have entered the adult stage, with considerable knowledge base and life experience. We should not require college students not to look at evil things, not to speak evil words. College chinese textbooks should not only select Confucian classics but also select other various classics. Now the college chinese textbooks have considerable improvement in this regard. Previously, very few of them selected the Taoist classics, but now, Lao Tzu, Chuang Tzu's works have been used in many versions. However, very few Legalist, Buddhist classics have been selected in college chinese textbooks. In fact, the Buddhist classics as "non-Bodhi tree, nor stand mirror, there is nothing, where can the dust alight" and "body is the bodhi tree, heart is a mirror set, and always wash or wipe, not let dust alight" can also be incorporated in college chinese textbooks. Chen Yinque praised Wang Guowei when Wang died "spirit of independence, freedom of thought." It should not do to develop this point with no concept of inclusive ideas and free thinking.

Openness principle also requires the college chinese textbooks must absorb modern multimedia technologyand audio-visual teaching technology to increase the amount of information. The national quality textbook " college chinese " by Mr. Wang Bugao as chief editor has electronic CD-ROM as well as supporting website. A lot of other materials have those supportings as well. These measures adapted modern college students' feature of loving to use the computer. Such materials also impelled college chinese teachers to learn multimedia technology, to make courseware, to increase the amount of information in the classroom. This is the only way to meet the needs of modern students.

References

[1] Lin, H.: "Chinese fever" touch the world China. Observation and Reflection (8) (2009)

[2] He, E.: Thirty years of college chinese (EB / OL), `http://www.eyjx.com/eyjx/1/ReadNews.asp?NewsID=5131`

[3] He, E.: Mother tongue obligations, no doubt. College Chinese education and Research. Nanjing University Press (2009)

[4] Wang, B.: On the meaning and role of mother tongue for the students' growth. College Chinese Education and Research. Nanjing University Press (2009)

[5] Wang, B.: On the double super-concept in the compilation of college chinese textbooks. College Chinese Education and Research. Nanjing University Press (2009)

[6] Huang, Y., Ju-jie: On the integration of Information technology and college chinese. Scientific Information (6) (2009)

[7] Sang, D.: College Chinese Teaching under the new situation: Thinking and Exploration. Journal of Jiangxi Institute of Education (4) (2010)

The Secondary Hash MM Algorithm Based on Probability Factor

Yan Niu and Lele Huan

SouthLake, Wuhan. 430068, Hubei, P.R. China
Computer School, Hubei University of Technology
ny@mail.hbut.edu.cn, 490525226@qq.com

Abstract. Chinese Words Segmentation is a very important section of Chinese information process. The speed and the accuracy of Words Segmentation will affect the results of information processing directly. Dictionary of secondary hash structure mechanism based on the probability factor is proposed through the analysis of the traditional Chinese Words Segmentation's algorithm. This segmentation mechanism combined with the original maximum matching algorithms those main points is to consider cutting accuracy, and also increased segmentation's accuracy while the new Segmentation mechanism improve the searching speed in the dictionary. Theoretical analysis demonstrates that the algorithm's speed and accuracy is better than the traditional method.

Keywords: Probability factor, Probability factor difference, Secondary hash.

1 Introduction

Rapid growth in the information and showing skyrocketing trend gradually of today, Chinese information on the Internet increased rapidly. How to make computers understand human language better and faster access to Chinese information is already a very important research topic. Chinese word segmentation is a critical factor in many information processing fields.

Segmentation is a string of consecutive sequence of words divided into word sequences according to certain rules. Due to the different Chinese and English, Chinese have no spaces between words. Only the paragraphs, sentences have a simple separator in Chinese article. From the end of the century have been the new segmentation algorithm is proposed constantly, but no matter which method, all need to use a lot of time to calculate the possibility of splitting the word statement. And then according to statistics or grammar rules processing out of the word segmentation, get the best results.

2 Introduction of Word Segmentation

Now commonly used algorithm is the maximum matching algorithm:

Assuming that the longest word in the words table formed by i characters. Each time intercepts i characters from the sentence opening, make it matches with the word

A. Xie & X. Huang (Eds.): Advances in Computer Science and Education, AISC 140, pp. 313–318.
springerlink.com

in words table. If in the words table has these *i* characters composted word, it matches successfully. Takes these *i* character as a segmentation of words from the head of the sentence out. Then intercepts *i* characters again from the sentence in opening of the remaining part.

Repeat the process until the sentence is cut exhausted. If you can not find a word in the words table to match with the current *i* characters, delete one character from the end of the *i* characters. Use the remaining *i-1* characters to match the words table to look up word. If it matches successfully, takes the *i-1* characters as a segmentation of words from the head of the sentence out again. If this match defeat, deleting one character from the end of the *i-1* characters. The remaining *i-2* characters match the word in the words table until successful.

The flow diagram of Forward Maximum Matching algorithm is shown in Fig. 1.

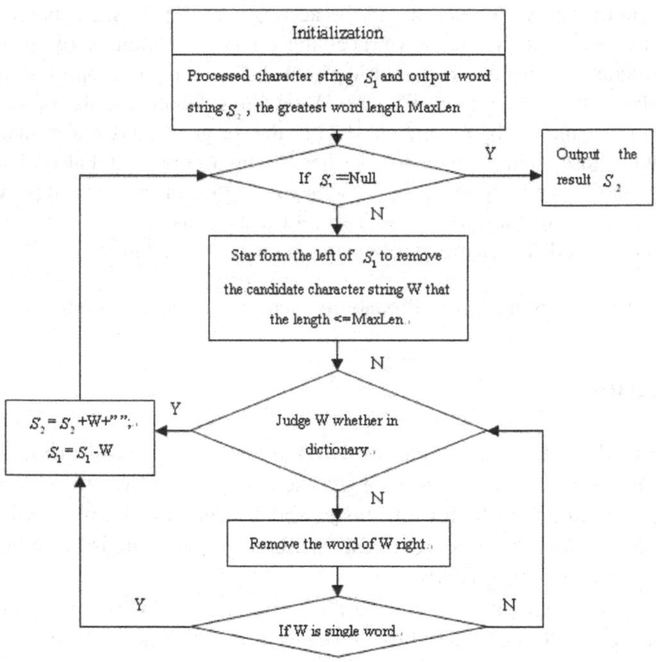

Fig. 1. The flow diagram of Forward Maximum Matching algorithm

3 The Second Hash Based on Statistical Segmentation Algorithm

General hash word segmentation only improves the efficiency of search, but nothing points to improve the accuracy of the word. This article inside the hash table to add a probability factor p and a probability of difference factor p_{max} mark the used probability of each word. According to the probability of word for segmentation, segmentation may obvious enhance the precision of the segmentation. Present's segmentation is based on the long word first, but sometimes goes wrong. We do not want to get the results of the participle. After we add the probability factor, take into account the short word, the basic process is as follows:

Supposition the string is $S = C_1C_2C_3 \cdots C_n$. It will carry on the participle to *S*. Assuming C_1C_2 is a word, the used probability is P_2.

The $C_1C_2C_3$ also is a word. Probability is P_3, if $P_3 - P_2 >= 0$, then select the word $C_1C_2C_3$, if $P_3 - P_2 < 0$, have two cases: If $P_2 - P_3 < P_{\max}$, select the word $C_1C_2C_3$, if $P_2 - P_3 >= P_{\max}$, select the word C_1C_2. This algorithm reflects the thinking of long-term priority and flexible, set a limit probability difference $P_{\max}$ when the short word's probability P_s is bigger than the long word's probability P_l, select the short word. This algorithm improves the segmentation accuracy.

Traditional several methods use the first character hash index algorithm, according to Chinese national standard area code, through hash operation direct localization Chinese character's position in first character Hasche table. Generally the used Hasche function is:

$$offset=(ch1\text{-}0xB0)*94+(ch2\text{-}0xA1) \quad (1)$$

offset is the Chinese word in the hash table in the first place, *ch1* and *ch2* are the machine codes of the character.

4 Design of Sub-dictionary

The traditional words table commonly use the plain text way to build words table, words table have no effective plan, causes the search waste massive time. This words table consists of 5 parts:

(1) The first character hash table, in accordance with international area code each Chinese character in the first character Hasche table has only mapping, which can be calculated by the formula (1).

Structure

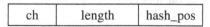

ch	length	hash_pos

ch is the Chinese character, *length* is the second character tables length, *hash_pos* is a pointer that point to the character word hash table.

(2) The second character hash table, In the words table, the mount of the first character's word inferior character decides the second character hash table the size, the mount of the first character's word inferior character and the hypothesis hash table packing factor decide the length of the second character hash table, we use *0.5* as the hash table packing factor, and use the commonly used chain address to solve the address conflict. Each table has a pointer to the first address of the list composted by second characters. In the structure of the second character hash table, *ch* is the second character, *first* is a pointer that point to the first position of the remaining characters in the words table, *last* is a pointer that point to the final position, *flag* express if only

have the word consist the two characters, *if_has* express if have the word consist the two characters, *next* is the address of the chain table constituted by second characters which produces the conflict, *p* is the used probability of the word constituted by the two characters.

ch	first	Last	flag	if_has	next	p

(3) Words index table: Because the length of words is variable, so we should select the appropriate variable-length storage, but also realize to the word random access. Word index table is structured as follows:

ch	word_pos	Length	p

word_pos is a pointer that point to the remaining characters of the word, *length* is the length of the remaining characters of the word, *p* is the used probability of word .

(4) Words table text: the ordered list of word's remaining characters.

(5) Probability factor array, Recorded probability of each word during each participle process.

The table determined by the number of Chinese character encoding standard. GB2312 Chinese character coding table containing one or two characters and a total of 7164 of certain symbols, so the first character hash table's length is 7164.

Second character hash function, hash function should as far as possible reduced conflict, more than this number of selection methods remains to be determined. The formula as follow:

$$offset =((ch1\text{-}0xB0)*94+(ch2\text{-}0xA1)) \bmod len \quad (2)$$

offset is the Chinese word in the hash table in the first place, *ch1* and *ch2* are the machine codes of the character, *len* is the length of the second character hash table.

4 Words Table's Search Algorithm

String $S = C_1C_2 \cdots C_{99}(n \in N)$ is assumed to be divided; L is the initial biggest length of word, which base on this article's biggest match participle algorithm, the steps of the search algorithm as follow:

Premise: Set integer *i=1*

(1) Takes the most left *L* characters, $C_iC_{i+1}C_{i+2}\cdots C_{L+i}$ *(n-L>0)*, then turn to (2).

(2) Query words table to the word led by C_i the maximum word length M_i, if $M_i < L$, the string of $C_iC_{i+1}C_{i+2}\cdots C_{L+i}$ is not a word, turn to (3); otherwise turn to (4).

(3) *i=i+1*, Turn to (1).

(4) According to C_i the internal code to calculate the offset of C_i in the code table, determine the position of C_i in the first character hash table. First determine

the *length* integer the first word hash table, if *length=0*, no words beginning with C_i , turn to (3), if *length>0*, according to the pointer in the first character hash table to find the position of the second character hash table, in the second character hash table look up the C_{i+1} , if can't find the character, turn to (3). Otherwise, judge *if_has*, *if_has=true*, put The probability of C_iC_{i+1} into the probability factor array *P[2]* which is in *P[]* (*P[0]=*P_{max}, *P[1]=0, P[1]* is a constant). Find string matching in the range of *first* and *last* in the second word hash table. If do not have matching string, $C_iC_{i+1}C_{i+2}\cdots C_{L+i}$ is not a word, we select C_iC_{i+1} as a word, and make *i=i+2*, then clear probability factor array *P[]*,,turn to (1), otherwise turn to (5). If *if_has=false*, turn to (3).

(5) According to the string's pointer to the words table to determine the string position in the words table, and start from this location to find whether there is a word match with string $C_iC_{i+1}C_{i+2}\cdots C_{L+i}$, and then record C_iC_{i+1} , $C_iC_{i+1}C_{i+2}$, $C_iC_{i+1}C_{i+2}C_{i+3}$ ... $C_iC_{i+1}C_{i+2}C_{i+3}\cdots C_{i+k}$ *(k<=L)* probability of $P_{C_iC_{i+1}}\cdots P_{C_iC_{i+1}C_{i+2}\cdots C_{i+k}}$. Put $P_{C_iC_{i+1}}\cdots P_{C_iC_{i+1}C_{i+2}\cdots C_{i+k}}$ in the position which in the probability factor array *P[2]...P[k]* corresponds. According to circulation computation, compare *P[j]* and *P[j+1]*, if *P[j]-p[j+1]>*P_{max}, select $C_iC_{i+1}C_{i+2}\cdots C_{i+j}$, and take $C_iC_{i+1}C_{i+2}\cdots C_{i+j}$ out ,make *i=i+j*, turn to (1), otherwise select $C_iC_{i+1}C_{i+2}\cdots C_{i+j+1}$, and take $C_iC_{i+1}C_{i+2}\cdots C_{i+j+1}$ out, make *i=i+j+1*, turn to (1).

5 Result Analysis

This article used the biggest matching algorithm on sentence segmentation, the search is based on the probability factor two Hasche search and the binary search to the surplus characters, this enhanced the participle precision and the seek rate. In this article probability factor and the probability factor difference has avoided the mistake which the pure long word first produces. This article only adopted to string first two word to hash search, making the whole word by hash search and searching efficiency will be further improved with the development of computer hardware and the expending of memory.

References

1. Wen, T.: Chinese automatic participle research development. Journal of Books and Information 5, 54–63 (2005)
2. Wen, T., Zhu, Q.: One kind of fast Chinese participle algorithm. Journal of Computer Project 3, 70–72 (2009)
3. Hu, X.: Biggest match law in Chinese participle technology application. Journal of Anshan Normal School 02, 30–36 (2008)

4. Wu, J., Jing, J., Nie, X.: One kind of fast Chinese participle dictionary mechanism. Journal of Chinese academy of Science Graduate School 05 (2009)
5. Ding, Z., Zhang, Z., Li, J.: Based on Hash structure reversion biggest match participle algorithm improvement. Journal of Computer Project and Design 12 (2008)
6. Zong, Q., You, J.: Based on the dual Chinese participle's highly effective searching algorithm. Books Intelligence 22 (2009)
7. Yan, W., Chen, W.: Construction of data and application algorithm course, background. Tsinghua University publishing house, China (2001)
8. Sun, B.: Modern Chinese text words and expressions segmentation technology, http://www.tinko.com/Lunwen/86087.htm

An Optimal Model of Selecting Players for Mathematics Competition

Ying Wei

Computer Department, Wuhan Polytechnic
Wuhan 430074, Hubei, P.R. China
915745208@qq.com

Abstract. In this paper, the estimating index system and optimization model for choosing players in Mathematics competition are proposed based on fuzzy mathematical theory and optimization theory. The example of choosing players for Mathematics competition show that the model proposed in this paper is feasible and reasonable.

Keywords: Mathematics competition, Selecting players, Optimization Theory, Fuzzy multi-attribute decision-making.

1 Introduction

The annual national college students' mathematics competition is important games in colleges and universities. On the other hand, not all people who want to take part in the competition can be employed due to the competition venues and other reasons. The coach cost a lot of energy in order to select excellent students represent school attends the national mathematics competition, such as the select problems of outstanding players. Design a index system and establishing optimization model for selecting the team member are very important.

In this paper, the team member selection problem for mathematics competition is studied, and the estimating index system and optimization model of choosing players for Mathematics competition are proposed based on fuzzy mathematical theory and optimization theory. The example of choosing players for Mathematics competition show that the model proposed in this paper is feasible and reasonable. The research offers valuable reference for developing extracurricular science and technology activities in colleges and universities.

2 Quality and Ability Analysis for Competition Players

College students' mathematical contest is a kind of open comprehensive challenging extracurricular technological activity. Therefore, the players also need have certain requirements. Searching out the key factors for players selection and then formulating a reasonable investigation plan are very important for improving math competition team member selection mechanism.

A. Xie & X. Huang (Eds.): Advances in Computer Science and Education, AISC 140, pp. 319–325.
springerlink.com

According to the principle of key performance indicators (Key Performance Indicator, KPI) and the practice of mathematical contest, the mathematics competition team member should have the following basic ability: (1) Good mathematics foundation, certain mathematical foundation of basic skills, solid background of mathematics knowledge. (2) Accurate understanding the problem, independent analyze the question, good mathematical thinking. (3) Problem solving skills, avoid arrogance for complex operation, good operation habit. (4) Positive thinking, a wealth of space for imagination.

The possible reasons of the above factors are as follows: (1) Mathematic is the main contents in Mathematics competition, almost all subjects of the competition have relationship with mathematical, therefore, having a solid foundation of mathematics is the necessary ability for the team member. (2) The test questions of Mathematics competition usually have the characteristics of comprehensive and difficulty, so good understanding analysis ability is very important. (3) Generally speaking, the solution processes of Mathematics competition test question is very complex and tedious, so good habits of the operation ability in mathematics competition is indispensable. (4) The content of Mathematics competition is diverse, so the wealth of space for imagination is a very key factor.

3 The Optimal Index System of Selecting Players

The index system is the whole formed by a series of connecting statistical index. It's a main factor for model evaluation. Therefore, index selection affects the model's scientific and rational. The optimal index system for mathematics competition member is composed of a set of independent and associated indicators which can be design by balanced scorecard. Through the above analysis, the key basic class index for level of the attended team are followed by these (includes qualitative indexes and quantitative ones): math background; competence indicators; understand analytical capacity; operation ability and space imagination; developing index; knowledge and so on. The optimization index system can be constructed by establishing the corresponding weights of each index, as shown in figure 1.

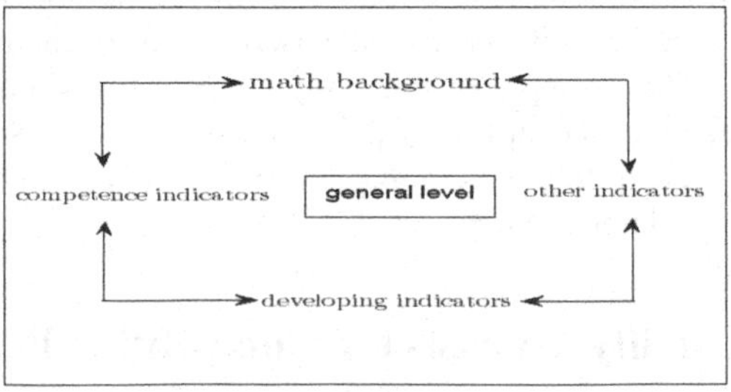

Fig. 1. The index system for mathematics competition member

4 The Model

The key ability of the attended team are followed by these: math background; understand analytical capacity; operation ability and space imagination. The specific index set to $C=\{C_1,C_2,\cdots,C_6\}$, where C_1 is test results; C_2 is understand analytical capacity; C_3 is knowledge, C_4 is operation ability, C_5 is space imagination, C_6 is other conditions. The weights of index set to $w=\{w_1,w_2,\cdots,w_6\}$. The set of players is $A=\{A_1,A_2,\cdots,A_m\}$. The initial data matrix recorded as $F=(f_{ij})_{m\times n}$ for studying the m players based on above six index. If the basic information of the matrix F is known, we can evaluate and sort the m players and choose the best players.

The basic idea of compromise type fuzzy decision[5] is: firstly, invented fuzzy positive ideal and fuzzy negative ideal based on the original sample data, where, the fuzzy positive ideal is maximum value of fuzzy index while fuzzy negative ideal is the minimum ones. Then, we can calculate the distance between optional object and fuzzy positive ideal and fuzzy negative ideal based on measurement tools of weighted Euclidean distance.

The membership of each fuzzy positive ideal is calculated. Generally speaking, the greater the membership means that the scheme more ideal. Its decision procedure is as follows:

Step 1: The original index data processing

First, all the qualitative、 quantitative index and weight are converted into triangular fuzzy numbers, the specific methods are as follows:

1) For qualitative indexes, the Bipolar scaling method can be improved to triangular fuzzy number proportion method, then, the qualitative index can be transformed into quantitative index, the transformation forms shown in Table 1.

Table 1. Method for qualitative index transformed into quantitative index

	(0,0,1)	(1,1,2)	(2,3,4)	(4,5,6)	(6,7,8)	(7,8,9)	(9,10,10)
cost type index	highest	very high	high	average	low	very low	lowest
benefit type index	lowest	very low	low	average	high	very high	highest

2) For quantitative indexes, the triangular fuzzy number also can be used. If a is an accurate number, the form of triangular fuzzy Numbers for a can be write as:

$$a=(a,a,a) \tag{1}$$

When all the attribute index convert into triangle fuzzy numbers, we can get the fuzzy index matrix $\tilde{F}=(\tilde{f}_{ij})_{m\times n}$.

3) The triangular fuzzy number for weight vectors

① If the form of weight vectors is quantitative, the formula can be expressed as:

$$\bar{w} = [(w_1, w_1, w_1), (w_2, w_2, w_2), (w_3, w_3, w_3), (w_4, w_4, w_4), (w_5, w_5, w_5), (w_6, w_6, w_6)] \quad (2)$$

② If the form of weight vectors is qualitative, the weight vectors can be converted into triangular fuzzy numbers based on table 1.

Second, fuzzy index matrix $\tilde{F}$ is normalized. For evaluation index j $(j = 1, 2, \cdots, 6)$, m players correspond to m fuzzy index recorded as $\tilde{x}_i = (a_i, b_i, c_i)$, $(i = 1, 2, \cdots, m)$. The normalization formula of $\tilde{x}_i$ is as follow:

$$\tilde{y}_i = \left(\frac{a_i}{\max(c_i)}, \frac{b_i}{\max(b_i)}, \frac{c_i}{\max(a_i)} \wedge 1 \right) \quad (3)$$

$\tilde{R} = (\tilde{y}_{ij})_{m \times n}$ is the normalized the fuzzy index matrix.

Then, the matrix $\tilde{R}$ were weighted to get the fuzzy decision matrix $\tilde{D} = (\tilde{r}_{ij})_{m \times n}$, where $\tilde{r}_{ij} = \tilde{w} \Theta \tilde{y}_{ij}$ $(i = 1, 2, \cdots, m \quad j = 1, 2, \cdots, 6)$.

Here, the *Bonissone* approximate product formula is used, that means that if $\tilde{w} = (a, b, c)$, $\tilde{y}_{ij} = (d, e, f)$, we can get

$$\tilde{r}_{ij} = \tilde{w} \Theta \tilde{y}_{ij} = (ad, af + ab - bf, ae + cd - ce). \quad (4)$$

Step 2: Determining the fuzzy positive ideal M^+ and fuzzy minus ideal M^-

Set

$$\tilde{M}^+ = (\tilde{M}_1^+, \tilde{M}_2^+, \cdots, \tilde{M}_6^+), \quad \tilde{M}^- = (\tilde{M}_1^-, \tilde{M}_2^-, \cdots, \tilde{M}_6^-) \quad (5)$$

where $\tilde{M}_j^{\ +} = \max\{\tilde{r}_{1j}, \tilde{r}_{2j}, \cdots, \tilde{r}_{nj}\}$ ($j = 1, 2, \cdots, 6$)is the maximum fuzzy set of fuzzy index with subscript j in $\tilde{D}$; on the other hand, $\tilde{M}_j^{\ -} = \min\{\tilde{r}_{1j}, \tilde{r}_{2j}, \cdots, \tilde{r}_{mj}\}$ ($j = 1, 2, \cdots, 6$) is the minimum fuzzy set of fuzzy index with subscript j in $\tilde{D}$.

Step 3: Calculate distance

The distance d_i^+ between player i and fuzzy positive ideal M^+ can be calculated as

$$d_i^+ = \sqrt{\sum_{j=1}^{6} [d(\tilde{r}_{ijL}, \tilde{M}_{jL}^{\ +}) + d(\tilde{r}_{ijR}, \tilde{M}_{jR}^{\ +})]^2}, \quad i = 1, 2 \cdots, m \quad (6)$$

The distance d_i^- between player i and fuzzy minus ideal M^- can be calculated as

$$d_i^- = \sqrt{\sum_{j=1}^{6} [d(\tilde{r}_{ijL}, \tilde{M}_{jL}^{\ -}) + d(\tilde{r}_{ijR}, \tilde{M}_{jR}^{\ -})]^2}, \quad i = 1, 2 \cdots, m \quad (7)$$

where $d(\tilde{r}_{ijL}, \tilde{M}_{jL}^+)$ means the Hamming distance between left fuzzy sets $\tilde{r}_{ijL}$ ($j = 1, 2, \cdots, 6$)and left maximum fuzzy set $\tilde{M}_{jL}^+$; $d(\tilde{r}_{ijR}, \tilde{M}_{jR}^+)$ means the Hamming distance between right fuzzy sets $\tilde{r}_{ijR}$ ($j = 1, 2, \cdots, 6$)and right maximum fuzzy set $\tilde{M}_{jR}^+$ (The meanings of $d(\tilde{r}_{ijL}, \tilde{M}_{jL}^-)$ and $d(\tilde{r}_{ijR}, \tilde{M}_{jR}^-)$ are similar).

Step 4: Fuzzy optimization decision

If player i belong to fuzzy positive ideal with membership μ_i, we can get

$$\mu_i = \frac{(d_i^-)^2}{(d_i^+)^2 + (d_i^-)^2} \qquad i = 1, 2, \cdots, m \tag{8}$$

Sort μ_i from large to small. The greater the value of μ_i valuation said the objects more optimal. So, we can determine the optimal results based on the order of membership.

5 Example

The example is that choosing 8 players from 16 candidate players. The index set is $C = \{C_1, C_2, \cdots, C_6\}$, where C_1 is test results; C_2 is understand analytical capacity; C_3 is knowledge, C_4 is operation ability, C_5 is space imagination, C_6 is other conditions. The weights of 6 index is set as $w = \{w_1, w_2, \cdots, w_6) = (0.5, 0.1, 0.1, 0.1, 0.1, 0.1)$. The initial indicator information of the 16 players is shown in Table 2 and we will chose 8 players from the 16 candidate players.

Table 2. Initial indicator information of the 16 player

student ID	test results	lectures number	Grade			
			understand analytical capacity	knowledge	operation ability	space imagination
S1	96	5	A	A	B	B
S2	92	4	A	B	B	C
S3	93	6	B	A	A	C
S4	81	3	A	B	A	C
S5	83	3	A	B	B	C
S6	83	5	D	B	A	B
S7	82	4	B	A	C	B
S8	84	4	B	A	B	B
S9	80	5	B	B	A	B
S10	79	4	B	D	A	C
S11	78	4	C	D	B	A
S12	77	5	B	A	C	A
S13	76	4	C	B	D	A
S14	74	2	B	D	A	B
S15	78	4	B	A	C	B
S16	76	5	A	B	B	C

1) We can process the initial indicator information in Table 2 based on the idea in Table 1. The full mark in test is 100 points and 1 point for attending a lecture. The quantitative form of A、B、C、D is show in Table 3.

Table 3. Quantitative form

Grade	A	B	C	D
Fuzzy number	(85,90,100)	(75,80,85)	(60,70,75)	(50,55,60)

2) The Indexes and their weight can be converted into triangular fuzzy numbers based on Table 3 and formula (1).

3) The fuzzy index matrix $\tilde{F}$ is normalized into $\tilde{D}$ based on formula (3) and (4). Then determining the fuzzy positive ideal M^+ and fuzzy minus ideal M^- based on formula (5).

$$M^+ = [(0.5,0.25,0.5),(0.1,0.01,0.1),(0.085,0.01,0.085),(0.085,0.01,0.085),(0.085,0.01,0.085),(0.085,0.01,0.085)]$$

$$M^- = [(0.3854,0.25,0.3854),(0.0333,0.01,0.0333),(0.05,0.01,0.05),(0.05,0.01,0.05),(0.05,0.01,0.05),(0.05,0.01,0.05)]$$

5) Calculating the distance d_i^+ between player i and fuzzy positive ideal M^+、d_i^- between player i and fuzzy minus ideal M^- and the membership.

The comprehensive level of the 16 player can be sorted as:

$$S_1 > S_3 > S_2 > S_8 > S_6 > S_9 > S_{11} > S_5 > S_{12} > S_4 > S_{16} > S_{10} > S_{15} > S_7 > S_{13} > S_{14}$$

Thus, we can choose the student with ID S_1, S_3, S_2, S_8, S_6, S_9, S_{11}, S_5.

6 Conclusions

In this paper, the optimization model of choosing players for Mathematics competition are proposed based on the methods and ideas in management, operations and mathematical. Combined the fuzzy mathematical theory and optimization theory, the model is analyzed and optimized. The example of choosing players for Mathematics competition show that the model proposed in this paper is feasible and reasonable. The main advantage of the model:

1) The structure and arrangement of the model are clear, it is a good way to solve the actual problem.

2) The qualitative index is converted into quantitative one to ensure the rationality of the model.

3) The method and ideal of this model can be used in other similar problem only by changing some parameters in the model.

References

1. Han, Z.G.: Practical Operations Research. Qinghua University Press, Beijing (2007)
2. Rao, C.J., Zhao, Y.: Multi-attribute Decision Making Model Based on Optimal Membership and Relative Entropy. Journal of Systems Engineering and Electronics 20, 537–542 (2009)
3. Wang, G., Wang, M.S.: Modern Mathematical Modeling Method. Science press, Beijing (2008)
4. Dong, W.Y., Liu, J., Ding, J.L., Zhu, F.X.: The Optimum Technology and Mathematical Modeling. Qinghua University Press, Beijing (2010)
5. Rao, C.J., Peng, J.: Fuzzy Group Decision Making Model Based on Credibility Theory and Gray Relative Degree. International Journal of Information Technology & Decision Making 8(3), 515–527 (2009)
6. Xiao, P., Li, Y.B.: The application fuzzy set theory in human resources performance evaluation. Business Studies 40(10), 25–27 (2010)

Gestalt Psychology Theory Based Design of User's Experience on Online Trade Websites

Peng Bian and Nairen Zhang

School of Mechanical and Vehicular Engineering
Beijing Institute of Technology
Beijing, P.R. China
camus10047@163.com

Abstract. As the popularity of Internet technology in public society, people shift to carry more and more business activities on Internet. E-commerce has now become one of the hottest Internet applications. But because of bad interaction design, some online trade websites show unfriendly results to user's experience. This paper combined with the theory of gestalt psychology which belong to cognitive psychology, analysis the interact behavior and experience feeling of online trade consumers toward websites. Summed up consumers' psychological model, have a full discussion of integrated design elements of online trade websites. Then summarize some design methods and strategies in line with the laws of psychology. Trying to put forward some original points of view toward the design of online trade websites.

Keywords: gestalt psychology, user's experience, interaction design.

1 Introduction

Nowadays with the development of Internet technology and the popularity of software and hardware equipment, the online living applications field also has a wide range of development. We can purchase clothing, shoes and other daily necessities from network; we can purchase books, CDs and other cultural items from network; we can purchase mobile phones, cameras and other electronic products from network; we can also booking tickets and hotels, purchase of intangible services from network, etc. Online trading is more convenient and efficient; it has an incomparable superiority towards traditional trade manners.

However, despite of so many advantages, many Internet users have never been tried online trading because of lots of reasons. Such as they don't understand the process, worry about their personal information will be leaked, consider the process of register and login is too complicated, etc. For ordinary Internet users, online trading is a field full of unknown, it requires users have a lot of special knowledge.

But these barriers could be solute actually by good interaction design. Because of bad interaction design, some online trade websites show unfriendly results to user's experience. This paper combined with the theory of gestalt psychology which belong to cognitive psychology, analysis the interact behavior and experience feeling of online trade consumers toward websites. Summed up consumers' psychological

A. Xie & X. Huang (Eds.): Advances in Computer Science and Education, AISC 140, pp. 327–332.
springerlink.com

model, have a full discussion of integrated design elements of online trade websites from two different aspects: logical structure design of websites and visual elements of web pages. Then summarize some design methods and strategies in line with the lows of psychology. Trying to put forward some original points of view and suggestions toward online trade websites design which is differs from computer technical field but could be helpful supplement.

2 The Development of User's Experience on E-commerce Website

E-commerce development in China has been very rapid in recent years. A group data from the Ministry of Commerce of the People's Republic of China shows: China's e-commerce trading amount to 491 billion dollar in 2008, concluding online shopping reach to 19 billion dollar; China's e-commerce trading amount to 594 billion dollar in 2009, concluding online shopping reach to 40 billion dollar; and China's e-commerce trading amount to 704 billion dollar in 2010, concluding online shopping reach to 80 billion dollar.[1] The growth of trading amount has a very quickly increase. Now in China there are the world's largest B2B e-commerce trade platform Alibaba.com and the Asia's largest C2C e-commerce website Taobao.com; the world's largest B2C e-commerce website Amazon.com and the world's largest C2C e-commerce website EBay.com have also entered Chinese market for years. Therefore in China the competition between e-commerce websites is very fierce.

Because the user's experience is very critical for the development of online trade websites, many scholars had put their energy into the research work of user's experience design on online trade websites. We can say that this is a brand new topic in the Internet era of which the practice develops faster than academic research. The result of Moesliger's research demonstrates that user's experience is component of practical experience, sensory experience and emotional experience.[2] The studies of Robert Rubinoff about user's experience on e-commerce websites should be measured from brand, content, function and usability.[3] Florien describes the credit framework of e-commerce websites from three aspects: graphic design structure, navigation design and content design.[4] Dorothy Lefore of A&T state university in North Carolina proposes some recommendation about application of gestalt theory to guide the visual design of web pages. It achieves some useful results. Garrett, who is a structure and usability engineer, believes it should design websites from strategy, scope, structure, skeleton, surface (5S) aspects according his lots of year's experience.[5]

3 Gestalt Psychology Theory

Modern cognitive psychology is sciences which focus on the systematic research of human cognitive and behavioral rules. Gestalt psychology believes that we should research experience of the phenomenon.[6] It is one of the main streams of western modern psychology. Experience in the observation of phenomena to maintain the original appearance of the phenomenon, but can't to analysis as sensory elements. In other word, the experience of phenomenon is a whole gestalt, so it was named gestalt psychology. Gestalt psychology brings the subjective consciousness and experience of the artist and audience into the research of graphic design.[7] The concept of "gestalt" is comes from but not limited to the research of visual field, even not limited to feeling area. Its application range is far beyond the limits of sensory experience. Its research target may include the process of learning, memories, aspirations, emotions, thinking and movement.

View from the gestalt psychologists, everybody including child and uncivilized people are organized the understanding of things according to the following principles.[8]

Graphics and Background----In a certain configured field, some object come to the fore to format graphics, some object relegate to the back to format background. General speaking, the more differentiated between graphics and background, the graphics could be more outstanding to be our perception objects.

Proximity----Some parts of short distance or close to each other are easy to compose a whole thing.

Tendency of Complete and Closed----The perceptual impression is easy to present the most complete form with environment. The things belong to each other are easy to form part of a whole thing. And the things not belong to each other are likely to be isolated.

Similarity----The parts of similarity are easy to compose a whole thing.

Conversion----Gestalt can experience a wide range of change without losing its own characteristics.

Simplicity----People often base on experience from the past to simplify what they perceived unconsciously.

4 User Behavior Analysis and Psychological Model

Compared with the traditional mode of shopping at mall or supermarket, people shopping on network are also basically follows the psychological model of sequence, "select goods---fill in delivery address---cash on delivery". The user's online shopping experience can be divided into five stages: view and browse, decide to purchase, implementation of the purchase, wait for receiving goods and finally, after receiving.

User needs to take a series of interact activities with the website at every stage above mentioned. At the stage of view and browse, users are attracted to log on the online shopping websites and wander casually to browse the goods. And then users enter the two key stages of shopping experience: decide to purchase and implementation of purchase. At the stage of decide to purchase, users need to take some activities like searching goods, comparing goods, etc. At the stage of implementation of purchase, users need to take some activities like finding purchase entrance, registrations, account confirmation, and fill in mailing information, log in online banking or paying in other mode. After users confirm the trading is successful, users will monitor the stage of commodities trading and delivering while waiting for the arrival of goods. After receiving users may be satisfied to show the goods, or may be not satisfied to return or change the goods. Or users most possibly do not make any feedback on the websites. The user behavioral process model is concluded in the following figure.

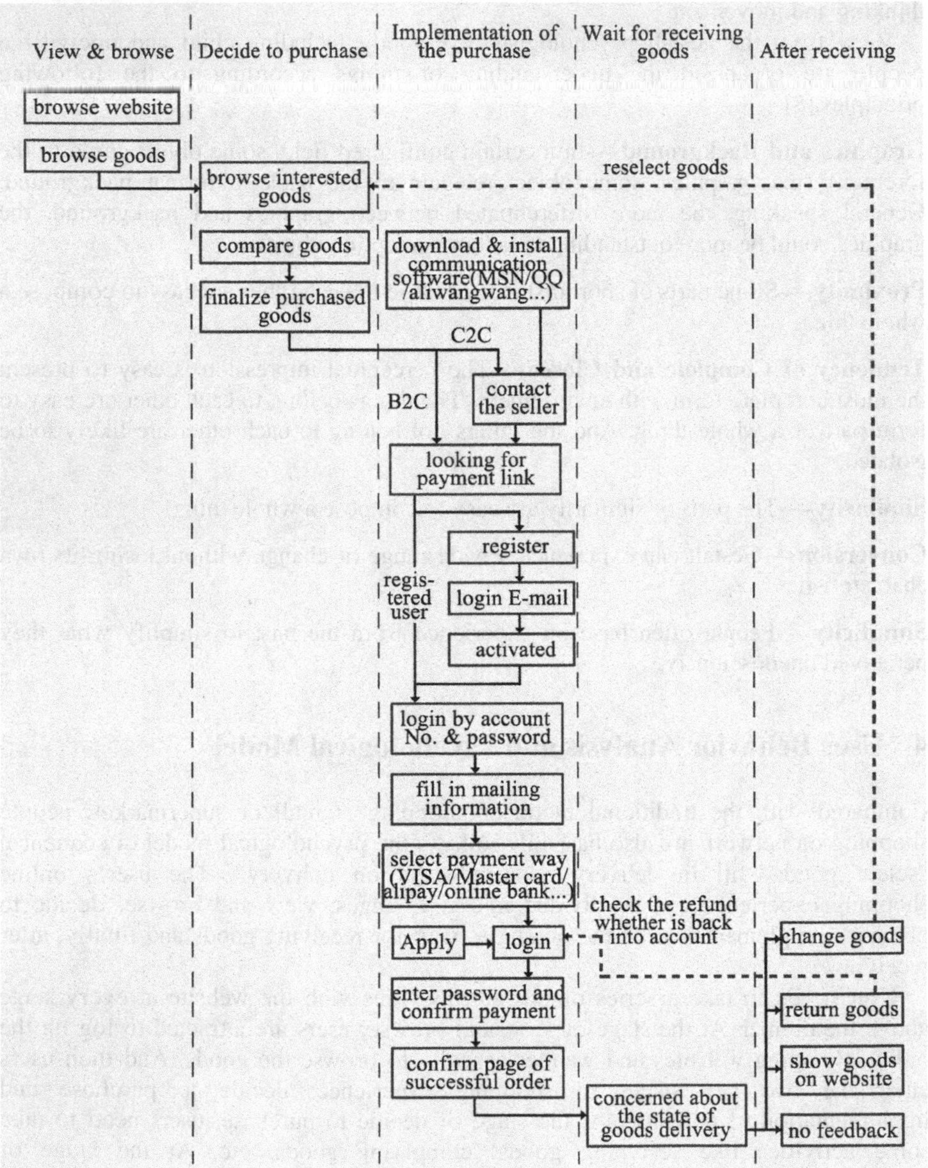

Fig. 1. The user behavioral process model

5 Application of Gestalt Psychology in Web Interaction Design

Based on the above mentioned model of user behavior, we analysis every stage of the shopping experience by applying the basic principles and laws of gestalt psychological theories. This in turn can be analyzed from two perspectives: websites structure design and visual elements design of web pages.

5.1 Websites Structure Design

1. Logical structure of the websites should be able to be understood by users. The designed online shopping process should consistent with user's psychological model which have already being established. That means to respect the user's existed gestalt psychological experience. If the websites are logical clear and category accurate, it will finally improve the efficiency of users.

2. Always describe the current status. In a step by step process, users need to know in what step it is now and how many steps later. We suggest that there should be tips of what status it is now in any web pages and any times.

3. Clearly tell users currently all possible operations. Be specific to each tip of the operation and explanation of the situation.

4. Do the job instead of users as possible as we can. If it is not avoided, help users to fill in blank or copy information if it have to.

5. Lists the information plates like "Recently Viewed...", "Users Bought This Goods Also Bought..." and "Recommend Goods You Maybe Interested..." on the web pages. According to analysis the users browse history, we should supply these information for them to provide better service.

6. Continuously release the information plates like "Price Reduction", "Promotion", "Group Purchase" and "Special Offer" to stimulate consumers' desire of buying.

5.2 Visual Elements Design of Web Pages

The concept of "gestalt" is comes from the research of visual field, so when we look at the visual elements design of online trade websites through the visual principles of gestalt psychology, we bound to find some new design ideas.

1. Apply the rule of "whole is greater than sum of parts" into web pages design. That means we have to around a vision center to create an overall visual image. Have a uniform style throughout the websites design to maintain the same effect. And ensure all sub-sections and links in a prominent personality while maintaining same style.

2. Apply the rule of "proximity" into web pages design. "Proximity" rules tell us, users are easy to put text or graphics which location nearby together as a whole thing. It can be used for the design of images, buttons, navigation bars and color applications.

3. Apply the rule of "tendency of complete and closed" into web pages design. People use to understand uncompleted graphics and text according their past experience. So we should try to avoid incomplete images because users will spend their extra time to wonder the meaning of images.

4. Apply the rule of "similarity" into web pages design. People tend to classify similar objects as a type of same thing. Human visual route is also tending to move with similar design. The rules can be used for the design of graphics, animations, buttons and color applications, etc.

5. Apply the rule of "simplicity" into web pages design. When people see a complex graphic, they will often simplify it as the form they can understand. So in the design of web pages, there is no need to over-concerned about graphic whether is exquisite or not. We should concern about they whether are in the habit of user living.

6 Conclusion

According to the basic theories and laws of gestalt psychology, we summed up consumers' psychological model, have a full discussion of integrated design elements of online trade websites from two different aspects: logical structure design of websites and visual elements of web pages. Then summarize some design methods and strategies in line with the lows of psychology. Based on these design methods, it should enable users to obtain better experience result when they are shopping online. Of course there should be improvement and development of this paper, for example new interactive technologies which different from traditional "keyboard, mouse and monitor" emerging will impact the user's trading experience. It will be our focus point in future further research.

References

1. China Electronic Commerce Research Center: 2010 China E-commerce Market Data Monitoring Report, Hang Zhou, 10–15 (2011)
2. Moeslinger, S.: Technology at Home: A Digital Personal Scale. CHI 97 Eletronic Publications: Formal Video Program (1997)
3. Rubinoff, R.: How to Quantify the User Experience (2004), `http://www.sitepoint.com/print/quantify-use-experience`
4. Egger, F.N.: Trust me, I'm an online vendor: Towards a Model of Trust for E-commerce System Design. In: Human Factors in Computer Systems, CHI 2000 (2000)
5. Garrett, J.J.: The Elements of User Experience: User-centered Design for the Web. New Riders Press, Indiana (2002)
6. Zhang, H.: The Research on the Design Strategy of Visual Effects of Network Curriculum Guided by Gestalt Theory. Paper of Master Degree: Northeast Normal University (2006)
7. Wen, B.: The Practice of Gestalt Psychology in Graphic Design. Package World 03, 122–126 (2011)
8. Gestalt Psychology, `http://baike.baidu.com/view/73571.html?wtp=tt`

Exploration and Practice on Construction of Curriculum Group of Numerical Analysis

Datian Niu, Xuegang Yuan, Jia Jiao, and Wei Zhao

College of Science, Dalian Nationalities University, Liaoning Dalian 116600
{niudt,yuanxg,Jiaojia,Zhaow}@dlnu.edu.cn

Abstract. To culture applied and cross-disciplinary talents with high comprehensive quality, initiative spirit and strong practical capability, we establish a curriculum group of numerical analysis in terms of the characteristics of Nationalities University and the developing direction of Information and Computing Science. Firstly, we introduce the arrangement of curriculum group and the information of faculty and staff briefly. Secondly, for the characteristics of curriculum group of numerical analysis and the present problems, we make a list of the exploration and practice on three aspects, namely, theoretical teaching, experiment teaching and network teaching. In particular, we carry out the teaching in mixed-level classes on the course of numerical analysis, the bilingual teaching of code-switching type on the curriculum group, the innovative experimental project and the study website of numerical analysis.

Keywords: curriculum group of numerical analysis, teaching in mixed-level classes, code-switching, study website.

1 Introduction

Many practical problems to be solved in scientific and engineering computing are often complicated and compositive. Generally, these problems can not be solved analytically due to their inherent complexities. Therefore, they are commonly solved numerically. The general processes for solving the complicated problems by computer can be came down to the following steps: 1. Abstract and simplify the practical problem and build the mathematical model; 2. Propose a numerical method to solve the model by its features; 3. Program by some computer language or mathematical software; 4. Debug and run the program, obtain the computational results, numerically simulate the practical problem by using the computational results.

With the development of science and technology, scientific computing, theoretical analysis and scientific experiment have been three foundation of scientific exploration. The curriculum group of Numerical Analysis mixes these functions together exactly.

Numerical Analysis (also called Numerical Computing Method or Numerical Method) is a comprehensive curriculum that introduces the basic theories and methods of scientific computing, it is also a combining product of applied mathematics and computer science. In recent decades, with the development of computer and computing technique, numerical methods for solving all kinds of mathematical problems have

A. Xie & X. Huang (Eds.): Advances in Computer Science and Education, AISC 140, pp. 333–338.
springerlink.com

been widely used in the fields of science and technology, and new computational cross-disciplinary branches, such as computational mechanics, computational physics, computational chemistry, computational biology, are constantly emerging. Furthermore, they also infiltrate into life sciences, military, economic and other fields of non-traditional mathematical applications [1]. That is to say, the theories and methods of Numerical Analysis play an increasing important part in high technology and traditional technology.

Since 1980s, in China at many colleges and universities, Numerical Analysis has been a compulsory course of undergraduates of Mathematics. At the moment, it is also the compulsory course of postgraduates of engineering major.

With the increasing power of computer and the rapid updating of computing technique, "Numerical Analysis" is always the forefront of exploration and reform in the teaching process due to its features. In [2] the authors introduced the history and the actuality of the development and reform of Numerical Analysis, summarized the suggestions and strategies of reform, and gave some discussion and thought of the reform. Other introductions about Numerical Analysis can be referred to [3-5].

2 Arrangement of Curriculum Group of Numerical Analysis

With the progress of times, the goal of talent training has been changed to culture applied and cross-disciplinary talents with higher creative spirit and practical ability. The major of Information and Computing Science is a major that combines theoretical foundation and practical application, and has remarkable characteristics in mathematical education. Dalian Nationalities University (DLNU) pays much attention to the frontier, the practicality, and the cross of the major in the construction process. The courses arrangement reflects the hierarchical structure of three-dimensional characteristics [6]. In order to make the students of Information and Computing Science grasp the fundamental theory and the necessary specific knowledge and understand the new trends in today's technological development, we build the curriculum group with Numerical Analysis to be the leading course. The group aims at the students who have intensely interest in scientific computing and get further education in this area. After several years of exploration and practice, we adjusted the main courses and their periods of the curriculum group. Now the main compulsory courses are: Numerical Analysis II (48 lectures and 16 experiments), Theory and Method of Matrix Computing (32 lectures and 16 experiments), Optimization Method (32 lectures and 16 experiments). Meanwhile, we set some elective courses such as Approximation (24 lecture and 8 experiments) and Algorithm Design and Analysis (32 lectures and 16 experiments). Each course is the continuation of Numerical Analysis and has its feature.

3 Explorations and Practice of Construction of Curriculum Group

Aiming at the features of the curriculum group and the existing problems [2-5], after several years of practice, by absorbing the reform achievements of the foundation of mathematics curriculum and by adding the modern teaching methods and philosophy, we make some meaningful explorations and practice on theoretical teaching, experimental teaching and network teaching.

(1) Theoretical Teaching

The fundamental courses of the curriculum group are Mathematical Analysis, Advanced Algebra and Ordinary Differential Equations. In order to study Numerical Analysis well, students must master these fundamental courses. In the process of theoretical teaching, we mainly list the important works in the following parts.

(i) Combining Theory and Application

The curriculum group of Numerical Analysis has strong features in theory, application and complexity. Therefore, it is difficult to study this course. So, before teaching each method, we first introduced some simplified mathematical model of the practical problems such that students can realize the applicability of the problems; then we taught the basic idea and the method for solving these problems, especially emphasized the history, the thinking process of the method; next we analyzed the merits and demerits, convergence, stability and error analysis of the method by combining the numerical experiments; finally, we introduced the applicability of the method. By these processes, we cultured the ability of students to abstract the mathematical model from practical problems, and to apply and analyze the algorithms.

(ii) Modernization of Teaching Stages

Before popularization of multimedia teaching, the traditional teaching mode was "teacher + (textbook + blackboard + chalk) + students". Since the formulas are too plentiful to remember and many algorithms are designed for specific problems, it will take up too much time on writing and drawing in blackboard and will cause less interaction between teacher and students. The emergence of multimedia teaching methods has affected on mathematical teaching deeply and widely, and it raises a series of revolution on teaching method, teaching mode and teaching concept. However, it also exists many drawbacks. By considering the above problems synthetically, we did not use multimedia courseware only. Inversely, by combining courseware with blackboard-writing, we shown definitions, theorems and figures on PPT and focused attention on basic concepts, basic theorems, construction of algorithms and implementation of algorithms by using blackboard-writing.

(iii) Teaching on Different Levels

In DLNU, about 50% students of Information and Computing Science are ethnic minority. Most of them come from underdeveloped regions (especially ethnic minority regions). The cultural qualities of students differ greatly and show a polarization. Except for Numerical Analysis, the foundation course of the curriculum group, other courses are limited or elective courses of Scientific Computing branch. Therefore, we carried out the teaching on different levels, the actual steps are: taught basic contents mainly in class; open the extracurricular classroom to answer students' question, encourage students' mutual help; for excellent students, encourage them to join the teachers' personal innovation workroom, contact the teachers' projects, and build good foundation of carrying out scientific research and scientific and engineering computing.

(iv) Bilingual Teaching of Code-switching Type
For students of non-English major, the main function of English language is to learn professional knowledge, to read scientific literatures, to take scientific research, and to serve for future work. However, the English teaching in many colleges is disjointed with the studying of professional knowledge. Many words have their special meanings in mathematics. To this end, we carried out code-switching teaching for all courses of curriculum group of Numerical Analysis to make students grasp the important professional words and phrases of mathematics and the translation skill in mathematical environment. In order to culture students' interest and motivation of learning professional English, we bring professional English test into the final exams. We organize a students' extracurricular scientific English translation group, and guide the students to translate original English textbooks and papers by teachers' help.

(2) Experiment Teaching

Numerical experiment is an important part of teaching process of Curriculum Group of Numerical Analysis, and it is a main approach to combine theory and practice. The development of numerical methods is inextricably linked with the updating of the computer science. Especially, the development of mathematical softwares, such as Matlab, Mathematica and Maple, brings the vitality into the construction of the Curriculum Group. Carrying out relative numerical experiments of numerical methods by using mathematical softwares and their visual functions can deepen students' understanding of numerical algorithms, improve their ability to choose a suitable algorithm, program and analyze computational results. According to students' level and the textbook used, we design three types of experiments: basic experiments, synthetic experiments and innovative experiments.

(i) Basic Experiments
Basic experiments are the main part of the experiment teaching. Students use Matlab (or C) to implement basic algorithms, compare their convergence and stability, grasp the cognition and comprehension of construction idea, basic principle and condition of the algorithms, strengthen the programming skill.

(ii) Synthetic Experiments
The contents of synthetic experiments are classical problems with background of practical applications. Students simplify and abstract the practical problems, build the relative mathematical models, solve the problems by choosing some suitable numerical methods, programs, carry out the numerical computing and simulation. The goal is to improve students' ability to program and analyze, construct and solve the mathematical models, evoke students to participate in the mathematical contest in modeling, culture the students' ability to do scientific research.

(iii) Innovative Experiments
Innovative experiments aim at the students who have great interest in scientific computing and want to get further education on this aspect. The experiments are grounded in teachers' innovation workroom. By experiments, students can know teachers' research work systematically, improve interest in scientific research, and

learn specifically, culture ability to carry out research independently, build good foundation for future scientific research. Meanwhile, those who want to go to work can also have the ability and experience to solve practical problems.

(3) Network Teaching

The rapid development of modern information technology and the popularity of the Internet hit the way of people's work, learning and life unprecedentedly. Specially, they provide abundant resources for college education, break the limitation of using textbooks as a unique approach to learn knowledge for students, and give a powerful support on changing the traditional teaching mode. Applying the modern information and network technology to teaching reform of mathematics major, taking advantage of modern education technology and improving traditional teaching methods have been the important task for teaching reform of mathematics major.

Against this background, in order to provide extra information resources and communication channels, we make a learning website for Numerical Analysis. In the website, except for information of the faculty and staff, course planning, teaching calendar and multimedia coursewares, we specially set up some subjects including achievement area, library and links. The achievement area includes several sub-forums, such as computational methods, guide examples, wonderful thinking and algorithm implemention. Each sub-forum provides three levels, namely, basic, synthetic and innovative contents to students to learn and consider. In library forum, we list some English and Chinese textbooks and monographs relatively to Numerical Analysis. To make students know teaching and research information of Numerical Analysis of other universities, we give the links of other universities whose courses of Numerical Analysis have been quality courses above provincial level.

4 Conclusions

After several years of construction, we have accumulated some experience and lessons of the construction of curriculum group of Numerical Analysis. We realize deeply the construction of the curriculum group is a long-term engineering. With the persistent efforts of the members, the curriculum group plays an important role in the culturing of the students of Information and Computing Science. In each year, there are many students becoming master students of famous universities. Students' employment rate always lies in the front rank of those of DLNU. Contracted companies include top 500 companies such as IBM, HP, Accenture. They also include some famous inland software companies such as Huawei, Neusoft, CSS and Huaxin. In the aspect of some students' contest, the students have won the first, second and third prizes of national and international mathematical contests in modeling many times.

Although we obtained some achievements of the construction of curriculum group of Numerical Analysis, there are still some works to do for us, such as further optimizing the curriculum system and contents, improving teaching materials, constructing the learning website, further improving the teaching level and academic level of the teachers, providing higher quality resources for students and improving the quality of education.

References

1. Li, Q.Y., Wang, N.C., Yi, D.Y.: Numerical Analysis, 5th edn. Higher Education Press, Beijing (2008) (in Chinese)
2. Du, Y.S.: Teaching Reform of Numerical Analysis: Overview and Thinking. Collage Mathematics 23(2), 8–15 (2007) (in Chinese)
3. Wan, Z., Han, X.L.: New Viewpoints and Practices for Reforming the Teaching of the Course: Numerical Analysis. Journal of Mathematics Education 17(2), 65–66 (2008) (in Chinese)
4. Zhang, Y.H., Chen, X.Q.: Reform Exploration of course of Numerical Computing Method. Collage Mathematics 19(3), 23–26 (2003) (in Chinese)
5. Wu, Y.J., You, C.H., Ding, Y.F.: Inheritance and Reform of Numerical Analysis. Higher Education of Science 8(1), 46–49 (2000) (in Chinese)
6. Ge, R.D., Sun, X.L., Liu, M.: Reform of Information and Computing Science Major Organization. Journal of Mathematics Education 17(1), 99–102 (2008) (in Chinese)
7. Wang, L.D., Yuan, X.G., Liu, Y.T.: Problems of Computer-aided Instruction Mathematics. Journal of Mathematics Education 17(2), 97–99 (2008) (in Chinese)
8. Xi, Y.R., Yin, R.: Instructional Design of Website for Learning Special Topics. E-education Research 117(1), 34–38 (2003) (in Chinese)

College Chinese and Humanities Education

Zhonghua Li

Department of Humanity and Social Sciences
Henan Institute of Engineering
Zhengzhou, China
lindalee2009@yeah.net

Abstract. The reason why college Chinese has attracted the attention of the state educational departments and the whole society lies in the fact that it plays an important role in quality education. It is an important way for University students to learn human knowledge, to help students understand human thought, to master human approach and human spirit. In order to improve college students' humanistic qualities through college chinese, teachers should reflect the idea quality education in his teaching. Students should be told to draw nutrients beneficial to the quality of education from the excellent cultural tradition, and college chinese teachers should pay attention to students' different personality.

Keywords: College Chinese, quality education, human knowledge, human thought, humanistic methods, human spirit, cultural traditions.

1 Introduction

It has been thirty years since the recovery of college Chinese course,. In the beginning of the restoration of college Chinese, scholars emphasized that it's a tool. Kuang Yaming, Su Buqing and other old educators said students lacked basic oral and written communication skills, so they strongly recommended to restore college Chinese. After the Cultural Revolution, Students' language ability was really very poor, they badly needed to continue to "catch-up." However, at present, it is difficult to sum up the necessity of college Chinese with the reason of "making up". If we just stress college Chinese as a tool, we have not only belittled college Chinese, but also misunderstood quality education. Today's college Chinese attracted not only the concern of the whole society, but also the concern of the state educational departments. What's the reason? The fundamental reason is that college Chinese has been playing an important role in quality education. We can see that the reason why the State stressed college Chinese is the important role it has played in quality education.

2 The Role of College Chinese in the Humanistic Quality Education

First, college chinese an important way for students to learn human knowledge.

Human knowledge is all the knowledge within the field of humanities, which includes historical knowledge, political knowledge, legal knowledge, art knowledge,

A. Xie & X. Huang (Eds.): Advances in Computer Science and Education, AISC 140, pp. 339–344.
springerlink.com

philosophical knowledge, religious knowledge, moral knowledge, etc.. All these categories have corresponding courses, which students in secondary schools can learn these courses, students in universities can still choose a variety of elective courses for further learning. However, these courses are not a substitute for college Chinese. There are two reasons. First, college Chinese trains college students the skills in Chinese language and literature reading, appreciation, and understanding. Only through the study of college Chinese, can students understand and grasp the beauty of Chinese language, the long and splendid history of chinese literature. Second, the knowledge students learn through College Chinese has the most essential human characteristics. Now the history course, philosophy course and so on are enveloped in the "scientism". The only course that are not enveloped in “scientism” is college chinese which includes the best of ancient Chinese literary classics. Scientism has made tremendous contributions for human development. But its negative effects can not be ignored. One of the negative effects is that it has given too much pressure on humanism. The humanism included in College Chinese is to some degree the opposite of “scientism" and "money fetishism. It worries about human life, value and significance. It does not focus on practical useness, but pay attention to human spiritual pursuit. In this materialistic age, the human knowledge in College Chineseis playing a great role to cover the limits of scientism, to resist commercialization and practical tendency.

Secondly, College Chinese can help students understand human thought.

This is the use of college chinese for college students in the ideological dimensions. Compared with students' capacity, technology and expertise, it is more important to cultivate students to develop independent thinking, to become a person who does not drift with the tide the known universities in west countries played a major role in making west countries strong and in society's progress. This role of promotion is realized firstly by promoting science and technology, and secondly by the humanistic philosophy. Therefore, in the famous Western universities, there is a balance between humanities and social sciences and the construction of science of techonology specialties. Humanities are not ignored and science and technology specialties are not over emphasized. In modern period, many Japanese students went to Europe to learn, many of them engeged in liberal arts, Itobowen returned from Germany and served as Prime Minister of Japan. But most of Chinese students going to the West tolearn science and technology, students who learned the arts were not given important positions after returning. Yan Fu was such a person. Therefore, Western philosophers predicted that China would sink and Japan would rise. After the founding of the PRC, there was a obvious tendency of paying attention to science and technology but light weighing humanities incollege and university construction. This is one of the major reasons which led to the "Cultural Revolution". If Chinese universities had been independent academic centers which not only provide the state of science and technology but also provide the state with philosophy of ruling the country, then such a tragedy as the "Cultural Revolution" would have been avoided.

College Chinese have a strong national character. today's world is getting smaller. Every nation must be connected with other nations in the global village to develop their own. In such a information, high-tech age, students must be familiar with the world, to understand the global trend. At the same time, students must also understand their own country. Only understand their motherland in the past deeply, can students grasp the present, predict the future, can they maintain their nation's cultural identity in the interaction with other nations, can they be clearly aware of their strengths and weaknesses, can they inherite our ancestors' ideological essence and create a a better future for the Chinese nation. College Chinese is deeply rooted in the soil of national culture. it conveys the nation's thoughts and feelings, contains the nation's spiritual and cultural values. In the study of national literary classics, college students will accept the subtle Chinese world view, values, modes of production, lifestyles and ways of thinking, receive the profound educationof national traditions and national consciousness. Only in this way, can students not be captured by national arrogance or national nihilism, can they become outstanding citizens of the world who is self-confident but without any self conceit.

Many hows and whys contained in College Chinese are also truths of universal value. the people-oriented thoughts of Confucius and Mencius, their thoughts of integrity, their thought on learning attitude "I know as I know, I do not know as I do not know", the ideas on the importance of people's moral self-cultivation and self-perfection in"university", the ideas of erudition deliberative, discernment in 'moderate', and so on, are truths of universal value. Chinese thoughts belong to the Chinese nation as well as the world. Like all ancient civilizations of the world, Chinese civilization has both its essence and its dross. Through the learning of the humanistic ideas in College Chinese, students can improve thinking ability, distinguishing ability, and their ideological level.

Thirdly, College Chinese can help students master humanistic approaches.

Humanistic approach is different from the scientific method. Scientific approach emphasizes accuracy and general applicability, but humanistic approach emphasizes experience, emotion, and practice. It emphasizes the specific and individual nature of different cultures. In this period scientific doctrine enveloping everything, many people believe that science can solve all problems. This is a mistake. In fact, science and the humanities are the two wings of the human spirit, loss of any, the human spirit can not soar in the vast universe of the spirit. Therefore, the the human way of thinking students learn from College Chinese can make up the deficiency of students' thinking, benefit students all their life. From this perspective, the quality requirements for College Chinese teachers is very high. Mr. Xu Zhongyu stressed that College Chinese needs to select the most knowledgeable teachers. This is very reasonable. The College Chinese teachers in well-known universities in China before liberation were famous professors You Guo-en, Feng Yuanjun, Zhong Jingwen, Guo Shao-yu, Zhu Dongrun, Zhu Ziqing, Lu Shuxiang and so on. They were all scholars of profound erudition. Mr. Xu Zhongyu himself has been teaching College Chinese personally for many years. College Chinese teachers should teach the students not only literary classics, but also research methods in literature. When Mr. Wang Bugao taught College Chinese at Tsinghua University, he not only stimulated students' interest in learning but also developed student's study habits and their basic research methods. All the College Chinese teachers should not only teach, but also lead the students to do research. This is the only way to make the end of students' College Chinese learning become a new starting point to learn their mother tongue, to enable students to learn their mother tongue positively instead of learning their mother tongue passively. Only so, can we say that the real purpose of College Chinese was achieved.

Finally, College Chinese can help students master the human spirit.

Human spirit is the core of quality education, the base of people's worldview and values. Human spirit has the following four basic principles: First, the principle of Subjectivity. We should look at people as subjects when dealing with the relationship between human beings and objects, between people and society, between man and nature. Second, the purpose principle. We should look at man as the ultimate purpose in the activities to understand and transform the world. All human practice is to meet people's needs. Third, humanitarian principles. We should recognize people as above all things. We should recognize that the spirit is more important than substance, that human values are more important than the value of all things, that the value of life is above all else. Fourth, the principle of equality. In interpersonal exchanges, we should think that people should respect each other, should preserve human dignity. Any insult to human dignity of others is not allowed.

In comparison, teaching students cultural knowledge and helping students master the cultural methods in College Chinese teaching are relatively easier. The most difficult thing is to make students to follow the human spirit. this is not easy to bear fruit. Teachers should through word and deed lead students to cultivate themselves to like real people, to follow the human spirit in accordance with the requirements of real people. If colleges and universities can nurture such talents, help students become comprehensively developed socialist builders and successors who have ideals, morality, culture, and discipline is the key to the implementation of quality education.

The realization of College Chinese to cultivate students' human spirit is through the promotion of students' emotion. After learning Dai Wangshu's "Rainy Lane" and "Reeds" in the "Book of Songs", the students can comparatively analyze the feelings of the two poems, and feel the sameness of the two poems: the authors' self-pity, sadness, helplessness, and loneliness, their search for beauty and hope. Teachers can also compare "that man" in the Reeds and the "Lilac Girl" in the "Rainy Lane", to guide students to understand the two hazy realm of fantasy, to understand the deeper implication in these two poems. Chinese classic literature also tells us a lot of beauty, pure love stories, let us understand that "love" is not just that simple bouquet of roses, a few luxurious meals, several eloquent love letters, a few sweet words, and a simple clause "I love you". Love also shoulders a heavy responsibility and the guardian of loneliness and the taste of loneliness. "Lady Xiang" and "Everlasting Regret", Li Shang-yin's love poems express the people's desire for pure love, their yearning for a better life. The pursuit of love, the loyalty and firmness shook everyone's mind, so that their humanity has been sublimated.

3 How to Achieve Quality Education in College Chinese Teaching

College Chinese teachers should effectively improve students' human quality through College Chinese teaching. In order to improve human quality of students, College Chinese teachers should make efforts in the following aspects:

First, teachers should teach for the purpose of quality education Teachers should combine teaching with the real life. In teaching, teachers should guide students to set up correct life view and world view, so that students are yearning for outstanding qualities such as dedication, courage, brave, sacrifice, selflessness, dedication to the mission. Teachers should cultivate students' qualities of honesty, integrity, kindness, humility, resolute and so on.

Teachers should strive to create a culture conducive to quality education in college chinese teaching. Language is the most important communication tool as well as the most important carrier of culture. Contained in the College Chinese textbooks is the selected cultural works of China since ancient times, expressed in beautiful language. The author bears the noble spiritual realm, a strong personality, healthy aesthetics in their works. In teaching the teacher should not forget humane quality training. Teachers should not lecture on the text. Teachers should train students healthy quality, noble ideas, and high morality, promote patriotism and national pride, improve students' aesthetic sensibility of taste, which is conducive to the formation of strong cultural atmosphere favorable to quality education.

School language teachers should have a passion. In some language classrooms, teachers taught wryly for the completion of teaching tasks. Such a class is placid and will not impress the hearts of students. Remembering the teaching state of Mr. Liang Qichao in the period of the Republic of China is good for today's college chinese teachers. Mr. Liang Qichao "sometimes break out, sometimes laugh, sometimes heave a deep sigh, sometimes cry, when his emotion turns good he laughs and opens his mouth big. After lecture, Mr. Liang sweated like very happy". Such a class the students would never forget in all their life. Passion is needed in teaching. Teaching need teachers to devote all our energies and dedication. Only if the teacher love life and cause, can he inspire students to love life and learning.

Language teachers should be good at creating circumstances compatible with the teaching context to stimulate students emotional experience. When teaching Zhou Zuoren's "a little black boat", teachers can allow students to listen to Ma Sicong's Violin Concerto "Nostalgia" which is created in 1937. the music has a sense which can deeply impress the students, because many of them are likely to be the first time to leave home and go out to school. The same feeling that they are immersed in will make it much easier to understand the article which contains the same feelings the students have when parting with their own home and their loved ones. Their understanding of the article will become much more profound.

Secondly, college chinese teachers should tell students again and again that they should learn beneficial nutrients from the fine cultural traditions for the improvement of cultural quality. Reality is the continuation of history. A person whithout any knowledge on the past will not the present. He who does not know the nation's history has no understanding of people. It is unlikely for him to be a great contributor to the building of socialism with Chinese characteristics. China's ancient literature has a long history, many ancient books and records not only has superb fine art of literary language, and contains a broad and profound traditional Chinese philosophy, such as "Laozi" "The Analects of Confucius", "Zhuang Zi", etc. They will give useful lessons for the modern socio-economic and cultural fields. Japanese entrepreneurs would love to read "Laozi" "The Analects of Confucius", "Three Kingdoms" and other Chinese classics. From these classics they have gained a lot of interpersonal skills and business management insight. There are also some scholars who applied Confucian ruling thinking to modern security management. This indicats that traditional culture is not a pile of paper, but has rich connotations. In language teaching, teachers should enhance the students' understanding of our ancient cultural excellence, so that students can improve the cultivation of classical literature, learn more about the splendid culture of ancient China. Teachers should give students skills of self enlightenment, enable students to be a qualified good citizen when they go to work in the future.

Again, College Chinese teachers should pay attention to student's personality. At the same time, teachers should guide students to associate their own feelings and their own life when reading works. So the students can resonate with the authors. Because the meaning of literary works is created by the readers involved. The meaning of the works is finished only if the reader has finished his creative process. Only if the readers take part in the meaning of the works, can the works be alive. Teachers must not simply replace the student to explain the works. They should guide students to read works individually, to explain the works with their own experience. Reading is a process for students to know the society, to know themselves, to know the reality. When students compare the works with their own life, their own fate, with themselves, they will have intense emotion. Thus, in the long-term reading students will have a sense of language and beauty, their spiritual world will be enriched, and the students' language proficiency, language literacy and cultural tastes would be improved.

References

1. Yu, J.: About College Chinese teachers and Some Thoughts. China University Teaching (3) (2007)
2. http://www.eyjx.com/eyjx/1/ReadNews.asp?NewsID=2984
3. Cao, H.: Language: cultural composition. Language teaching communication (7-8) (2004)
4. Xu, Z., Qi, S.(eds.): College Chinese. East China Normal University Press (2005)
5. Chen, H., Lee, S.-S.: Native Higher Education: From Crisis to transfer. China's Higher Education (3), 34–36 (2008)
6. Wei, Y.: Construction of college chinese courses and practice. Journal of Hunan University of Arts (6), 5–7 (2006)
7. Qi, S.: Speech on ZHC's Shanghai forum (EB /OL). National Vocational Chinese Proficiency Test wetpage (September 2007)

Distributed Detecting and Real-Time Advertising System of Equipment Failure Based on MANET

Xia Qin and Tian Yi

[1] School of Electrical Engineering, Jiu Jiang University, Jiu Jiang, China
13701885490@qq.com
[2] School of Media Arts Suzhou Science and Technology College, Suzhou, China

Abstract. In this paper, we propose a real-time advertising solution based on MANET for equipment failure information, and provide a fairly detailed description of the development of a distributed sensing and visualization system of equipments failures. The system consists of clients, middle nodes and central server, and includes such subsystems as failure information generator, failure information transfer, routing protocol, distributed sensing and visualization system based on geographical information system (GIS). Experiments demonstrate that the system can advertise the failure information in a timely way and provide rich inquiry functions and real-time situation display.

Keywords: MANET, Failure Declaring, GIS.

1 Introduction

The lack of a fixed field environment network infrastructure, mobile devices can constitute a mobile ad hoc networks for communications, remote transmission equipment failure information. Perception of mobile equipment failure distribution and visualization systems is to solve the field environment, equipment failure information transmission, management and display system developed, the equipment maintenance and protection of information systems an important part of [1]. Distribution of mobile equipment failure can be perceived and real-time visualization system perceived battlefield equipment failure, and fault information via mobile wireless ad hoc networks spread to service center for repair centers provide access and management capabilities, and real-time battlefield situation display.

2 Relate Work

U.S. troops and equipment to protect information technology practice leader and loyal monk. In recent years, the U.S. wartime equipment and technical support means showing a new trend. First, the artificial intelligence technology used in working condition monitoring and equipment fault diagnosis [2]. By sensors or special measuring device detection equipment working state, the analysis by the expert system to predict potential equipment failure may be, actively implement preventive and targeted maintenance. The second is the use of information networks, the

A. Xie & X. Huang (Eds.): Advances in Computer Science and Education, AISC 140, pp. 345–352.
springerlink.com

implementation of remote technical support [3]. "Remote technical support" system is to protect the U.S. Navy an important part of equipment, it is to be installed in the ship's remote technical support (maintenance) prototype fiber-optic local area network connection with the ship, through an antenna system and the Navy's "Smart links "connected to the ships at sea from the shore to provide maintenance and logistics data. "Remote technical support" has been in the "Lincoln", "Stennis' aircraft carrier battle group installed the software, is planning to install a new aircraft carrier battle groups in the other version of the system [4].

Currently, based on GPRS network in the civilian aspects of remote fault system has been applied in many research fields, such as distributed micro-power station based on GPRS remote fault diagnosis system [5], GPRS-based remote diagnostics system and forklift GPRS-based paving Machine Remote fault diagnosis system. GPRS (General Packet Radio Service) is developed based on existing GSM network, a data communications network that wireless packet-switched technologies to provide end to end wireless IP connectivity [6]. Remote fault diagnosis system to on-site equipment failure state information is encoded [7], sent through the GPRS network to the remote diagnostic expert systems, expert systems by remote staff to view relevant information about the devices, and again through the GPRS network service information or feedback modify the parameters [8].

3 Systems Funcation and Structure

According to the battlefield environment features to build networks based on wireless ad-aware and real-time equipment failure declaration system, need to complete the definition of failure to collect the information, select and improve the wireless ad hoc network protocol, the classification of fault information display and query geographic information-based visualization system work.

3.1 System Require

Mobile ad hoc networks based on equipment failure sensing and real-time systems need to declare the following functional requirements.

1. Distributed Sensing
As the battle area, equipment and more, which requires a distributed sensing system design, sub-level perception, will be relatively small range of equipment to the middle of a server to collect information and then forwarded to a higher level server. Such a structure to facilitate the collection of information on fault and fault management of equipment [9].

2. Real-time Announced
Rapidly changing battlefield conditions, to grasp the situation on the battlefield to grasp war, victory has a crucial role. The normal operation of equipment on the battlefield, whether the decision of the combat effectiveness of the play, so when equipment failure requires declaration of fault information in real time, to win in the shortest possible time to repair the device and restore combat effectiveness. This requires the system design to protect the transmission of information in real time.

Real-time three aspects, namely failure to collect real-time information, and second, real-time information transmission failure, the three fault information is displayed in real time.

3. Transport Layer Protocol Design
Design transport layer protocol, requires the ability to accurately reflect the status of battlefield equipment, such as equipment failure information, location information to the maintenance and management personnel concerned with the other information.

4. Failure of Information Transmission
In the battlefield environment, without a fixed wireless base stations, equipment, mobile speed, the device between the network topology change, which requires us to choose the battlefield environment suitable for wireless routing protocols to ensure the smooth transmission of fault information.

3.2 Network Structure and Node Functions

Mobile equipment failure distribution system consists of visual perception and the client, intermediate server, the server consists of three parts. Figure 1.

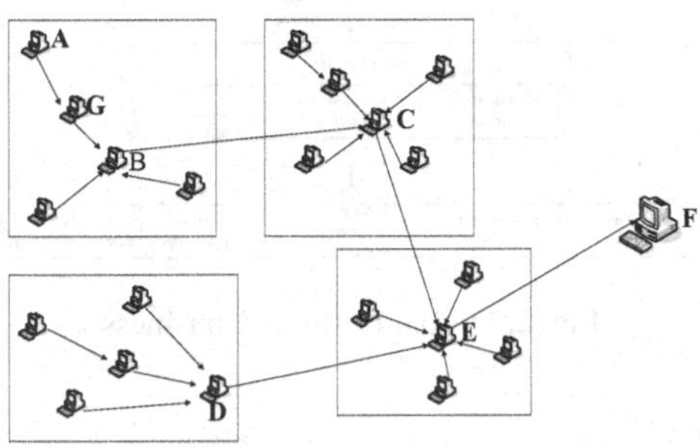

Fig. 1. System topology in MANET

In wireless ad hoc networks, including faulty equipment generates nodes, such as the A nodes; packet transmission node, such as the G node; intermediate server nodes, such as B, C, D, E node; server node centers, such as the F node.

(1) Client
System client is installed on the device, its main function is to collect failure information and the information sent to the wireless ad hoc networks. Collecting client information, including: device type, device number, fault type, fault levels, equipment and facilities managers the latitude and longitude in which you want to send a custom message. Client information is collected, it will add it in the packet sent to the wireless ad hoc networks.

(2) Intermediate Server
Intermediate server is to play the role of cluster head, a small range of client data collection to the intermediate server, and then there is an intermediate server forwards the data to the server. Intermediate server device node, node and server in one routing node. It provides fault information classification queries.

4 System Implemention

This section describes the various modules of business processes and implementation of functional classification.

4.1 Business Process

System's overall business process is the client acquisition fault information, to generate a fault message, the message transmitted by the wireless mobile ad hoc networks to the intermediate server nodes, the intermediate server nodes to provide fault information and fault messages query forwarding, packets failure intermediate server forwards to the server, providing fault information display, fault information classified information, maintenance information management, geographic information display and show the statistical distribution function of failure, the system's business process shown in Figure 2.

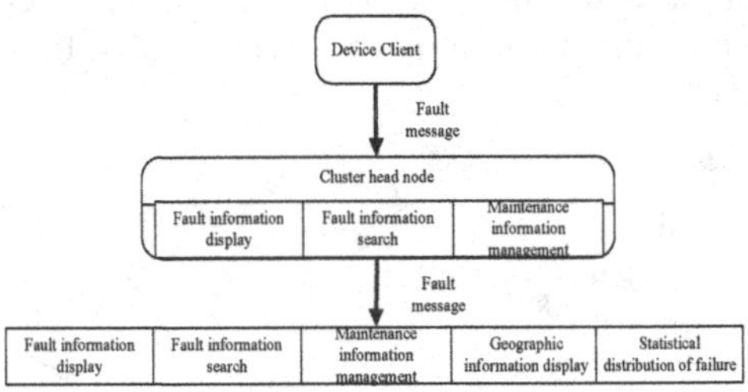

Fig. 2. System flow chart business

4.2 System Module

Fault information collection system is divided into modules, the routing module, data storage modules, data query and display modules, display modules situation. Module structure shown in Figure 3.

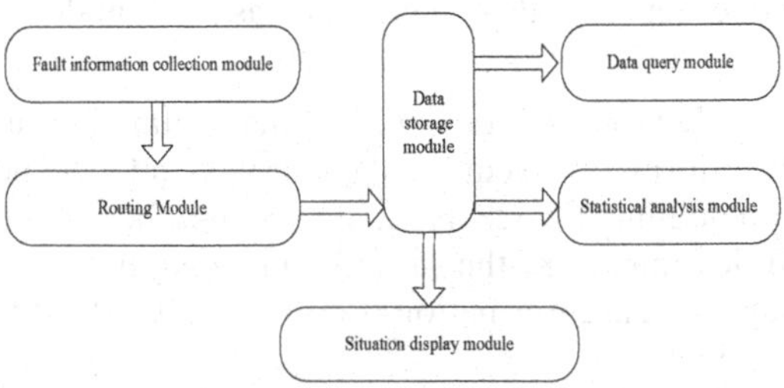

Fig. 3. Block diagram

1. Fault Information Collection Module
Fault information collection module's main function is to collect failure information, fault information generated message. Fault information packet includes the device

type field, type field equipment failures, failure severity field failure time field, the field equipment to send location information to customize information fields and so on.

2. Routing Module
Choose the type of wireless mobile ad hoc networks is the failure information transmission protocol and correctness of real-time protection [11]. We select the OLSR [12] protocol as a system of routing protocols, it has a transmission delay, higher packet delivery ratio characteristics.

OLSR protocol implementation process, the link state information forwarded by radio to be selected as those generated by the node, thereby reducing the flood control network, the amount of information, to achieve a further optimization. While only MPR nodes forward the broadcast node or choose to forward the broadcast nodes choose to link state information between the transmission. Running OLSR protocol for distributed network computing part of the link state routing information, they make changes to the shortest hop routing. Based on the above optimization, the OLSR protocol than the traditional link-state has a higher efficiency. OLSR is mainly applied to large and intensive wireless network [13].

3. Data query and display module
Data query and display the function module is provided depending on the classification of the query. The main classification of queries, including: query by device type, by fault type queries, according to the maintenance of state inquiries.

5 Experimental Results and Analysis

In order to verify the information transmission system and situation display capabilities, we conducted a simulation and design of a small wireless mobile ad hoc networks simulate actual application environments.

5.1 System Simulation

Using ns2 (ns-2.30) as a simulation platform for the routing of the transmission capacity of the system simulation, the simulation parameters shown in Table 1.

Table 1. System simulation parameters in Table

Range of network topology	1000m*1000m
The number of nodes	50
Pause Time	0、50、100、200、400、800
Node maximum speed	18m/s
Routing Queue Interface	50
Reference point group mobility model	Set maximum number of nodes 9, the minimum number of nodes is 3.
Simulation time	1800s, 800s stability of the system to skip time
Cbr data identification bit	Data flow nodes 60%, 512 packets

Main simulation evaluation of the data packet delivery rate and end data packet delivery delay. Data packet delivery ratio reflects the robustness and reliability of transmission protocol, it is successfully received and the total number of data packets issued by the ratio of the number of data packets;-end data packet delivery delay reflects the average transfer agreement data, delay characteristics, it is the end to end packet delivery delay and successfully received the sum of the number of data packets. The results of data curves shown in Figure 4 and Figure 5.

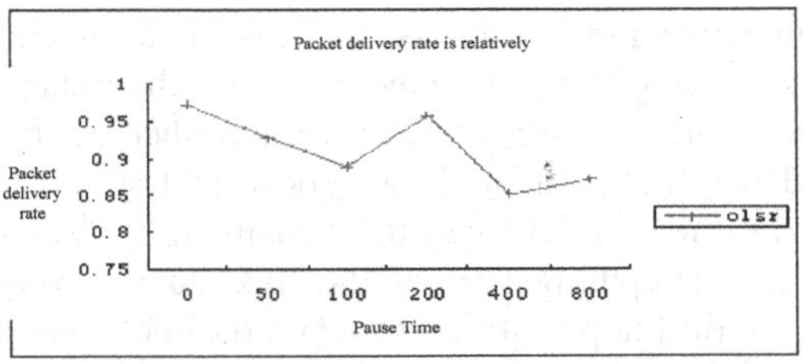

Fig. 4. Node 50, under different pause time packet delivery rate

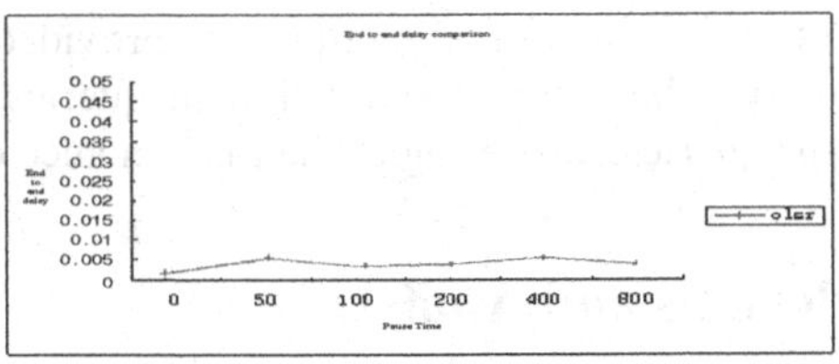

Fig. 5. When node 50 end to end delay under different pause time

5.2 System Physics Experiments

Small wireless mobile ad hoc networks by the two desktop computers, three laptop computer components, each computer has a wireless card. One desktop computer simulation intermediate server nodes and the central server node, the node faulty equipment notebook computer simulation.

We set the client node, the three client nodes on the move failed random time information packet. Intermediate server and the central server on the various functional modules, test systems to transmit information and information display capabilities. End to end communications test parameters shown in Table 2.

Table 2. Parameters physics experiments

Range of network topology	50m*50m
The number of nodes	5
Node maximum speed	3m/s
Simulation time	600s
Simulation data generated	600

Of 600 data transmission delay statistics, obtained an average delay of 2.13ms, to ensure data transmission in real time. Geographic information system situation display device shown in Figure 6.

Fig. 6. Geographic Information trend exhibited

Physical experiment results show that small wireless mobile ad hoc networks of communication within the mobile node with the client, the network topology changes, but the client sends the fault information packets can be intermediate server and the central server to receive to. System data storage module, the data query module, statistical analysis module, display module work situation.

5.3 Analysis of Experimental Results

Route of transmission capacity simulation results showed that with the node mobility increases, packet delivery rate of decline. The main reason is as node mobility increases, so the connection has been established off the possibility of increased again, resulting in data transmission failure, thus the packet delivery rate of decline.

Set of nodes in the speed range, with the node mobility to enhance the transmission delay between nodes has increased, but the increase is very small. This is because each node in the OLSR protocol is stored with all network routing information, packet transmission can be transmitted directly without the need to re-find the route, so the increase in node mobility limited circumstances, the system transmission delay increases very small.

6 Conclusion

In this paper, mobile battlefield environment declaration of the actual fault information needs, designed and implemented based on wireless mobile ad hoc networks of mobile devices and real-time distributed fault-aware systems declared. We selected the experimental transmission of environmental information for the battlefield as a system of OLSR protocol routing protocol, designed to truly reflect the equipment failure information transmission protocol, and through the fault information and geographic information systems combined, the system has real-time sensing equipment failure , real-time data transmission, equipment, real-time situation display functions. Transmission system simulation and system routing physical experiments show that the system can meet the battlefield environment, equipment failures and transmission of information disclosure requirements to the system design purposes. Next we will further improve the fault reporting models, improve the wireless mobile ad hoc network protocols, make the system more suitable for the battlefield environment, perception, and device information declaration.

References

1. The SNIA IOTTA Repository, http://iotta.snia.org/
2. Bairavasundaram, L.N., Goodson, G.R., Pasupathy, S., Schindler, J.: An analysis of latent sector errors in disk drives. In: ACM SIGMETRICS, pp. 289–300 (2007)
3. Schroeder, B., Damouras, S., Gill, P.: Understanding latent sector errors and how to protect against them. In: 8th USENIX FAST (2010)
4. Schroeder, B., Gibson, G.: Disk failures in the real world: What does an MTTF of 1,000,000 hours mean too you? In: 5th USENIX FAST (2007)
5. Mi, N., Riska, A., Smirni, E., Riedel, E.: Enhancing data availability through background activities. In: 38th Annual IEEE/IFIP DSN (2008)
6. Schwarz, T., Xin, Q., Miller, E.L., Long, D.D.E., Hospodor, A., Ng, S.: Disk scrubbing in large archival storage systems. In: IEEE 12th MASCOTS (2004)
7. Anderson, E.: Capture, conversion, and analysis of an intense NFS workload. In: Proccedings of the 7th Conference on File and Storage Technologies, pp. 139–152 (February 2009)
8. Kent, K., Souppaya, M.: Guide to Computer Security Log Management. Recommendations of the National Institute of Standards and Technology, USA (2006)
9. Leung, A., Pasupathy, S., Goodson, G., Miller, E.: Measurement and analysis of large-scale file system workloads. In: Proceedings of the USENIX 2008 Annual Technical Conference (June 2008)
10. Forte, D.V.: SecSyslog: an Approach to Secure Logging Based on Covert Channels. In: IEEE Proceedings of the First International Workshop on Systematic Approaches to Digital Forensic Engineering (SADFE 2005). IEEE Computer Society (2005)
11. Sambasivan, R.R., Zheng, A.X., Thereska, E., Ganger, G.: Categorizing and differencing system behaviours. In: Hot Topics in Autonomic Computing (June 2007)
12. Gulati, A., Ahmad, I., Waldspurger, C.P.: Proportionate Allocation of Resources for Distributed Storage Access. In: USENIX FAST (February 2009)
13. Shen, Y.-L., Xu, L.: An efficient disk I/O characteristics collection method based on virtual machine technology. In: 10th IEEE Intl. Conf. on High Perf. Computing and Comm. (2008)

Strengthen Study Motivations of College Students via Learning Environment Improvement

Xin Zhang[1,2] and Libo Huang[2]

[1] School of Humanities and Social Science
[2] School of Computer
National University of Defense Technology
Changsha 410073, China
{xinzhang_pst,truthseeker}@163.com

Abstract. Current college students are no longer a scarce resource and it is becoming increasingly prominent phenomenon they can't find jobs after graduation. This is because of not only the oversupply of graduates but also the inability of some students. Meanwhile, it has become a serious problem that college students are lack of study motivations. This leads to low efficiency in the class, improper use of spare time, academic fraud and other phenomenons, which will seriously affect the quality of personnel training. So how to create a good learning environment to enhance college students' study motivation and lay solid foundation for employment becomes a question that worth exploring. This paper presents some suggestions by improving learning environment, which try to solve the study motivation problem of college students: improvement of in-class efficiency and efficient utilization of after-class time.

Keywords: college student, learning environment, motivation, employment.

1 Introduction

As shown in Table 1, from year 2004 to 2010, the number of graduate students is raised from 2.8 million to 6.31 million, which increased 2.3 times in seven years and will continue to increase. Unlike 70s and 80s, current college students and graduates are no longer a scarce resource, on the contrary, it is becoming increasingly prominent phenomenon they can't find jobs after graduation [1].

Table 1. Number of Graduate students from year 2004 to 2010 in China

Year	Number(million)	Growth rate (%)
2004	2.80	32
2005	3.33	18.9
2006	4.13	24
2007	4.95	19.9
2008	5.59	13
2009	6.10	9.1
2010	6.31	3.4

A. Xie & X. Huang (Eds.): Advances in Computer Science and Education, AISC 140, pp. 353–358.
springerlink.com

The MyCOS Human Resource Information Management Company [2] classifies the majors whose unemployment rates rank top ten in the country for two consecutive years as risk majors. In 2009, the risk majors include Law, Computer Science and Technology, Accounting, English, International Economics and Trade, which were once popular majors.

Current college students' unemployment can be broadly divided into two categories: structural unemployment and voluntary unemployment. Structural unemployment is caused by the unreasonable distribution, including labor supply and demand in the professions, regions, etc, reflected in the number of graduates crowded on the popular companies for job opportunities, while nobody interested in companies with poor location and low pays. Voluntary unemployment refers to the cases that college students are not satisfied with the job and want to give up the chance, continuing to seek other jobs.

All graduate students expect to find jobs with high paying, good location, and jobs are better if they can be relevant to students' expertise or interest. But the number of such jobs is far less than the number of candidates, so the employer organizations selects students according to their learning ability, communication skills and other aspects of the candidates. Students with low ability or skills will face failure in the selections.

In order to stand out in the candidates, students needs to fully develop and exercise themselves in universities, and lay a good foundation for employment. However, parts of the current students go to college just to meet the requirements of the employer diploma, regarding university as a springboard, rather than a stage to develop their own capacity. Lack of motivation leads to low efficiency in the class, improper use of spare time, test cheating, academic fraud and other phenomenons. From the 1980s onwards, the lack of motivation for college students has been studied and has now become the focus of many discussions. But the problem remains exist, causing widespread concern in the education community [3].

Four-year training of students in universities involves a large number of human resources. Once a student graduates from school and faces unemployment, human resources will be wasted. Over time, the performance of students in the university will have a great impact on their future developments. Although the growth of students are more relying on their own initiative, but how to improve the class efficiency and make good use of spare time, the universities can still make a difference. It is a challenge problem worthy of study that how to create a better learning environment to enhance the students' motivation and to lay the foundation for their employment.

This paper will introduce two ways try to solve the study motivation problem of college students: improvement of in-class efficiency and efficient utilization of after-class time, as will be discussed in detail in the following sections.

2 Improvement of In-Class Efficiency

It is a common phenomenon that the class efficiency is low because many students always fall asleep, play games, receive calls, and talk to each other. As students have different learning abilities, some students may still make up after class, but other students can not pass the examination. Experienced students may know that it is difficult to fully compensate for the loss of content in the classroom because part of the course content is difficult to represent in the form of words.

However, when parents and teachers criticize the students for the class inefficiency, they should not ignore the other negative aspects such as not good enough learning environment, lack of professional interest and inadequate supervision of students. To solve these problems, we propose following solutions.

2.1 Improve the Hardware Construction

To build up a comfortable learning environment, it is suggested that school could provide air-conditioning in classrooms and dormitories. For example, in the hot summer, if the classroom is not air-conditioned, hundreds of students are crowded into a classroom, sweating, which is hard for them to concentrate listening lectures.

In the case of adequate teaching resources, introduction of small class teaching (SCT) method is recommended for several important courses [4][5]. SCT can avoid some obvious drawbacks like that teacher is difficult to communicate with each student at large class; it is hard to ensure lectures effect for students in the back; and it is difficult to mobilize the enthusiasm of students. SCT can have a small size of students, generally 20 ~ 25 and can change the form of seats according to the characteristics of the course. This will help the communication between teacher and students, and mobilize the enthusiasm of each student. Based on the SCT, these students can be further assigned to different groups according to their scores. Each semester students exchange between the various levels of groups based on the study effect, which can not only stimulate the enthusiasm of students, but also play individualized results [6].

SCT is currently piloting in a number of colleges and universities and gets good results. For example, Computer Department of the Air Force Aviation University takes SCT at different levels in their computer basis course, resulting in resources sharing and personality highlighting. The excellence rates and pass rates are drastically increased than before.

2.2 Provide Opportunities for Re-selecting Major

It can be seen from the many survey data that some students lack of learning motivation because of dissatisfaction of their majors [7]. That is to say, there are large differences between their majors and personal interests, or even majors are just their shortcomings so than even if seriously listening to classes, performance improvement is very slow. It would be a good thing if the university can provide students the opportunity to choose major after entrances, such as encouraging students to listen interdisciplinary courses, and after 1-3 months re-selecting their satisfactory majors.

2.3 Improve the Capabilities of Teachers

Universities should strengthen the training of teachers and enhance the overall level of the teachers. Though some teachers have good performance on academic research, they lack of teaching experience. Even if there is a good idea, teacher can not share with students in the classroom. Such scripted teaching would not achieve the teacher's preaching, tuition, and FAQ's role. It can be often seen that students with the same class listening to some teachers in high spirits, while to some other teachers in sleepy mode. As the dominant role of class, the importance of teacher can be imagined, especially in SCT, requiring teachers to better grasp the classroom and carry out various forms of teaching, or they will greatly reduce the advantages of SCT [8].

In addition, for courses with adequate teachers, we can arrange to open several different classes rather than one class for the same course, enabling students to freely choose a teacher. This can effectively stimulate teachers' sense of responsibility and improve their teaching level, rather than complaining about their students.

3 Efficient Utilization of After-Class Time

It is very important for student to grasp the spare time beside eight hours in class. Over time, spare time will gradually widen the gap between different students. Therefore, universities should pay attention to the supervision and management of spare time, build a better platform for students, and help them lay a good foundation.

3.1 Enhancing the Supervision of Students

Many students believe that study in college is an easy thing, so they do not pay attention to study at the beginning. Until they discover that they can not keep up with the study progress, they are repentant and loss their study inspiration. Therefore, universities should organize students to learn the training program from the beginning, letting them have a clear plan for four-year university study.

To increase students' sense of crisis, we can strictly implement the elimination system for selecting the best. For example, regardless of test situation for a course, the last 5% of the students are regarded as failed and their scores are published online. Thus enhanced transparency can prevent some students from placing their hopes in other ways and make students work hard not to be eliminated.

Another important aspect of enhancing supervision is to strengthen the management of students by counselors. These counselors are responsible for all aspects of student management including studying, daily life, psychological. The university could combine the rewards of counselors with the progress of their students, which can make them more responsible. Counselors need to find out the basic situation of their students, understand their psychological state, organize the students in order in the classroom, and know their credit situation at various stages, so that counselors can provide timely help.

3.2 Strengthen the Psychological Counseling

Current college students have big psychological pressure. It is very important for their growth to detect and ease the psychological problems of students. A positive and optimistic student is easier to communicate with other people and motivated to study; while a depressed student is often interested in nothing and tends to abandon himself to despair when confronting difficulties.

Since students have different psychological crux, so there is a need to create a professional counseling organizations to solve the problems students meet. For example, establish "Heart of Bridge" counseling centers, adhering to "What's your trouble, your problem, just come here!"

3.3 Work-Study Normalization

According to incomplete statistics, 70% of college students desire to work-study, 47% of the students take part in work-study program for the money, but their motivations vary: some students have economic difficulties and want to reduce the family burden; some students want social contact as soon as possible and exercise personal capacity; some students would like to contribute to charity through work-study and so on. However, most of these students choose work-study positions irrelevant to study: 50% of students choose a tutor; 40% of students choose the simple odd jobs such as leaflets or waiter, only 10% of the students are engaged in a job that using their knowledge learned from universities [10].

In fact, in order to truly grasp the university for four years, students should invest a lot of time. But some students failed to distinguish between primary and secondary, making the time occupied by things that are unplanned or work-study with non-technical content. This would result in that students do not have enough time to study.

To better address these issues, the university could set up a special work-study organization. It can contact the employers with a high popularity among students and organize the students using the summer vacation time for practice. This can not only meet the aspirations of students, but also make the employers more familiar with our students so that students can be more competitive in the job hunting. In addition, universities could encourage students to participate in some academic projects and provide material assistance according to their contributions. This method not only allows students to have some income, but also mobilize their enthusiasm for study.

3.4 Establishing Platform for Communication

Currently, many universities invite some well-known domestic or foreign experts to come to school for lectures. However, the role played by a single class is not immediate. It is recommended that universities with adequate funding can invite experts in related fields for one-month intensive training. Through daily communication with students and reinforcement learning, it will receive better results. In addition to "come in" model, universities can also use the "send out" model to stimulate students' motivation for learning. Universities can establish friendly relationship with some well-known foreign universities, and then send elite students abroad for joint training. This can not only help students open up their views, but also helps universities learn each other.

4 Conclusion

College students are the top priority of national education. As a cradle of high-quality students, the university should create a better environment to stimulate study motivations of college students and to lay the foundation for employment and further study. This paper discusses two ways that try to solve the study motivation problem: improvement of in-class efficiency and efficient utilization of after-class time.

Through the learning environment improvement, study motivations of college students could be strengthened.

Acknowledgements. We gratefully acknowledge the supports of the National Basic Research Program of China under the project number 2007CB310901 and the National Natural Science Foundation of China under Grant No. 60803041 and No. 60773024.

References

1. Xu, X., Discussion, A.: of the Unemployment Problem of Graduates in China. Journal of Yangzhou University (Higher Education Study Edition) 8(6), 48–50 (2004)
2. http://www.mycos.com.cn/
3. Li, N., Zhao, Z., Wang, Y.: Discussion on Factors of Insufficient Study Motivations for College Students. China Electric Power Education 4, 166–168 (2011)
4. Han, Y.: New Views on Superiority of Smaller Classes Teaching. Journal of Yichun University, 199–206 (2005)
5. Xue, G.: A Comparison of Studies of Class Size Reduction in China and the U. S. Comparative Education Review 7(25) (2004)
6. Zhang, L.-W.: Implementation of Teaching Reform Regarding at Different Levels in Small Classes. Computer Education (13), 50–52 (2010)
7. Yu, L., Liu, X., Wu, H.: Creative Research on Development of Study Motivation for College Students. Legal & Economy (227), 116–117 (2010)
8. Tao, Q.: Teacher's Professional Development for Small Class Teaching Based on the Experience of America. Development and Research of Education, pp. 50–53 (2008)
9. Hou, W., Shi, S., Zhang, J.: Course and Solution Research of Inefficient Study Motivation for Local College Student. Education and Career (26) (2010)
10. Pu, H., et al.: Research on Current Status and Related Solution of College Student Work-Study. Higher Education Research 27(4), 84–87 (2010)

The Mediating Role of Entrepreneurial Efficiency on the Relationship between Human Resource Management and Corporate Entrepreneurship

ChunLi Liu and HongLi Wang

School of Management
Zhejiang University
Hangzhou, 310012, China
lclmn@hotmail.com

Abstract. Now firms are facing greater pressure and they have to be sensitive and act quickly to grab the opportunities. Corporate entrepreneurship involves recognizing the opportunities and taking actions to take advantage of them. It is widely agreed that the capability of CE is an important criteria to evaluate enterprises. And human resource management is regarded as an important factor to decide the level of corporate entrepreneurship. Although it is widely accepted that human resource management influence corporate entrepreneurship, it is still not clear how human resource management influence corporate entrepreneurship. The paper offers a new perspective, that is to study the mechanism using enterpreneurial efficiency, and specifically this paper focuses on small and medium enterprises.

Keywords: entrepreneurship, entrepreneurial efficiency, human resource management, small and medium enterprises.

1 Introduction

Nowadays, firms are facing more and more competitive business environment, not only because their customers are changing tastes quickly, but also their competitors keep launching new products and improving manufacturing methods. Therefore, in order to satisfy the customers and grow the firm, they have to be able to explore the potential needs of their customers and develop new products quickly and effectively. Now, quite a lot of firms have already realized the importance of improving the capability of corporate entrepreneurship, and they spare no efforts to achieve this. Due to the fact that all the actions which can promote corporate entrepreneurship are initiated and complemented by employees, it's vital to select the right person, train them effectively and motive them to take these actions. Until now, many papers have discussed the relationship between human resource management and corporate entrepreneurship. This paper aims to discuss the existing relevant papers, and based on previous study, this paper suggests we could study the mechanism between HRM and CE using the concept of entrepreneurial efficiency.

A. Xie & X. Huang (Eds.): Advances in Computer Science and Education, AISC 140, pp. 359–364.
springerlink.com

2 Literature Review

2.1 Corporate Entrepreneurship

Since the concept of corporate entrepreneurship is put forward, there has been much argument on what it actually is, and nowadays researchers has generally reached an agreement that corporate entrepreneurship is not a single-dimension concept but a multi-dimension concept, that it can indicate entrepreneurial actions in an existing firm, and also can refer to starting a new business. Regarding to the specific definition of corporate entrepreneurship, there are mainly two streams: Lumpkin suggests that corporate entrepreneurship means the willingness of a firm to take entrepreneurial actions to improve firm performance; Brikinshaw points out that corporate entrepreneurship should be the entrepreneurial actions a firms has taken instead of the willingness to do it [1], as he believes the willingness will not inevitably turn into actions. Now, the second explanation is the dominant one since it can show the actual capability of corporate entrepreneurship in a firm than the first explanation. Some scholars believe corporate entrepreneurship should contain five dimensions: innovation, risk-taking, proactive, venturing and strategic renewal [2]; Some scholars insist corporate entrepreneurship should be composed of proactive, venturing and strategic renewal; When it comes to the definition of corporate entrepreneurship in China, Song Lin points out that corporate entrepreneurship means a firm develops new products or services to increase their profitability and competitive advantage when faced intense competition and constantly-changing environment, and it mainly refers to the innovation and improvement in the research, manufacturing and marketing departments. Besides, he suggests that corporate entrepreneurship should contain innovation, risk-taking, value creation, strategic renewal and proactive [3]. As is showed above, although different scholars explain the concept of corporate entrepreneurship in a slightly different way, we can see their explanations have quite a lot in common. Specifically, innovation refers to firm behaviors such as the creation and introduction of new products, production processes and organizational systems. Venturing refers to firm behaviors such as entering new businesses through the creation or purchase of new business organizations. Strategic renewal refers to firm behaviors such as transforming the firm or revitalizing its operations by changing the scope of its business or its competitive approach. This paper would take this definition of corporate entrepreneurship: innovation, venturing and strategic renewal.

2.2 Human Resource Management

The capability of corporate entrepreneurship in a firm is often showed through its innovation, flexibility and the speed of response to changes, and all these features rely on the knowledge, capabilities and sense of risk-taking of the employees in the firm [7]. This shows the important role human resource management is playing when it comes to improving the capability of corporate entrepreneurship in a firm. Scholars have mainly five frameworks to describe the human resource management system of a specific firm: some scholars propose the system should include employee influence, the process of human resource management, compensation design and job design; some scholars think the system should contain employee election, compensation design, performance appraisal and employee training; some scholars believe it can be divided into seven categories: job design, employee recruiting, employee training, compensation design, employee communication and change management; some scholars point out the system can be understood by job analysis, the overall plan of human resource management, recruiting and selection, performance appraisal, employee training, security conditions and the relationship with labor union; there are also some scholars who suppose the human resource management can be interpreted as employee election, employee training, performance appraisal, compensation design, job design, employee participation and the sense of security of employees. Besides, the framework of high performance human resource management is gaining more and more popularity nowadays. It contains three categories: appraisal and rewards, people flow and employment relations. Specifically, the first category is composed of compensation design and performance appraisal; the second category includes staffing, training, mobility and job security; the last category shows the job design and participation. As is listed above, scholars use these frameworks to understand and describe the human resource management of a specific firm, laying the foundation for discussing the relationship between human resource management and corporate entrepreneurship.

2.3 The Relationship between Human Resource Management and Corporate Entrepreneurship

In the past decades, scholars in the field of human resource management have demonstrated that the success of a firm relies on employees' constantly-improving job performance [2]. The relevant research can be mainly put into two categories: the first category examines if human resource management influences the capability of corporate entrepreneurship; the second category explores the mechanism how human resource management influences corporate entrepreneurship. Under the first category, Ralf Schmelter shows that employee recruiting and employee training has a positive impact on corporate entrepreneurship while the relationship between compensation design and corporate entrepreneurship is uncertain. Besides, others scholars have also demonstrated the contingent correlation between compensation design and corporate entrepreneurship [5], and furthermore, they propose the assumption that this relationship may be moderated by the feature of firm task and employee perceived organization support. Block and Ornati have done empirical research and found that compensation design have little impact on corporate entrepreneurship. Laursen and Foss have demonstrated that team collaboration and empowerment can promote corporate entrepreneurship. Besides, research shows that employee recruiting, training, performance appraisal and compensation design can promote corporate entrepreneurship [7]. Under the second category, that's how human resource management influences corporate entrepreneurship, Zhe Zhang points out that human resource management will promote organizational citizenship behavior, and therefore employees are motivated to take more tasks, volunteer to help each other to finish work, and take initiatives to innovate, in the end, the corporate entrepreneurship will be improved. Also, some scholars explore the mechanism how human resource management influences corporate entrepreneurship from the viewpoint of perceived organizational support [9].

3 Discussion and Proposal

3.1 Introduction of Research Model

When I look back on the existing research on human resource management and corporate entrepreneurship, I realize most research is based on large firms, and only several papers focus on small and medium firms. However, the only several papers on small and medium firms are based on foreign firms. As the small and medium firms have a lot of different features compared with their counterparts in foreign countries, we cannot declare the conclusions which have been demonstrated will be also valid in Chinese small and medium firms. All in all, I would say it's necessary and important to discuss the relationship between human resource management and corporate entrepreneurship in small and medium firms in China. When I examine this relationship, I decide to introduce the concept of entrepreneurial efficiency since only when human resource management practices can radically influence how people feel and how people act it can at last effect the level of corporate entrepreneurship. Now, I propose my model how I will do further research on the relationship between human resource management and corporate entrepreneurship. That is to examine the relationship between human resource management and corporate entrepreneurship in small and medium firms in China, take entrepreneurial efficiency as a mediator, and take firm size, firm age and ownership as control variables.

3.2 Human Resource Management in Small and Medium Firms

There is no world-wide definition of small and medium firms. ILO concludes that Scholars often use criteria such as the number of employees, the yearly output, the value of assets and so on. Due to different social features, customs and other relavant factors, different countries believe small and medium entreprises are not always the sa me, such as enterprises with about 1000 people in South Korean are regarded as small and medium enterprises, but they are taken as large companies in other countries, like in Germany. Within the same country, the definition does not stay the same, instead the definition changes gradually, such as the definition of small and medium firms in China has been modified many times since the economic environment always changes. Nowadays, the definition of small and medium firms in China is no more than 50 people for non-manufacturing firms [9].

Ram says that informality means the way how employees act and make decisions is influenced by the habits thy see and learn, the unwritten down rules . Many scholars have pointed out that informality is the fundament feature of small and medium firms. Informality can be interpreted in two ways: there is not witten-down regulations to guide employees' actions and decision-making process; employees don't agree with each other regarding to how they feel about the firm and how to go ahead with their wok [6]. Regarding to the reason why small and medium firms show the feature of informality, some scholars discuss the effects of organizational atmosphere, some scholars believe government regulaitons; there are also scholars who explain the differences from the viewpoint of firm resources and management rather than discussing the external factor, such as Cunningham and Rowley declare that it's because small and medium firms tend to have uncertain strategy and lack professionals to management human resource management [9].

Due to the fact that small and medium firms have distinctive features compared to large firms, it's not proper to go on using the frameworks which are developed to describe human resource management in large firms, and it's necessary to adapt a system designed especially for small and medium firms. Collins and Allen developed a framework to describe the human resource management in small and medium firms. It includes three parts: employee selection strategy, employee management strategies and employee motivation and retention strategies. Specifically, employee selection strategies discusses selecting employees based on fit to the company versus fit to the job; employee management strategies discuss managing employee performance through employee involvement versus tight controls; employee motivation and retention strategies discuss motivation employee commitment through creating a family atmosphere versus individual monetary incentives, the incentives include: sponsoring company social events so employees can get to know each other, offering employees profit- or gain-sharing pay, creating a strong social environment at work and so on. We are going to adapt this framework to describe human resource management in this paper.

3.3 Entrepreneurial Efficiency

Entrepreneurial efficiency evolves from the concept of self-efficiency. Self-efficiency means how much one believe himself or herself when confronted with one specific. We can see from the concept thant self-efficiency is a subjective concept, besides it is influenced by the feature of work. Therefore, the level of self-efficiency is associated with the specfic context and work. Here the concept is defined on individual level, and then some scholars extend the concept to organizational level. And it then means how much one believe the organization he or she is in can finish the job. When the concept of self-efficiency is used in the field of corporate entrepreneurship, it means how much one employee believe the organization when it comes to corporate entrepreneurship. In this paper, we discuss the relationship between human resource management and corporate entrepreneurship, and I think when companies adopt proper human resource management practices, the level of entrepreneurial efficiency will be enhanced, and then employees are encouraged to work acitively to promote corporate entrepreneurship.

4 Conclusion

This paper has reflected on the relationship between human resource management and corporate entrepreneurship. We have carefully illustrated the importance of promoting corporate entrepreneurship, and looked back on the existing research in this field. Based on the research, this paper suggests a new model for further research: that is to examine the relationship of human resource management and corporate entrepreneurship in small and medium firms in China while taking entrepreneurial efficiency as the mediator. It is still not persuasive to put forward this model, empirical research is needed to demonstrate it.

Acknowledgment. I would like to thank Zhejiang University and those people who helped us during this period of time, especially professor Xuejun Chen.

References

1. Cooke, F.L. (How) Does the HR Strategy Support an Innovation Oriented Business Strategy? An Investment of Insitutional Context and Organizational Practices in Indian Firms. Human Resource Management 49, 24–49 (2010)
2. Hayton, J.C.: Promoting Corporate Entrepreneurship through Human Resource Management Practices: A Review of Empirical Research. Human Resource Management Review 15, 15–21 (2005)
3. Schmelter, R.M.: Boosting Corporate Entrepreneurhip through HRM Eractices: Evidence from Germany. Human Resource Management 49, 27–49 (2010)
4. Bowen, D.E., Otroff, C.: Understanding HRM-Firm Performance Linkages: The Role of the "Strength" of the HRM System. Academy of Management 29, 203–221 (2004)
5. Schuler, R.S.: Forstering and Facilitating Entrepreneurship in Organizations: Implications for Organization Structure and Human Resource Management Practices. Human Resource Management 25, 24–25 (1986)
6. Tocher, N., Rutherford, M.W.: Perceived Acute Human Resource Management Problems in Small and Medium Firms: An Empirical Examination. Entrepreneurship: Theory & Practice 33, 455–479 (2009)
7. Zhang, Z., Jia, M.: Do High-Performance Human Resource Practices Help Corporate Enterpreneurship? The Mediating Role of Organizational Citizenship Behavior. Journal of High Technology Management Research 19, 11–19 (2008)
8. Zhang, Z.: Using Social Exchange Theory to Predict the Effects of High-Performance Human Resource Practices on Corporate Entrepreneursip: Evidence from China. Human Resource Management 49, 23–49 (2010)
9. Li, C.X., Rowley, C.: Small and Medium-sized Enterprises in China: A Literature Review, Human Resource Management and Suggestions for Future Research. Asia Pacific Business Review 16, 319–337 (2010)

M-Learning Activity Design for College English Courses

Lin Hu[1,*], Honghua Xu[2], Shuying Zhuang[1], and Xiaohan Zhang[2]

[1] Jilin University of Finance and Economics, Changchun, China
[2] Changchun University of Science and Technology, Changchun, China
{huhu315,honghuax,shuangyingz,xiaohanz}@126.com

Abstract. Through introducing mobile devices into conventional classes, this paper analyzes the revolutionary feature of mobile learning contexts. On the basis of analyzing the teaching contents of college English, this paper constructs an actual m-learning example for college English to increase the efficiency of English class teaching, to extend the depth, scope of the teaching as well as the participation of students. While promoting knowledge learning, it can help cultivate ability.

Keywords: design model, conversational framework, mobile learning, m-learning activity.

1 Introduction

As knowledge sharply accumulates, it is quite urgent to study anytime anywhere. Early in year 2000, Sharples who is an expert in mobile learning in University of Birmingham pointed out that advanced mobile technology was promoting the transformation from e-learning to m-learning(1). In recent years, with fast development of mobile communications and growing increase in the property of mobile devices, our mobile devices such as cell phone, PDA etc. can meet the demand for study, which provides the probability of lifelong study. Therefore, how to combine mobile devices with traditional teaching activities has become leading edge and hot issue in academic world at home and abroad in this century.

2 Revolutionary Feature Analysis of Mobile Learning

Papert has ever said," When everyone could afford a pencil, learning method would accordingly change." Similarly, when every student owns a mobile device that has strong ability of multi-media information processing, our teaching will experience a thorough revolution. Traditional teaching system consists of teacher, student and teaching content. Teachers reach the target of knowledge transmission through imparting knowledge. The relation among the three elements in traditional teaching system is relatively loose. The approaches of their interrelatedness are imparting, receiving and communication between teachers and students.

---

* Corresponding author.

A. Xie & X. Huang (Eds.): Advances in Computer Science and Education, AISC 140, pp. 365–370.
springerlink.com

Basic elements of teaching consist of teacher, student, teaching content and mobile device. As for the teaching content, mobile device is an implementation tool, which can realize the most optimized content presentation. As for the teacher, mobile device is a teaching organization and implementation tool. Whereas, for the student, mobile device is cognition tool, which can not only help obtain knowledge but also help improve cognitive ability of students as well as may increase the participation of students.

Traditional teaching structure is interconnected and interrelated among teachers, students, textbooks and teaching media (2). The teaching essential quality after introducing mobile devices has transformed from "teacher-centered structure" to "teacher-led, learner-centered" teaching structure.

3 Structure Design for Mobile Learning Activity

As is said above, introduction mobile learning devices into teaching activities makes substantial change to teaching structure. Therefore, it is quite necessary to redesign teaching activities which are inserted mobile learning devices. According to activity theory, learning activity is the best analysis unit during the process of teaching design. Learning is internalized activity. Learners reach learning targets by fulfilling a series of teaching activity tasks. Hence, we must put emphasis on learning activity design.

3.1 Traditional Teaching Design Model

At present, it is widely believed that teaching design process includes five essential elements, i.e. ADDIEL (analysis, design, development, implementation, evaluation) (3). To be specific, under the circumstances of traditional school education, teaching design should start with demand which includes demand analysis of learning, content analysis, learning target illustration, learner's analysis, laid-down strategy, choice and usage of teaching media as well as evaluation of teaching design.

The teaching design of traditional design model is mainly applied to formal learning environment. Although this model takes learner's feature analysis into consideration, there is clear mark of imparting knowledge, which apparently adapts to traditional teaching environment.

3.2 Mobile Learning Activity Design Model

The mobile learning design process should be filled with ideas and methods of teaching design and at the same time should make full use of mobile technology as well as emphasize mobile learner's learning experience (4). The common model of comprehensive teaching design is the core of the design. We believe that mobile learning activity design should include demand analysis, focusing on learners, learning context design, providing necessary technical contexts, restraint conditions' analysis and learning support design. These six elements should be complementary and together set up mobile learning activity design model which is shown in Fig. 1.

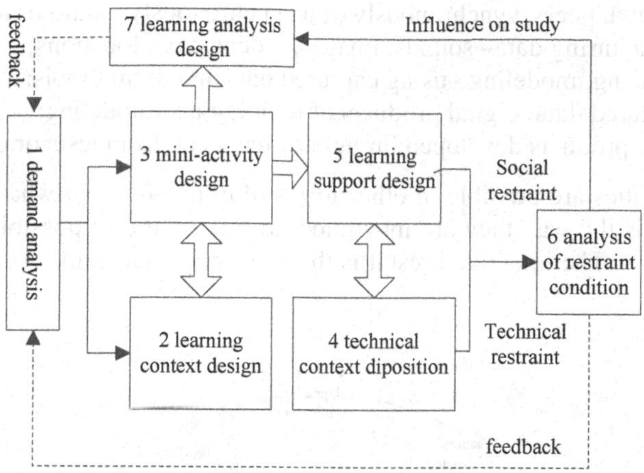

Fig. 1. M-learning activity design model (also called MLADM model)

Demand analysis is the first procedure of mobile learning activity design whose aim is to judge whether particular learning task needs unfolding through the form of mobile learning. Mobile learning design must satisfy the actual quality and special demand of a learner. No personalized design of " one-to-many" is proved to be a failure. After demand analysis, then come designs for learning contexts and mini activities. Mobile learning context consists of many elements such as background, user, target, event, activity etc. Mini-activity is made up of one or many sub-activities to improve learner's knowledge and skills. During the steps of fulfilling learning tasks, the results of these mini-activities can be assessed through the performances of learners. Then come the stages of providing necessary technical contexts and learning support service, during which support service must be taken into consideration. Restraint conditions should overemphasize analyzing social restraints and technical restraints. Social restraint mainly refers to the obstacles which impact on interaction and collaboration. Whereas, technical restraint mainly refers to the systems and tools which lack of necessary technical support or using skills. At last, as analysis of restraint conditions may influence the final learning result, so it may directly impact on learning assessment. The constant feedbacks between learning assessment and restraint analysis promote front-end demand analysis to be continuously refined and perfected.

4 Mobile Learning Activity Design for College English

From mobile learning activity design model, mobile learning activity design is the most important part. It is key to have a successful teaching. Mobile digital devices may offer a wide range of learning activities that could be supported through mobile digital tools and environments:

- Exploring–real physical environments linked to digital guides;
- Investigating–real physical environments linked to digital guides;

• Discussing–with peers, synchronously or a synchronously, audio or text;
• Recording, capturing data– sounds, images, videos, text, locations;
• Building, making, modeling –using captured data and digital tools;
• Sharing–captured data, digital products of building and modeling;
• Adapting–the products developed, in light of feedback from tests or comments.

All these activities are possible in other forms of e-learning, but what may be critical to m-learning is the way they are integrated, to bring the best possible support to the learning process. The integrated result is the conversational framework as is shown in Fig.2.

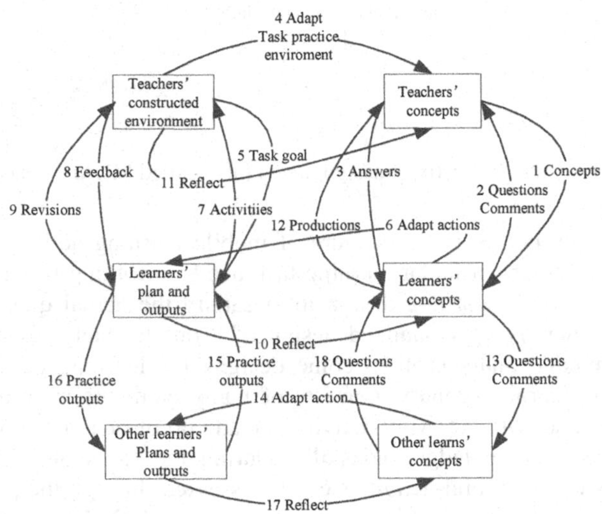

Fig. 2. Conversational framework for formal learning

The form of the Framework defines a dialogic process between 'teacher' and 'student' on two levels, the discursive level, where the focus is theory, concepts, description-building, and the experiential level, where the focus is on practice, activity, procedure-building. Both levels are interactive, but at the discursive level the interaction will take a communicative form – the teacher describes, i.e. the teacher decides what is to be 'framed' (Kress & Pachler, 2007), the student asks questions, the teacher elaborates, the student states their own idea or articulation of the concept (i.e. their conceptual resources are 'augmented' in Kress and Pachler's sense). At the experiential level, the interaction is adaptive, where the student is acting within some practical environment to achieve a goal, and experiences the results of their actions as changes in that environment, enabling them to see how to improve their action. The interaction at the experiential level benefits from the student adapting their actions in the light of the theoretical discussion. The interaction at the discursive level benefits from the students' reflection on their experiences. Similarly, the teacher's construction of a suitable learning environment benefits if it is adapted to their students' needs, and their explanations at the discursive level will benefit from reflecting on their students' performance at the experiential level. The whole process is the same for every teacher-student pair, but also links students with each other, by the same interaction type of communication at the discursive level. At the experiential level, the feedback between peers takes the form of shared comparisons of their outputs from actions on the environment.

The Conversational Framework is designed to describe the minimal requirements for supporting learning in formal education. It can be interpreted as saying that, on the basis of a range of findings from research on student learning, if the learning outcome is understanding, or mastery, the teaching methods should be able to motivate the learner to go through all these different cognitive activities. In that sense it should be able to act as a framework for designing the learning process.

In order to introduce it into college English teaching activity, I would like to take Unit6 " The Last Leaf" in college English book three as an example to design an mobile learning activity based on conversational framework.

Traditional teaching activity design is as follows:

1) Teacher may provide some related background materials about the text including information about the author, O'Henry and several famous short novels and may ask some questions about the materials.(1,2,3)
2) Teacher instructs students to finish certain learning tasks and divides students into several groups, reading the text, summarizing the main idea of the whole text and main idea for each part, discussing to find out the similarity among O'Henry's novels and taking notes.(4,5)
3) Each group may study itself important language points appearing in the text by means of learning tools such as dictionary. In this process, teacher may provide necessary help and instruction. (5,6,7,11)
4) Teacher makes a conclusive speech, presenting his/her own idea which students can refer to.(1)

By contrast, a typical mobile learning activity design is as follows:

1) Teacher may provide some related background materials about the text including information about the author, O'Henry and several famous short novels and may ask some questions about the materials. All the materials teacher provides are downloaded to their mobile devices or uploaded to a designated web page to make students read them anytime anywhere. (1,2,3)
2) Teacher instructs students to finish certain learning tasks and divides students into several groups, reading the text, summarizing the main idea of the whole text and main idea for each part, discussing to find out the similarity among O'Henry's works and taking notes. Then teacher asks students to upload the discussing result to the designated web page. Simultaneously, teacher may upload the sample answer to the web page for reference. (4,5)
3) Each group may study itself important language points appearing in the text by means of learning tools such as dictionary. In this process, teacher may provide necessary help and instruction. In order to better grasp important language points, each group can launch challenging activities to other groups on the shared internet message board and other groups must accept challenges. This way students can fully grasp and understand language points of the text. (5,6,7,8,11,13,14,15,16,17,18)

4) Later in the seminar, students report on what they have learned, uploading them to the designated web page for reference. Teacher may ask questions and requires answers. (1,2,3,10,12,13,14,16,17,18)

From traditional classroom activity design, students can only get knowledge by means of taking notes. At the same time, teacher doesn't set up other motivations to inspire students to study. As long as they take notes and bring it back to the class, that is Ok. They may share their knowledge during learning process but it is not convenient to obtain other students' learning results. In the design for mobile learning activity, mobile devices help teacher set up a cooperative and competitive system and at the same time promote the final learning result. In traditional design, teacher makes a summary at last which maybe partially include what the students' ideas but unavoidably it mainly includes teacher's ideas and knowledge. Whereas, in the mobile learning design, students hold the initiative of final summary. Teacher uploads students' learning results to the designated web page after arranging, editing and correcting. In this process, teacher may only add some extra knowledge which doesn't mentioned by the students. From the latter design, we can see that students; mobile learning practices are greatly enriched. Just as the converstional framework shows us, the main reason is that mobile devices help students obtain relative materials and data with ease in particular learning contexts.

Hence, from above analysis, conversational framework provides us with a strong method for designing learning activities. It will be far-reaching if it is introduced into classroom teaching.

5 Conclusion

This paper makes a trial research to introduce mobile learning into college English classroom teaching. Although this field is still brand new, this research will be definitely focused on by more and more scholars for the features and developing tendency of mobile learning. Introducing mobile learning will revolutionize college education with teaching efficiency greatly improved, classroom's depth and breadth extended. Finally students may enthusiastically attend teaching practice. So while promoting learning knowledge, ability and quality can be cultivated.

References

1. Sharples, M.: The Design of Personal Mobile Technologies for Lifelong Learning. Computers and Education 34, 177–193 (2000)
2. Yu, S., Chen, L.: Practical Meaning of Teaching Structure. Audio-Visual Education Study (2), 21–27 (2005)
3. Huang, R., Salomaa, J.: Mobile Learning—theory, current situation, tendency. Sicence Press, Beijing (2008)
4. Aderinoye, R.A., Ojokhela, K.O., Olojede, A.A.: Integrationg mobile learning into nomadic education programmes in Nigeria: Issues and perspectives. International Review of Research in Open and Distance Learning 8(2), 1–17 (2008)

A Novel Approach to Online Image Education Resource Retrieval by Semantic Network and Interactive Filtering

Xianlin Peng[1], Ling Gao[1], and Yuwen Ning[2]

[1] The Modern Educational Technology Center, Northwest University, Xi'an 710069, China
397448497@qq.com, gl@nwu.edu.cn
[2] The Information Network Center of the Fourth Military Medical University, Xi'an 710032, China
ningyuwen@163.com

Abstract. Image resources play an important role in education. With the rapid development of the Internet techniques, it is an effective way to get large amount of education resources through web search. While it is hard for the current keywords based retrieval methods to reach people's query attention quickly and accurately. In this paper, a novel approach to online image education resource retrieval is proposed. First, a semantic network is constructed to help users to input more precise query words to the Baidu search engines. Then, a novel interactive junk images filtering algorithm is developed with the combination of clustering and the hyperbolic visualization, to filter junk images from Baidu's search results effectively. The experimental results show that the developed method can acquire image education resource from web effectively.

Keywords: Image Education Resource, Retrieval, Semantic Network, Interactive Filtering.

1 Introduction

The development of teaching resources is the basis of information technology in education, and image education resources play an important part of the whole teaching resources. Currently, we are faced with the situation of a serious lack of available material resources since it is time consuming and high cost for material production, which has seriously affected the information technology in network teaching[1-2].

Since there are large amount of education resources on web, more and more attentions are paid to web education resources retrieval. While traditional web search engines, such as Baidu, Google, Yahoo. etc., mainly utilized the keywords-based query techniques, which often return too much junk images and so users could not reach their needed images quickly. The main reason is that it is hard for users to input several query keywords accurately enough to describe their true query intention and so it is hard for the search engines to evaluate it effectively [3-4]. Besides of this, the search results are set out one by one, which makes it inconvenient for user to further

A. Xie & X. Huang (Eds.): Advances in Computer Science and Education, AISC 140, pp. 371–376.
springerlink.com

search what they wanted. Therefore, it is needed to develop new algorithms to filter junk images from search engines effectively.

Gao etc. had developed an approach to filter out junk images from Google search results [4]. The returned images are divided into the relevant sets and the irrelevant sets, and the irrelevant images are filtered as junk images. And the users' query intentions and personal preferences are considered to filter out the junk images more effectively. But when the returned images are divided to two groups with SVM, some needed images maybe filtered out without user's evaluation.

In this paper, a novel interactive approach to image retrieval is suggested. A semantic network is constructed to represent the relation among various image concepts, and to help users input more precise query word; the junk images from the Baidu search results are filtered out interactively until the user's requirement is reached.

2 Proposed Approach

The block diagram of the proposed approach is shown in fig. 1.

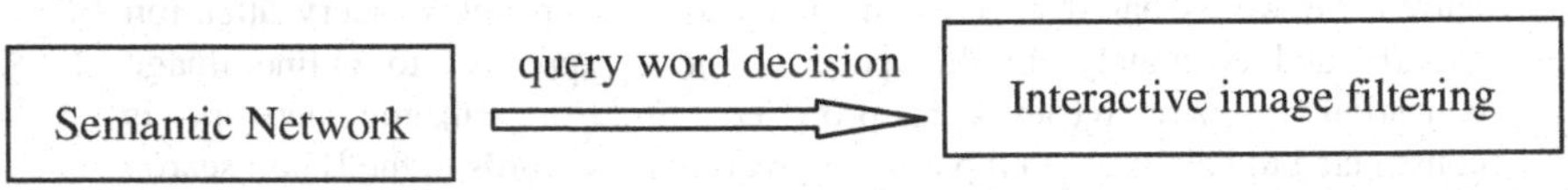

Fig. 1. The block diagram of the proposed approach

2.1 Semantic Network Construction

In our work, semantic network is constructed with the combination of the semantic similarity (see fig. 2), and the similarity between two image concepts is measured [4]:

$$c(c_i, c_j) = as(c_i, c_j) + bp(c_i, c_j) \tag{1}$$

Here, c_i and c_j denotes two image concept respectively, $s(c_i, c_j)$ is their wordnet -based similarity, $p(c_i, c_j)$ is their co-occurrence probability, and a and b is the weight [4].

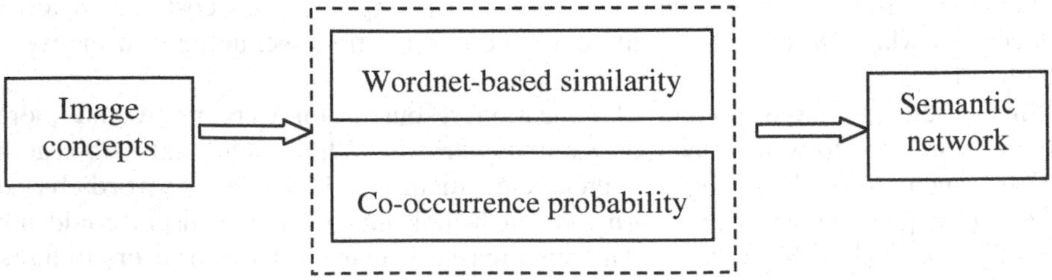

Fig. 2. The block diagram of semantic network construction

$$s(c_i, c_j) = -\log \frac{length(c_i, c_j)}{2D} \tag{2}$$

Here, $length(c_i, c_j)$ is the distance of two image concept in wordnet, D is the depth of wordnet.

2.2 Interactive Image Filtering

The suggested image filtering method include the following steps: First, large amount of images are obtained by the Baidu search engine with query words; then, the returned images are clustered by the Affinity Propagation (AP) algorithm; then, hyperbolic visualization is adopted to display these images for user to assess the relevance between the query intentions and search results; finally, the junk images are filtered out interactively until the user's requirement is reached (see fig. 3).

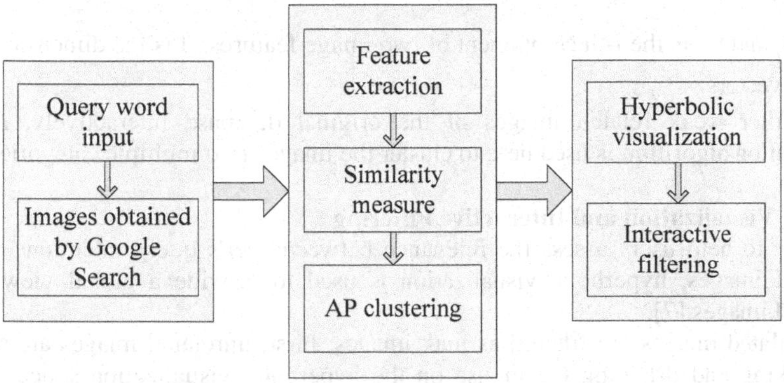

Fig. 3. The block diagram of the proposed image filtering method

2.2.1 Initial Image Collections Obtained by the Baidu Search Engine

The keywords such as dog, pig and cat are used respectively as query words to the Baidu search engines, and large amount of images are returned for each keyword. The frontal 100 images for each query word are selected automatically as the initial set of images for further process. It is well known that some returned images are not what we want. These unwanted images are called junk images and need to be filtered out.

2.2.2 Image Clustering with the AP Method

First, the local visual Gabor features are detected. A 2D Gabor wavelet transform is defined as the convolution of the image $\mathbf{z}$ as follow [5]:

$$\Gamma_{\mathbf{k}}(\mathbf{z}) = \int\int I(\mathbf{z}')\psi_{\mathbf{k}}(\mathbf{z}-\mathbf{z}')d\mathbf{z}' \tag{3}$$

Here Gabor filter is defined as [5]:

$$\psi_{\mathbf{k}}(\mathbf{z}) = \frac{\mathbf{k}^T\mathbf{k}}{\sigma^2}\exp(-\frac{\mathbf{k}^T\mathbf{k}}{2\sigma^2}\mathbf{z}^T\mathbf{z})(\exp(i\mathbf{k}^T\mathbf{z}) - \exp(-\frac{\sigma^2}{2})) \tag{4}$$

Here $\mathbf{z} = (x, y)$ and $\mathbf{k}$ is the characteristic wave vector. The mean value and variance of Gabor coefficients are calculated to form the feature vector $h(z)$.

The similarity between two images *x* and *y* is defined as follows [4]:

$$K(x, y) = \prod_{i=1}^{p} e^{-\chi_i^2(h_i(x), h_i(y))/o_i} \tag{5}$$

Here $\sigma_i = [\sigma_1, ..., \sigma_m]$ is the set of the mean value of the χ^2 distances of all the feature vectors, *p* is the features' dimension. The χ^2 distance is defined as [4]:

$$\chi^2(u, v) = \frac{1}{2}\sum_{i=1}^{d}\frac{(u_i - v_i)^2}{u_i + v_i} \tag{6}$$

Here u_i and v_i is the *i*-th component of two image features, d is the dimension of the feature vectors.

To filter weak related images of the original database interactively, Affinity Propagation algorithm is used here to cluster the images into multiple categories [6].

2.2.3 Visualization and Interactive Filtering

In order to help users assess the relevance between user's query intentions and the returned images, hyperbolic visualization is used to provide a global view of the returned images [7].

Unrelated images are filtered as junk images. First, unrelated images are selected by clicking and dragging the mouse on the hyperbolic visualization space; then, a small number of images that are highly similar to each selected images are filtered.

If there are still unwanted images, AP clustering is used again to the remaining images and the above interactive filtering procedures are repeated.

3 Experimental Results

The proposed method is realized by the Visual C++6.0 platforms and its flowchart is shown in Fig.1. In order to evaluate the proposed method, a small online image education resources retrieval system with 20 kinds of animals is implemented. The semantic similarity is calculated based on the animal sets in the MIT's LabelMe database and the semantic network is constructed.

In the experiments, 20 keywords such as horse, cow, cat, tiger, snow, panda, sheep etc. are selected randomly, and then 5 most similar keywords are selected automatically for users to decide more precise query word(see fig.4) with the semantic network, and then returned images from search engines are filtered finally(see Fig.4).

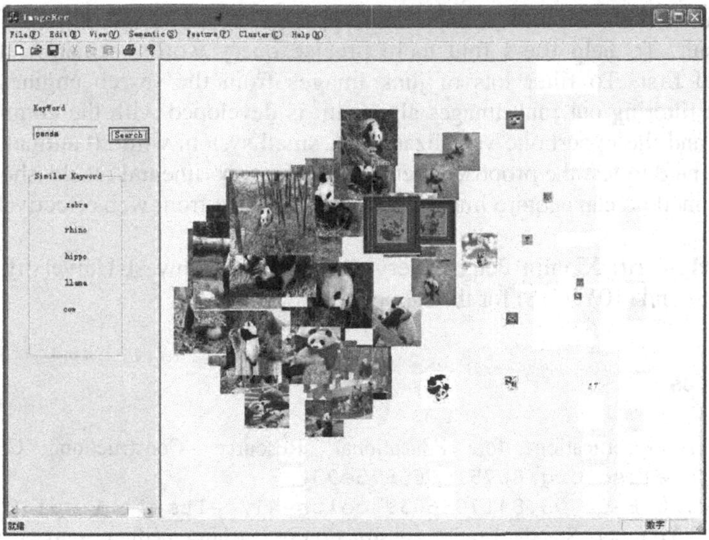

Fig. 4. Result of our supposed method

Fig. 5. Result of the Baidu search engine

To illustrate the validity of our method, result of the Baidu search engine with word 'panda is shown in Fig.5. It can be seen that the suggested method can reach user's attention more precisely.

4 Conclusions

A novel approach to online image education resource retrieval method is proposed in this paper, which is based on the semantic network and interactive image filtering, and which can both help user to input more precise keywords and reach more precise search result. To help users find more precise query words, semantic network is constructed first. To filter lots of junk images from the search engines, a novel interactive filtering out junk images algorithm is developed with the combination of clustering and the hyperbolic visualization. A small system with 20 animal keywords is implemented to test the proposed method and the experimental results show that the developed method can acquire image education resource from web effectively.

Acknowledgment. Xianlin Peng is very grateful to Northwest University Graduate Innovation Fund (10YJC18) for the support.

References

1. Technical Specification for Educational Resource Construction, CELTS-41.1, http://celtsc.org/80751c665875e93/folder.2006-04-03.8417036039/celts-41/celts-41-1-wd1-0.pdf
2. Yang, Y., Yong, Q.: Standardization construction of teaching media resource system based on web. Chong qing Technol. Business Univ. (Nat. Sci. Ed.) 122(1), 83–86 (2005)
3. Vendrig, J., Worring, M., Smeulders, A.W.M.: Filter image browsing: Interactive image retrieval by using database overviews. Multimedia Tool and Application 15, 83–103 (2001)
4. Gao, Y., Peng, J., Luo, H., Keim, D., Fan, J.: An Interactive Approach for Filtering Out Junk Images from Keyword-Based Goolge Search Reuslts. IEEE Trans. on Circuits and Systems for Video Technology 19(10), 1851–1865 (2009)
5. Ioan, B., Ioan, N., Ioannis, P.: Global Gabor features for rotation invariant object classification. In: Intelligent Computer Communication and Processing, ICCP (2008)
6. Frey, B.J., Dueck, D.: Clustering by passing messages between data points. Science 315(5814), 972–976 (2007)
7. Lamping, J., Rao, R.: The hyperbolic browser: A focus+content technique for visualizing large hierarchies. Journal of Visual Language and Computing 117, 33–55 (1996)

A Study on the Establishment of Interpretation Teaching Model Based on Multi-media Network*

JinBao Cai[1], Jing Xiao[2], and Ying Lin[3]

[1] Jiangxi University of Science and Technology, Ganzhou City, Jiangxi Province, P.R. China, 34100
[2] Gannan Normal University, Ganzhou City, Jiangxi Province, P.R. China, 34100
caijinbao@yahoo.com.cn

Abstract. The traditional interpretation teaching model has it own advantage that it is convenient for the teachers to play a role as a good controller of the classroom teaching activities. But it is not helpful for the students develop the interpretation skill from all-round aspects and the correctness percentage of the terminology, message completeness and accuracy in the process of interpreting is also not satisfying. This paper, based on the questionnaire survey on the traditional interpretation teaching model and empirical analysis of the teaching effect of the interpretation teaching model based on the multi-media network technology, is aimed at proving that the interpretation teaching model based on the multi-media network technology can be more helpful for the students to develop their interpretation skill than the traditional one.

Keywords: Traditional Interpretation Teaching Model, New Interpretation Teaching Model Based On Multi-media Network Technology.

1 Traditional Teaching Model of Interpretation

Most interpretation teaching activities at the universities in China are held in the language lab. The teaching contents, teaching method, textbook and teaching target are as follows:

It can be clearly seen that the teachers take what he or she likes, such as interpretation theories and their own experience, as the main content in the process of teaching activities, and take topics speech or role play as main teaching means. This kind of traditional teaching model has its own advantages that every single student has opportunity to participate in the interpreting activities and the teacher can effectively control the classroom teaching activities. But it has obvious shortcomings: first of all, the students can not actively attend the interpreting activities, and the students are only asked to do interpretation in turn; Secondly, under this teaching model, the students can be dispatched only several minutes to participate in the interpreting activities, and their attention may focus on outside-classroom affairs when they are not in turn. Thirdly, this teaching model is lack of interactive communication between the teachers and the

* This research is funded by The Education Department of Jiangxi Province. The title of the teaching reform project is "The Practice and the Comparison Research of the Translation & Interpretation Talents Training Model between Chinese and Japanese Universities. The authorization number is JXJG-10-6-44.

A. Xie & X. Huang (Eds.): Advances in Computer Science and Education, AISC 140, pp. 377–383.
springerlink.com

Table 1

NO.	TEACHING (BOOK/TARGET /CONTENTS/ METHOD/PROCESS)	CONTENTS
1	Text Book	One book published in China is chosen
2	Teaching Target	To grasp basic interpretation skills.
3	Teaching Content	Topic Training or Interpretation Skill Training
4	Teaching Method and Teaching Process	(1) The teachers read the articles in the textbook, or play CD for once or twice; (2) The students are required to note down what the teachers read or what the CD plays; (3) Then, the students are asked to interpret what they have noted down; (4) And then, the teachers will comment on what the students have interpreted

students. All in all, the most deadly defect of the traditional teaching model is that it is lackof live scene. Although the teachers in interpreting teaching have designed different scene and theme, the students can only establish corresponding situations by their imagination. Simulation training is lack of practical challenge. Its result is that the students cannot develop their reaction speed and thinking rate, the other non-linguistic skill training is also be restricted.

2 The Establishment of Interpretation Teaching Model Based on Multi-media Network

2.1 The Interpretation Teaching Concept and Model Based on the Multi-media Network

The interpretation teaching concept based on multi-media network means to establish student-centered teaching model. This model requires that the teachers should not act as the controller and dominator in the classroom anymore, but play a role as helper, promoter and director of the students. This is good for helping students develop self-study ability and provide students with a real language environment and wide range of language material so as to construct an interactive platform of communication and study between the teachers and the students.

The interpretation teaching model based on multi-media network is to depend on modern information technology, especially on the network technology. Under this model, the teaching resources of different medium can co-operate with each other and the learners' sense is activated and the learning effect is promoted. This resoles the main problem in the traditional teaching model which is lack of language learning environment. The adoption of multimedia technology and network can create for students with a comparatively real situation and provide students with fresh and rich teaching contents. Thus the students can obtain real interpretation skill after the students are trained and experience under this teaching model.

2.2 The Establishment of Interpretation Teaching Model Based on Multi-media Network

This research mainly focuses on resolving the following problems: ***firstly,*** interpretation multi-media teaching model integrated watching, listening and speaking as a whole, which provide students with a comprehensive training platform. The multi-media equipment has the function of a large storing capacity, repeatable program with changeable contents and data system management, which can provide the students with a network platform with endless studying resources. Thus the students can grasp different interpreting skill by repeatedly practicing different interpreting skill. Meanwhile, the students can record the source language material and what they themselves retell so that the students can improve the studying efficiency by after-class comparing, analyzing and summing up. Actually, the students can better understand the teaching content by muliti communicating and discussing base on the related technology of multi-media; ***Secondly***, under the environment of muliti-media, the students can freely choose different time, different place and different partners to do more practice according to their own English proficiency. They are not constrainted by the time and place of the traditional teaching model. They can also resolve the problems through feeding back to the teachers by by by the means of Email, message board and so on. The students can not only have a lot of interpretation video material and improve their comprehensive ability of interpretation, but also can enhance their intercultural communication consciousness and ability by watching the video of different cultural communicating; ***Last but not the least,*** the students can view and emulate the operation process of interpretation at the simulation environment created by multi-medial technology, which can help the students get intuitive experience of interpretation. All in all, with the interpretation teaching model based on the multi-media network technology, the teachers can not only carry out personalized teaching, the shortcomings of the traditional teaching model can also be overcome.

The specific teaching process is as follows:

(i) How to make interpretation preparation is the first skill taught by teachers. It includes long-term preparation and short-term preparation. Long-term preparation means that interpreters shall prepare every day for the potential interpreting tasks, which requires students to equip themselves with: (1) a strong sense of duty;(2)a high level of linguistic proficiency;(3) wide encyclopedic Knowledge;(4) A Good Mastery of Interpreting Skills. In contrast, short-term preparation is more direct and efficient. It refers to the job that can only be prepared shortly before the task is taken. Such preparation is highly necessary for the successful accomplishment of a certain interpreting task. Specifically, the interpretation learners should grasp interpreting skills.

(ii) Topic and topic background are handed out to interpretation trainees in advance in order to train interpretation learners with short-term preparation skill by teaching them how to prepare related documents and glossary lists through asking WH(what/who/when/where). Thus, skill training becomes the main task in class. The impact of language and background knowledge in class is weakened by pre-preparation.

(iii) In class, teaching efficiency is improved by group training based on the multi-media means. The trainees are divided into two groups: one group is acted as speakers and the other group is acted as interpreters. Under the simulation situation, the pre-designed speech is delivered by the speaking group whereas the interpreting group takes the interpretation task. At this time, teachers play a role in controlling the time and recording the students' presentation. Thus the students can experience the similar pressure of interpreting activities and improve the familiarity degree of interpreting activities. After that, by watching the recorded video, the students can correct their unsatisfying presentation including facial expressions, gesture, volume, logic of expression, speaking, eye contact, speed, message completeness, grammar, terminology, expression smoothness, intonation and pronunciation and so on. On the other hand, the students of speaking group can also benefit by thinking how to be a qualified interpreter on the position of speaker.

(iv) After class, the students can improve the interpreting skill through various means. First, the students can, based on the integrated platform of teaching and learning, file the relevant material of the topic including terminology, knowledge of the topic background and interpreting skills, which can be used by the interpreting trainees to do more studying after class. In addition, the integrated platform has been input many video of different topic presented by professional interpreters. The students can also improve the interpreting skill by watching these videos.

(v) The integrated platform of teaching and learning is also a communication platform between teachers and students. It forms a unified effective training system of in-class teaching and after-class learning. Based on the system, the students can be directed by the teachers in time and avoid ineffective practicing. In addition, the teachers can improve their teaching methods by the feedback from the students. Furthermore, based on the platform, the teachers can set up a file for the students by the presentation of the students and teach the students how to correct it.

3 The Questionnaire Survey and the Empirical Analysis of Interpretation Teaching Model Based on Multi-media Network

3.1 Questionnaire Survey of the Application of Multi-media Network in Interpretation Teaching

In order to implement the interpretation teaching model based on the multi-media network technology, the writer did a questionnaire survey to collect the data whether the students are interested in the internet and so on. 120 undergraduate English majors 2008 in Jiangxi University of Science and Technology are chosen to do questionnaire survey. The research methods such as *quantitative analysis, questionnaire survey, interview, observation and test* are adopted so as to comprehensively study the application of multi-media network in interpretation teaching. The students are mainly asked to answer the questions as follows:

Table 2

No	Questionnaire Survey
1	How many hours do you use internet every day?
2	Chinese website and English website, which is more helpful?
3	Do you ask help when you don't understand the meaning while you are surfing English website?
4	Can internet resources provide you with useful interpretation materials?
5	When is the internet more useful? Before or after class?
6	Is classroom teaching time enough for improving interpreting skill?
7	What arc the advantages of the traditional teaching method?
8	What are the advantages of multi-media network teaching method?

From Table (2), we can find that the questions are mainly set up from the perspective of the students' attitude to the teaching method based on multi-media network and assessment of teaching efficiency of internet teaching method. 120 students participated in the questionnaire survey.

Questionnaire statistical results are shown as follows. 88% students use internet at least 1-2 hours per week. 68% students believe that Chinese website is easier to find what they want but not English website. 63% students don't ask help when they meet difficulties while they are surfing English website, and they prefer to give up, which means that the teachers should recommend suitable website platform to direct the students how to use internet to look up for what they want, including related background knowledge and glossary. Only by this way can the students demonstrate their interpreting skill in the classroom for that over 50% students answer that the internet resources can provide with the most modern information and especially helpful for expanding vocabulary. In addition, 45% students believe that using internet resources before class is as important as it after class. Though about 50% agree that they don't need the teachers to guide them how to use internet resources, 90% students answer that classroom teaching time is too limited for them to improving interpreting skill. On the contrary, they mainly promote the interpreting skill by after-class self-studying because the online resources such as video and conference speech are more attractive than the tape recording played in the classroom.

However, 78% students believe that the interpretation teaching model based on multi-media network can motivate the students in the classroom and the simulation training is the most popular, 98% student don't believe that the network teaching way can completely replace the role that the traditional teaching method plays. The teachers' pronunciation, experience knowledge and personal demonstration also play an irreplaceable role. In addition, the teachers play a significant role in making attractive PPT and guiding the students how to make use of the internet resources. Moreover, the teacher can help the students improving their interpreting skill by establishing online communicating platform, online classroom and online forum.

3.2 Empirical Analysis of the Teaching Effect of the Interpretation Teaching Model Based on the Multi-media Network Technology

In order to check the teaching effect of the interpretation teaching model based on the multi-media network technology more objectively and practically, the writer has made a contrast test at Jiangxi University of Science and Technology. The test result can illustrate the effect of the teaching model. It can aslo better understand the advantages of the new interpretation teaching model. The test is as follows.

30 undergraduate English majors enrolled in 2008 at Jiangxi University of Science and Technology were selected out as the participants of the test. They have studied the interpretation courses for two semesters up to 36 learning hours. They were tested by finishing a interpretation task related to Boao Forum 2010 before and after being trained by the interpretation teaching model based the multi-media network technology. The students' presentation is as follows: (five aspects including grammar, terminology, message completeness, pronunciation & intonation and interpretation accuracy are chosen as the criteria.)

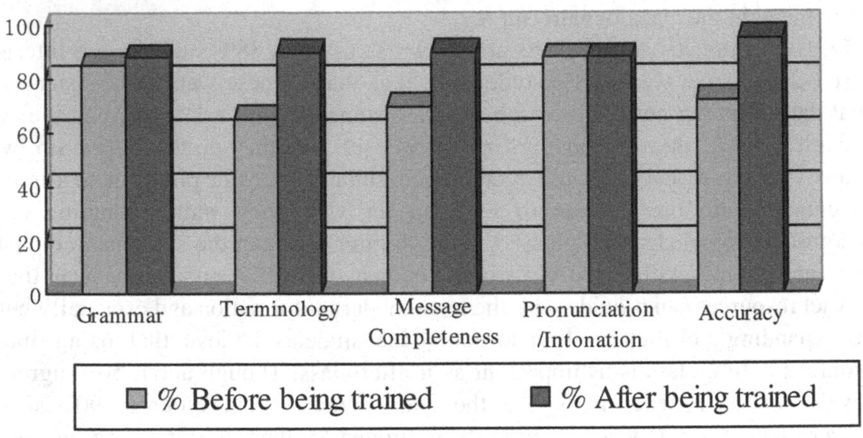

Fig. 1

NOTE: *Before being trained* means that before the students are trained by the interpretation teaching model based on the multi-media network technology;
After being trained means that after the students are trained by the interpretation teaching model based on the multi-media network technology.

From the Figure-1 above, it can be clearly seen that the correctness percentage in the field of terminology interpretation, message completeness interpretation and accuracy rises dramatically after being trained. Though the correctness percentage of the pronunciation and intonation is on the same level, the grammar mistakes decrease slightly. It demonstrates that the teaching effect of new teaching model is obvious.

4 Conclusions

To improve interpreting skill by using multi-media network technology is a complicated systematic engineering work. So the systematic support and guarantee is needed from different aspect. First of all, education technology should be supported by software and hardware and depended on teaching method. So the interpretation teaching model based on multi-media network technology need investment of hardware. Besides comparatively advanced teaching equipment, the corresponding software should be developed and good-quality multi-media PPT should be made. The PPT's features of modularization, intelligentization and networking helps the students finish the virtual interpretation task set up by the multi-media computer. Furthermore, they students can self-correct their mistakes by the computer's assessment. Secondly, the teacher should be good at the skill of multi-media network technology. Thirdly, the teacher should keep an eye on the development of the theory of interpretation teaching and continuously update the teaching theory. Meanwhile, the teachers should collect the newest rich language material which should keep pace with the students' psychological expectations, which is helpful for the interactive communication between the teachers and the students.

References

1. Nolan, J.: Interpretation: Techniques and Exercises/Multilingual Matters LTD (2005)
2. Jones, R.: Conference Interpreting Explained. St Jerome Publishing (2002)
3. Pochhacker, F.: Introducing Interpreting Studies. Routledge (2004)
4. Skrinda, A.: The challenge of language teaching: Shifts of paradigms (2004)
5. Gile, D.: Basic Concepts and Models for Interpreter and Translator Training. John Benjamins Publishing Company, Amsterdam (1995)

4 Conclusions

The [illegible] interpreting skills [illegible] multimedia network technology is a complicated systematic engineering [illegible] support and guarantee is needed [illegible]. First of all [illegible] should be supported by software and [illegible] and [illegible] the interpretation [illegible]

[illegible] should be developed and [illegible] and [illegible] should be made. The [illegible] students [illegible] interpretation [illegible] more [illegible] systematic [illegible] [illegible] teachers [illegible] and [illegible] the [illegible] should collect [illegible] the students [illegible] [illegible] and [illegible]

References

[illegible]

Retracted: Inter-domain Communication Mechanism Design and Implementation for High Performance

Tan Jie and Huang Wei

School of Information Science and Technology, Jiu Jiang University, Jiu Jiang, China
1370188549@qq.com

Abstract. Running multi-OS on a physical machine is the major method to improve the utilization of computer. With the widely use of virtualization technology in cloud computing, the efficiency of inter-domain communication becomes the key factor for performance of distributed applications, especially for some network-intensive applications. The communication synchronous mechanism used by traditional VMM is based on asynchronous signal provided by VMM and often leads to high latency, low performance. In this paper, we design and implement a communication mechanism named OSVSocket which uses inter-processor interruption(IPI) to synchronize and eliminate some useless packet check. We use shared-memory to reduce the time for data copying. Our prototype is implemented on a X86 VMM which is developed by ourselves. The experiment shows that OSVSocket has lower latency and higher performance compared with UNIX IPC.

Keywords: VMM, OSVSocket, Performance analysis.

1 Introduction

The virtualization technologies were first introduced in 1960s, and IBM 370 mainframe series stood for it's success in commerce. As the growing performance of computer hardware, the virtualization technologies become the efficient method to improve the utilization ratio of hardware resource. Running multi operating systems on a physical machine can not only rise the utilization of resource but also cut down the cost of management and get higher performance, be more reliability compared with running only one OS on a machine. The traditional distributed applications can be deployed on this platform easily and transparently. So when deploying some network-intensive applications, such as Internet servers, network file systems, the efficiency of inter-domain communication becomes a major challenge.

Although most of VMMs have provided some methods to meet the requirement of inter-domain communication, like Xen [1], which provides two ways for inter-domain communication, one is based on tcp/ip protocol, the other one is a new protocol named Xensocket[2] which is used to get high throughout, they all need VMM to intercept the communication, which leads to low performance and high latency. So the inter-domain communication protocol becomes a major bottleneck of the system when deploy some data-intensive applications on VMM. For example, the bandwidth of Xen is only 130Mbps, which is far lower than Gigabit Ethernet's

A. Xie & X. Huang (Eds.): Advances in Computer Science and Education, AISC 140, pp. 385–390.
springerlink.com

capacity. This situation is mainly due to the low performance, complicated stack of TCP/IP and the frequent page-flipping during communication. There are some papers eliminate the useless packet check to get transmit throughput of 3310 Mb/s [3].

2 Performance Analysis of TCP/IP Protocol

In order to transmit information on a long way and complicated network environment, Robert Elliot Kahn designed and developed TCP/IP protocol. The information should first be put into a socket buffer which is allocated by TCP/IP protocol and be constructed into a data package when user sends a message. And then TCP/IP will calculate the check-sum of this data package. After this, NIC will send the data package onto network using DMA. If user wants to receive message from network, TCP/IP will unpack the data package and check the check-sum to make sure that the receiving message is correct. If there are anything wrong happens during this process, the corresponding message will be sent again. A relational experiment shows that there is no data bit missed when transmit a message for 94 kilometers on a complicated network.

All the reliability of data transmission is due to the strict and complicated package checking. Unfortunately, while this technology makes a high reliability, it causes an extra high performance overhead. According to Annie P. Foong's analysis[7], the TCP/IP process procedure will take up 15% cpu time, and data copy and checking operation will even take 34% cpu time. Dave Minturn et. have analyzed the effect of TCP/IP to operating system[8]. Foong also analyzes the memory load caused by TCP/IP. In his paper, the memory ops can be divided into three types: reading data form user space into kernel space which causes 0.1% overhead, writing data to socket buffer which causes 1.1% overhead, and reading and sending data by NIC which cause 2.5% overhead.

3 Design and Implementation

A standard socket interface is provided by OSVSocket, with which user can use it easily as if using UNIX IPC. In this chapter, we will give the details of OSVSocket's design principles and implementation including the framework, buffer management, message notification and so on.

3.1 Design Principles

The purpose of TCP/IP is to ensure the correctness and reliability of data transmission on a vile network. So there are some complicated and time-consuming operations in TCP/IP protocol like we analyzed in chapter 2. We can assume that there is no error happens when we copy data in a physical machine, because of the stable environment and the reliability of memory's ECC check. So it will lead to low performance, if we use TCP/IP in inter-domain communication without any modifications. Therefore, in order to get high performance, OSVSocket must eliminate the unnecessary and time-consuming data package check and simplify the stack management.

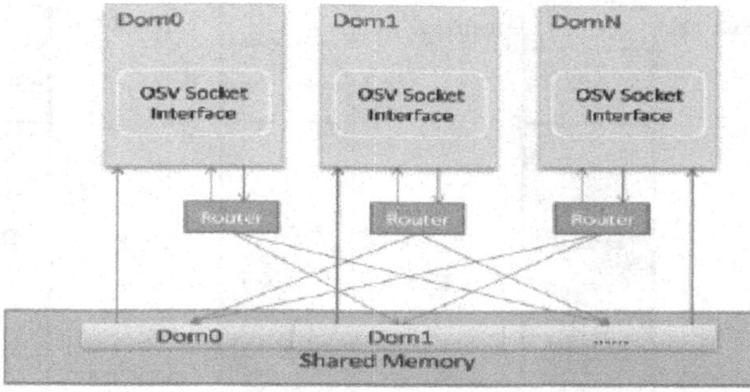

Fig. 1. The architecture of OSVSocket

3.2 System Architecture

Figure 1 shows the architecture of OSVSocket. Programmer can use OSVSocket interface to communicate with other domains. There is a router in every domain which is used to decide the message is sent to which domain. If the target domain is itself, the router will transmit the message to the corresponding socket, and if the target domain is others, the router will copy the message to the corresponding domain's shared buffer and construct the related data structure to describe the message and put it into the target domain's receiver queue. If the target domain does not exist, the router will discard the message and return an error code to the application. Every domain can only read his shared buffer and write others' shared buffer which ensures the isolation of privacy.

There is a daemon in each domain to scan the receiver queue and get the description of the message from which to get the message address in the memory. In the message description structure, there are information about the message's sender, receiver and so on, and detailed information is showed in chapter 3.6.

3.3 Message Buffer Design

The shared memory buffer is used by OSVSocket to transmit data. The shared buffer is reserved by VMM when start operating system, and OS will map the buffer's physical address to it's address space through a VMM invocation. Each domain has a buffer shared by others which can only be written by other domains and be read by itself.

As shown on the Figure 2, every shared buffer is divided into 8kb blocks, and labeled by bit-map. When a domain sends some message, it should first request a block to put the message into and mask the corresponding bit to indicate the block is in using in bit-map and then put the message descriptions into the corresponding domain's receiver queue. If it's necessary, notify the receiver domain. The receive daemon should scan it's receiver queue to get the message description with which to get the message's size, memory address and some useful information about the receiving message. It then constructs the related structure and put it into the corresponding socket's receiver list from which the application will get the message. After reading the message, OSVSocket will clean the corresponding bit of the message in bit-map to free the space.

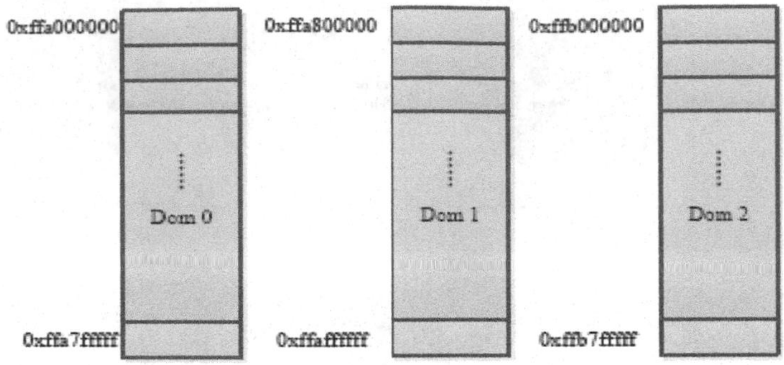

Fig. 2. Shared memory architecture

Through the above description, we can find that there are no VMM calls during the data transmission and only need just two times of data copy, one is from sender's user space to shared memory and the other is from shared memory to receiver's user space.

3.4 Address Structure

In order to reduce the difficulty of programming and maintain the programming style, OSVSocket uses the same address structure as IPV4. The network is 192.168.0.0, which is a private address. Every domain's IP address is it's domain id added by 2. For example, if the domain id is 3, the corresponding IP address is 192.168.0.5. As same as IPV4, 192.168.0.255 is used for broadcasting. The port number is used in OS to distinguish the different applications to ensure the uniqueness of message sending and receiving.

3.6 Related Data Structure

One important data structure of OSVSocket is it's sock structure named osv_sock as shown in figure 3. There is a standard sock included in osv_sock. Additionally, the target address, sender address and a message list are all members of osv_sock. The message list is constructed by osv_skb structure. In order to ensure sync, a spinlock is used by osv_sock. The message type, length, offset, address and some useful information are all putted into osv_skb.

The structure osv_recv_msg is used to describe the message in which the sender's address is recorded in. In addition to this, the target's port also can be found in osv_recv_msg. The member type is used to distinguish the different message type, that is to say, the message is a data message or is a connection requisition and so on. The msg_data is the address of a message, from which the OS can read message data directly.

3.7 API and Data Transmission

A standard Berkeley socket interface is provided by OSVSocket and programmer can use it like using UNIX IPC or TCP/IP interface. The difference is that, the protocol number is AF_OSV instead of AF_INET or AF_UNIX when create a socket. After that, when using bind(), listen() and accept() and other operations to bind a port, listen and accept a connection requisition, OSVSocket is the same as other protocol. Figure 3 illustrates the use of OSVSocket. After the connection is established, each other can invoke read or write to get or send message to others.

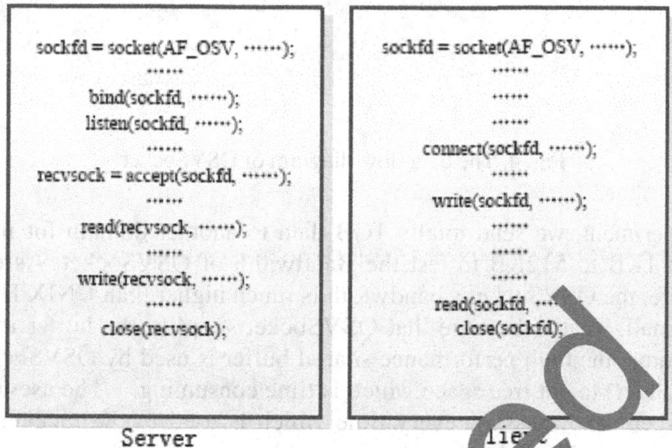

Fig. 3. The use of OSVSocket

A simple data flow diagram is shown in figure 4. When user invokes send()/write() system call to trap into kernel, the sender should first scan the bit-map of the target domain to get a free block. And then, using the copy_from_iovc() function to copy data to the target domain's shared buffer directly. If the target domain is not processing message, the sender also need to send an IPI to the target to notify the message arriving explicitly. The receiver daemon will traversal the message list to process the message received, and put the messages to corresponding socket's skb list. When user invokes receive()/read() system call to trap into kernel, the OS will use get_msg_from_skb() to get message from skb list, and then using copy_to_iovc() function to copy data to user space directly. After that, the OS will clear the corresponding bit in bit-map to free the block.

4 Experiments and Conclusion

We implement the prototype of OSVSocket on a X86 VMM named Trochilus which is developed by our own. Trochilus can run multiple operating system instances currently with no performance overhead. The Trochilus kernel allocates the resources to an operating system and protects from unauthorized accesses from others. While with the Trochilus micro kernel the many core based servers can be flexible and extensible as traditional VMM, also has good performance. This is quite suitable for the Cloud Computing. In this paper, we use a DELL T605 server which has two AMD Opteron processors and 16GB DRAM to test OSVSocket's performance. OSVSocket has higher performance compared with UNIX IPC and achieves the expected goals. Figure 4 shows the result of OSVSocket, UNIX IPC and memory copy performance comparison. We have not compared OSVSocket with Xensocket, because that Xensocket has lower performance than UNIX IPC.

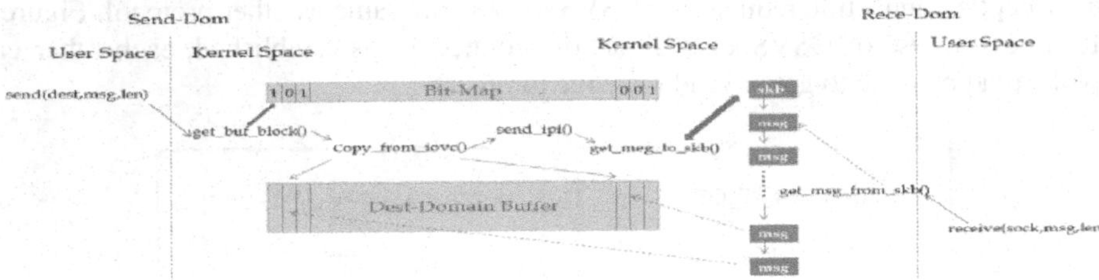

Fig. 4. The data flow diagram of OSVSocket

In this experiment, we send totally 1GB data to another domain for message size ranging from 1KB to 512KB to test the Bandwidth of OSVSocket. As can be seen from the figure, the OSVSocket's bandwidth is much higher than UNIX IPC when the message is small. That is because that OSVSocket simplify the buffer management. At the same time, the high performance shared buffer is used by OSVSocket to avoid invoking kmalloc() to get free space which is time consuming. The use of IPI avoids VMM call when send message every time which is used by XenSocket and makes OSVSocket more simple and efficient.

References

1. Barham, P., Dragovic, B., Fraser, K., Hand, S., Harris, T., Ho, A., Neugebauer, R., Pratt, I., Warfield, A.: Xen and the art of virtualization. In: SOSP 2003: Proceedings of the nineteenth ACM Symposium on Operating Systems Principles, pp. 164–177. ACM Press, New York (2003)
2. Zhang, X.L., Mointosh, S., Rohatgi, P., Griflin, J.L.: XenSocket: A high-throughput interdomain transport for virtual machines. In: Proceedings of Middleware 2007 (November 2007)
3. Menon, A., Cox, A.L., Zwaenepoel, W.: Optimizing net-work virtualization in Xen. In: 2006 USENIX Annual Technical Conference, Boston, Massachusetts, USA, June 2006, pp. 15–28 (2006)
4. Burtsev, A., Srinivasan, K., Radhakrishnan, P., Bairava-sundaram, L.N., Voruganti, K., Goodson, G.R.: Fido: Fast inter-virtual-machine communication for enterprise appliances. In: USENIX Annual Technical Conference, San Diego, CA (2009)
5. Wang, J., Wright, K., Gopalan, K.: XenLoop: A Transparent High Performance Inter-VM Network Loopback. In: Proceedings of the 17th International Symposium on High Performance Distributed Computing, Boston, MA, pp. 109–118 (2008)
6. Kim, K., Kim, C., Jung, S.-I., Shin, H., Kim, J.-S.: Inter-domain Socket Communications Supporting High Performance and Full Binary Compatibility on Xen. In: Proc. of Virtual Execution Environments (2008)
7. Foong, A., Huff, T., Hum, H., Patwardhan, J., Regnier, G.: TCP Performance Analysis Revisited. In: IEEE International Symposium on Performance Analysis of Software and Systems (March 2003)
8. Minturn, D., Regnier, G., Krueger, J., Iyer, R., Makineni, S.: Addressing TCP/IP Processing Challenges Using the IA and IXP Processors. Intel. Technology Journal (November 2003)

The Study on College Library Electronic Resources Integration Based on Internet

Min Ren

China College of Mining and Technology (Beijing), Beijing, 100083

Abstract. The purpose of integration is to improve the resource utilization, saving the users' time. To achieve the integration of electronic resources, we need to use a variety of retrieval techniques and methods of integrating these distributed, heterogeneous electronic resources into a unified environment, to form a new organic whole, and connecting those electronic resources seamlessly. The integrated electronic resources system should be able to provide users with a unified search platform where all processes are transparent to the user. The initial unified retrieval of integration of electronic resources focus on the multi-directional links among information, so as to solve the problem of information silos. With the construction of electronic library in-depth, service-oriented will become its main shape. The integration of electronic resources is not just integrating these distributed, heterogeneous resources together simply; what's more, it is the integration of services and resources together.

Keywords: Internet, college library, electronic resource integration.

1 Introduction

With the social development, scientific progress, especially popularization of information technology and computer networks, the library has undergone enormous changes. In the knowledge-based economy and information age with a rapid development, the desire for knowledge and information had changed accompanying by the development of times. As forms of knowledge and information become diversified and the content becomes integrated, people require a fast, timely and accurate access to the necessary information and knowledge. The dominant position of Library services in the traditional literatures faces with severe challenges. In this changing environment, the library has deeply affected, and electronic libraries came into being in order to adapt to this change. Electronic Library is built on the basis of information technology, using electronic technology and computer network to access, store, distribute information.

2 The Need for Integration of College Library Electronic Resources

In the network and electronic age, electronic resources occupy an increasingly proportion in the collection resources in library. Electronic resources are characterized

A. Xie & X. Huang (Eds.): Advances in Computer Science and Education, AISC 140, pp. 391–396.
springerlink.com

by easy of sharing and convenient retrieval. However, with the growing number and types of electronic resources, coupled with the differences between database application systems, the content of overlapping, knowledge with low association, diversity of data formats and storage methods, as well as the differences caused by retrieval language and methods, the effect of the using electronic resources is not good. One of the reasons is that when the user is using different database and network resources, he/she needs to enter different system and database with repetitive or even the same steps which cause a waste of time, over-consumption of equipment resources and the reduce of service efficiency.

Therefore, how to abandon the heterogeneity between the different databases to build a unified search platform by realizing organic integration of electronic resources to make one-step inquiry come true has become an important task of the construction of colleges and universities' electronic libraries as well as an important research topics and pressing issues in the area of information resources management. At present, electronic resources have become the most favorite resources of users. As a base for teaching and research services, the college library plays an irreplaceable role in feeding readers' needs. Making full use of it can serve the students and teachers better as well as saving money.

3 The Principle of Integration of Library Network Information Resources

Under the condition of the digitizing, due to the convenience of Internet delivery, the principle of being characteristics is even more important in the integration of information resources. In addition to following the general principles of the construction of information resources, such as the principle of practicality, the user supreme principle, the principle of systematic, novelty principles, some principles that can be suitable to the electronic networked environment should be adopted. Those principles are:

An overall principle means to maintain the integrity of academic information resources object. The integrated information resources system should cover the internal functions of the various subsystems, reflecting the intrinsic relationship between data objects.

Standardization principles include the standardization of data formats, description language standardization, the standardization of indexing languages, as well as the standardization of communication protocols, security technology standardization, and the standardization of database management software and hardware to ensure that information resources can be shared.

The principle is to make the optimization of information resources by using certain techniques and methods and to obtain the best organizational structure and weaving capabilities.

Security policy is to take the necessary security measures to ensure the security of information resources. Computer viruses, hackers, software bombs, the information garbage, storage devices and other problems have brought about tremendous security threat to the integration of information resources. We must pay attention to information resources security in the integration of information resources.

The principle of continuity refers to the development and continuing character of the integration of electronic resources. Only through continuous, systematic, and dynamic integration process, the electronic resources can play a sustained effect and show its vitality.

This principle refers to that in the integration process of electronic library network information resources, in accordance with the inherent sources of information resources for the user's information needs, information resources should be reasonably integrated to form a hierarchical, networked, three-dimensional nature of information resources system in order to complement each other and achieve the goal of sharing together.

4 The Content of Integration of Library Network Information Resources

From the development and utilization of network information resources, it can be seen that the network information resources can be divided into sub-zero documents (information), first information, second information, third information level and other levels which models the division of literature on the traditional documentation systems. And restructure and integration can be done in accordance with these levels. Zero Literature, is also known as gray literature. It refers to those literatures that are neither the public offering nor confidential documents among the white public offering literature and the non-public offering confidential documents. It mainly includes the government's administrative reports and science and technology reports, dissertations and internal publications of universities, scientific research institute, the manuscripts of experts and scholars, academic conference materials, social investigation reports and so on. These documents are of high quality and professional, as well as informative and covering a wide range, but it is also characterized by narrow reader ranges and low transparency. Zero literature can reflect the level of domestic and foreign research and technological developments and trends. It is the important information resources in China's reform and opening up and economic and cultural development and has become an integral part in literature information resources. Therefore, for this part of the literature, they should be classified and processed deeply, to establish the zero Literature Database, to realize remote query through computer networks, making it an important information resource available online.

The so-called first information refers to the electronic information that is being organized into hietmet through processing which derived from the rich information resources outside network. The first information has a wide, rich-contented and varied-typed resource. Because of this, first information is characterized by being of messy content, which requires us to select the information through getting rid of redundant useless information, choosing and organizing those of high practical value, high-quality, authoritative and reliable information after processing into the Internet. For the first information, we can organize them in the way of free text mode, super text mode and home web mode. Free text mode is mainly used for the construction of full text databases, the full-text information outside network that people created or collected is the one that be organized, which are the new resources to input network.

The way of Hypertext is to organize the information of related text on the network organically so that users can start from any one node to view and query information from a different angle, which is the most used organization way of the first information adopt. The home page is similar to the file Fonds Organization Act, which is used for the agency or personal information organization.

The second information is the product obtained after the processing, refining and enrichment of the first information. It is the accumulated instrumental literature after editing and publishing for facilitating management and use of the first literature. As the Internet does not have the function to transform a multitude of information resources pouring into network with a rapid growth into the specific information automatically as users required ,which requires us to apply the relevant knowledge , theory, principles and methods of information organizations to build Online information retrieval tool-- "information resources chain" of the organization, to control the free text, hypertext, and home web to enter the first information online. Currently, the organization of the second information on line is mainly the search engine. Search engines are a class of sites on the Internet. Its main task is to search for pit Fan b server information automatically, classifying the information and setting the index, and then store the contents of indexes to the database.

The current direction of search engines is a full-text search (FullTxetSearhc) engine, it can conduct a comprehensive site search for the page text, providing a new, powerful search function. You can retrieve directly according to the content of literature. And this engine supports comprehensive use of information resources in multi-angle and multi-lateral. Full-text search technology is the main technology base for the applications such as discovery of information, analysis and filtering information, the information agent and information security control, and so on. Search engines with Full-text search as the core technology has become the mainstream technology of the Internet age. Comprehensive, accurate and fast is the key indicators for measuring full-text search system. For restructuring and integration of the second information, first of all, we must select the desired targeted literature from the Internet, and to do a brief introduction and evaluation of its particular character and academic value after scientific analysis. Then according to the uniform format, set links under the names of literature for relevant articles or knowledge. Secondly, we must develop software which can automatically converted, automatic search, automatic bibliographic, automatic extraction of keywords of the web documents so that it can automatically identify and organize to pass it in the form of convenient web to the user.

Third, we should set up database mirroring, selecting the source database on a regular basis at home and abroad and making a mirror copy on the local server for users to make use of the latest external document resources repeatedly and timely. In addition, we should open up sections for two-way communication link between users and library website, allowing users to leave comments for us taking the initiative to analyze and track their needs and to provide the retrieved documents for the users by e-mails timely.

Different search engines, providing the different search approaches and implementations, in order to help users quickly and accurately find a suitable search engine in a short period of time" we can use the "Documents chain" alternative to second information in the web to form the third information to further improve the retrieval efficiency and network information resources development and utilization level.

5 Electronic Resource Integration Models

The integration of electronic resources is a systematic project which takes into account multiple factors; each of a college library should find a electronic resource integration approach suitable to its own features to improve the utilization of electronic resources effectively. Integration of College Library electronic resources can be divided into the following pattern:

This is the simplest model of the integration of a collection of electronic resources. This integration model provides dynamical link of variety of resources by using network hypertext link, allowing users to get additional information expanding services.

This integration regards bibliographic data collection as its core, which is one of the most basic ways of the integration of a collection of electronic resources. It can be divided into two modes--internal museum and external museum integrations, according to the source of resources. Internal museum integration integrate real and virtual collections together primarily through recording electronic documents URL in the MARC856, revealing and linking to full-text electronic documents in real collections. External integration, primarily through implementation of the Z39.50 protocol, make the aggregate of the OP AC databases on different platforms, integrating and generating national or regional joint Bibliography query system, so that users can retrieve the relevant library's OP AC through a unified interface.

This refers to integrating the search portal of the collection of electronic resources together to build resource navigation database, providing resource that can be access to according to its name alphabetically, subject category, resources, and so on. This integration model is not high-tech, easy to implement, and it is a currently widely used mean of college library. Electronic Resources navigation integrated into the overall navigation and sectional navigation. Overall navigation regards a variety of collections of electronic resources as a navigation object, and providing a unified search interface or links.

6 Conclusions

In today's era of electronic libraries, many libraries have increased its purchase of electronic information resources and inputs. It is not difficult for a library to have information resources and a fairly complete library automation system. The question is how to combine the different database systems organically to provide readers with a comprehensive and efficient information services. Information Resources Integration is a way of library developing from information service orientation to knowledge orientation. Integration of information resources can solve problems caused by geographical space, and readers can use the Internet in the library's search platform to retrieve their required information staying at home.

References

1. Wang, L.: Research on Integration Mode of Electronic Resource of College Library. Journal of Modern Information (02) (2009)
2. Guo, J.: Analysis of Electronic Resources Integration in College Libraries. Contemporary Library (03) (2008)
3. Xia, M., Qiang, Q.: Survey & Proposal on Academic Library Resource Integration in China. Journal of Academic Libraries (0 1) (2008)
4. Gong, Y., Song, C., Zhang, L.: Research on the Integration of Electronic Resources in College Library. The Journal of the Library Science in Jiangxi (02) (2007)
5. Yang, B.-W.: On Combination of Electronic Resources of Library Books of Colleges. Journal of Guizhou College for Ethnic Minorities(Philosophy and Social Sciences) (03) (2007)

A Multimedia Based Natural Forest Park for Ecological Education

Zhen Liu and YanJie Chai

Faculty of Information Science and Technology, Ningbo University, Ningbo 315211
liuzhen@nbu.edu.cn

Abstract. Applications of multimedia in ecological education for children are becoming interesting subjects. A virtual natural ecological park demo is set up with Web3D technology; the virtual park is based on real countryside scene near NingBo. In the virtual world, there are trees, water and animals. The design method of interactive animation is proposed. The program techniques of interactive animation are introduced, the design of sensor node and Lod node are essential to a large virtual scene. A user can explore the virtual park from different view and learn ecological knowledge in interactive manner.

1 Introduction

Ecological education for children is very important to our society, children should understand environment and interact with it. In china, the children in cities seldom wander in a real natural landscape; the parents are usually worried about possible dangers for children in a natural environment. In recent years, multimedia is becoming to be a new technical manner for natural exploration of human, it can make up the insufficiency of realistic eco-tourism in some degree, ecology resources can be visualized in multimedia, the ecological knowledge are vivid for children. As an attempt, we presents the method of setting up virtual forest park in web3d based multimedia, the virtual landscape is based on some of real country landscape near NingBo of China.

We select VRML as the tool for the programming [1]-[6], VRML integrates animation and sound together, it can create 3D multimedia software and it integrates animation, spatial sound, collision detection and scripting. The landscape is set up in 3DS Max, and can be converted into VRML file, we use bitmanagement company's VRML viewer, a user can explore a virtual landscape.

2 Design of Virtual Scene

We use 3DS Max to draw the 3D scene of the landscape geometry and exported to the models with a VRML file, some of interactive function are created by VRML helper objects in 3DS Max(see Fig.1). For example, in VRML language, if we open a door, we should use TouchSensor node is as the following:

A. Xie & X. Huang (Eds.): Advances in Computer Science and Education, AISC 140, pp. 397–402.
springerlink.com

```
TouchSensor{
    EnabledTRUE                 # exposedField SFBool
    isActive                    # eventOut SFBool
    isOver                      # eventOut SFBool
    touchTime                   # eventOut SFTime
    hitPoint_changed            # eventOut SFVec3f
    hitNormal_changed           # eventOut SFVec3f
    hitTexcoord_changed         # eventOut SFVec3f
    }
```

We can design a TouchSensor node in VRML helper button, and the above code is created automatically by 3DS Max. For example, if we can add a TouchSensor to open a door, the main code is as the following:

```
DEF Rectangle03 Transform { # Rectangle03 is the geometry model of the door

......
  DEF Rectangle03-TIMER TimeSensor { loop FALSE cycleInterval 1.667 },
  DEF Rectangle03-ROT-INTERP OrientationInterpolator {
    key [...... ]
     keyValue [...... ] },
    Shape {
      appearance Appearance {
        material Material {
          diffuseColor 0.6941 0.3451 0.102
        }
      }
      geometry DEF Rectangle03-FACES IndexedFaceSet {
        ccw TRUE
        solid TRUE
        coord DEF Rectangle03-COORD Coordinate { point [
          ......]
        }
        coordIndex [            ......]
        }
    }
  DEF TouchSensor01-SENSOR TouchSensor { enabled TRUE }
   ]
ROUTE Rectangle03-TIMER.fraction_changed TO Rectangle03-ROT-INTERP.set_fraction
ROUTE Rectangle03-ROT-INTERP.value_changed TO Rectangle03.set_rotation
  }
ROUTE TouchSensor01-SENSOR.touchTime TO Rectangle03-TIMER.startTime.
```

In a landscape, the irregular objects are realized by adding noise effect on a regular object in 3DS Max. Finally, an appropriate texture is used for representing the superficial vision effect. For example, the model of a stone is illustrated in Fig.2. A Tree is created in Billboard node, and a virtual animal is set up in CAT tool in Fig.3.

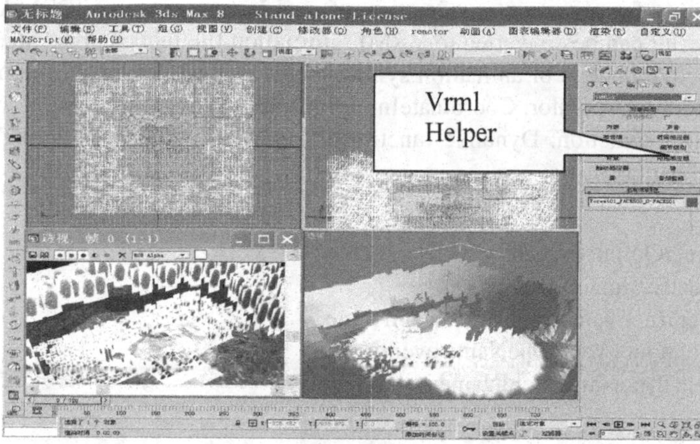

Fig. 1. Landscape of a virtual park in 3DS Max

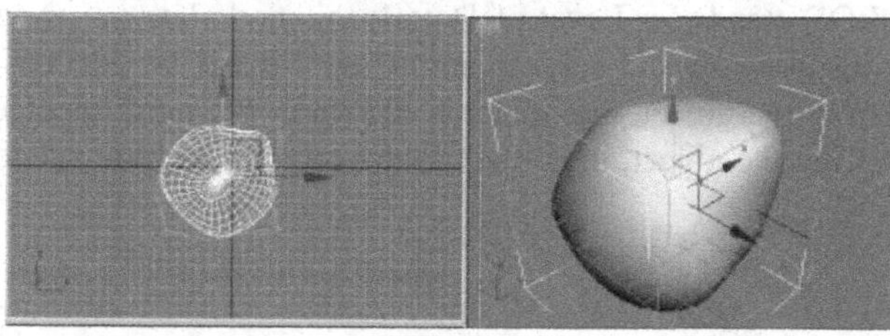

Fig. 2. Model of a stone by shape noise

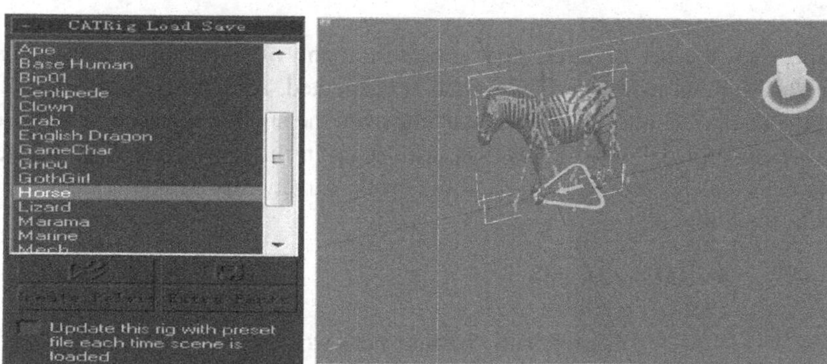

Fig. 3. Design of a virtual zebra in 3DS Max

3 Programming of Interactive Function

Dynamic effects of a virtual forest include the following several aspects:

(1) Establishing different viewpoints: the roam effect from different viewpoints is realized by viewpoint syntax in 3DS MAX.

(2) Animation of a object: the animation of a object is set up in 3DSMAX beforehand, and result can be converted to VRML animation file that can be edited by a text editor, some parameters of animation syntax (such as TimeSensor, PositionInterpolator, OrientationInterpolator, CoordinateInterpolatoretc.) can be repaired.
(3) Dynamic interaction: Dynamic function is realized by Script node in VRML, A Script is defined as flows:

```
Script {url[]
        directOutput FALSE
        mustEvaluate  FALSE
        eventIn  eventTypeName eventName
        eventOut eventTypeName eventName
        fieldTypeName fieldName initialValue
}.
```

The url specifies the program script to be executed. The program languages currently supported are JAVA, JavaScript, and VRML Script.
(4) Lod node: LOD stands for Level of Detail; this node lets a user specify alternative representations of an object by distance ranges. We use two level of Lod. If a user is not far away from the object, the object is drawn in the most detailed version. If a user is far away from the object, the object is drawn with a crude version. Lod Syntax is as flows:

```
LOD{
 center     0 0 0    #field SFVec3f
 level       []      #exposedField MFNode
 range       []      #field MFFloat
 }
```

level specifies the set of alternative representations of a graphical object, *center* specifies a 3D point to which the distance is computed. *range* specifies a set of distances, the distances must be ordered starting from the smaller distance, otherwise the results are undefined. Two VRML files are set up in this paper (see Fig. 4), the detailed is named as babmo-near.wrl , and the crude is named as babmo-far.wrl, the Lod code is as the following:

```
#VRML V2.0 utf8
LOD       {level [
          Inline {url "babmo-near.wrl"}
          Inline {url "babmo-far.wrl"}
                    ]
  range [80] }
```

If the distance from tourist to object is less than 80, the scenario will be rendered with detailed version of object.

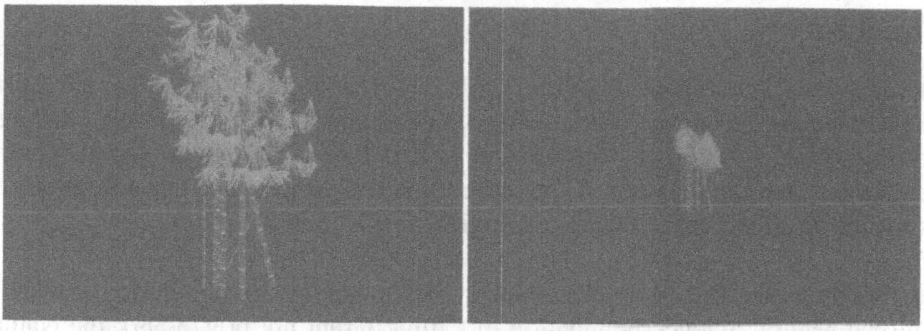

Fig. 4. The illustration of a virtual bamboo in Lod mathod

Fig. 5. The avatar watched a zebra in the park

Fig. 6. The avatar flew and watched the park

4 Conclusion

An internet virtual forest park is set up in this paper, and the demo system is realized by VRML language, a virtual tourist can roam the virtual forest park from different location viewpoint in internet explorer, the virtual forest park integrates all kind of multimedia information together. The method of this paper can offer a kind of new means for the development of natural ecological resource in the future.

Acknowledgements. The work described in this paper was co-supported in part by the National Natural Science Foundation of China (Grant no: 60973099), the Natural Science Foundation of Zhejiang Province (grant no: Y1091158), Zhejiang Public Technology Application Project(2011C23027).

References

1. Havaldar, P., Medioni, G.: Multimedia System:Algorithms, Standards, And Industry practices. Course Technology Press (2010)
2. William, R.S., Alan, B.C.: Understanding Virtual Reality, Interface, Application, and Design. Elsevier Press, Amsterdam (2003)
3. Geroimenko, V., Chaomei, C. (eds.): Visualizing Information Using SVG and X3D. Springer, London (2005)
4. Marrin, C., Campbell, B.: Teach Yourself VRML 2 in 21 Days, Sams.net (1998)
5. Patmore, C.: The Complete Animation Course. Quarto Publishing PLC (2003)
6. Kelly, L.M.: 3ds max 6 Bible. Wiley Publishing, Inc., Hoboken (2004)

Web-Based Interactive Animation for Children's Safety Education: From 2D to 3D

Zhen Liu and YanJie Chai

Faculty of Information Science and Technology, Ningbo University, Ningbo 315211
liuzhen@nbu.edu.cn

Abstract. Web Based animation include 2D animation (such as Flash animation) and 3D animation (such as WEB3D based animation), animation is a good tool for safety education of children; it can attract children to learn abstract safety regulations in an interactive manner. According to the psychological characteristic of children and constructivism learning theory, a method of developing web-based animation on safety education of children is presented. The design of cartoon's scenario, character's dangerous behavior and interaction are introduced. 2D animations are exaggerative and suitable for the young children, 3D animation are realistic and suitable for exploration in different views.

1 Introduction

Web based animations are popular on Internet[1]-[8], development of web based animation for children's safety education is a meaningful task. Watching case movies or learning safety regulations is traditional education manner for children. These methods are usually not effectively, they are not vivid and interactive for children. Based the investigation on accidents, there are some dangerous area in a child's life space. Based on topology psychology, behaviour is related to environment, dangerous behaviour occurs in a given environment[9]-[12]. If we design a traffic safety education animation, we can use interactive animation to visualize the dangerous behavior, if a use r select the behavior, its avatar will be injured. In the manner, a child user can understand what is the right behavior in a certain environment. We also design a storytelling for the animation, there are virtual characters and moving objects in a animation scene. Web based animation is a visualization tool for illustration safe walking in a traffic scene, children often like visual and interesting things by their psychological characteristic. Based on constructivism, a child safety knowledge is from the cognitive constructions for the world, animation is useful for the visual constructions among environment.

Web based animation can play a very important role in safety education of children. 2D animation has the expression force of exaggeration and can attract children to learn knowledge, and 3D animation has realistic images and interactions. Safety regulations are shown in a vivid web based animation, a child user can act as a character in cartoon animation and learn right action for dangerous situation.

A. Xie & X. Huang (Eds.): Advances in Computer Science and Education, AISC 140, pp. 403–407.
springerlink.com

Existing children's safety education methods need improving. Children's safety education in the 21st century should fully utilize web animation technology. In recent years, we develop some 2D and 3D web based animation for safety education, preliminary result shows the new method can enhance the learning interests of safety education effectively.

2 2D Flash Interactive Animations

We can use Flash to design 2D interactive animations for children's education, when a user selects more choice parameters by buttons, we can use gotoAndPlay function to realize animation's jumping on the axle of parallel time. For example, we can add action sentence on a button:

```
on (release) {
gotoAndPlay(200);
}// Film switches over to the 200th frame when the button is released
```

After finishing in a scenario, we can use loadMovieNum function to realize switchover of scenarios, For instance:

```
if( _ root == Number( _ root)) {
loadMovieNum("next_scenario.swf", _ root);
}
else {
_ root.loadMovie("next_scenario.swf");
}
```

We tested some of 2D animations for children(see Fig.1-2), we find that exaggeration effects in 2D animations can enhance children's interests greatly, a 2D animation scene is not like a realistic image, but children express more passion to 2D animations. In fact, children usually explore Internet, they are familiar with Flash animation. We can use exaggerated warning picture to visualize a safety regulation.

Fig. 1. 2D safety education animation of an electric fan

Fig. 2. Children like 2D animation very much

3 Web3d Interactive Animation

We select the traffic education as the example; traffic safety education for children is a very urgent task in china, teaching children to learn correct walking methods in traffic environments is very important. Based on the psychological characteristic of children and constructivism learning theory, 3D cartoon animation is a good tool for safety education of children, the traffic animations can attract children to learn abstract safety regulations in a vivid manner. We use Virtools to develop Web 3D animation, a user can act a avatar in the virtual environment.(see Fig.3-5).

We tested some of 3D animations for children, these 3D animations have no exaggeration effects, and a user controls the avatar just like playing computer game. Because boys often play 3D game, they like explore the virtual word bravely and make dangerous behavior, some boy control the avatar waking to a moving car. Girls like to control the avatar to walk in a right manner. 3D animations are more interactive than 2D animation; a user can learn more safety knowledge from a 3D animation.

Fig. 3. 3D animation for traffic safety education

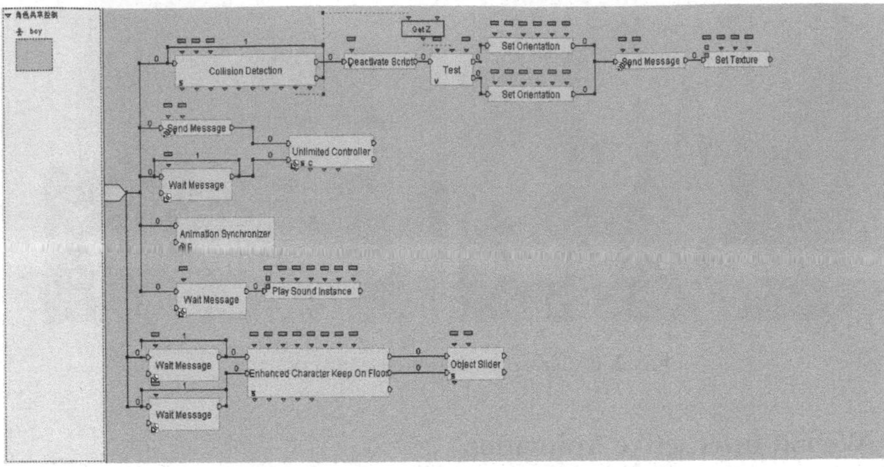

Fig. 4. The avatar's behavior scripts in Virtools

Fig. 5. We tested these cartoon animation in some school in Ningbo

4 Conclusion

Web-based animation will play an important role in the safety education of children; many abstract safety rulers can be visualized by animation. In this paper, a method of visualization of children's dangerous behavior is presented. The design of web based animation for children's safety education is proposed, 3D animations are more realistic than 2D animation, but 2D animations can have more exaggeration effects. Our demo system is developed in FLASH and Virtools; we have tested all these animation in some school in Ningbo, children very liked these animations.

Acknowledgements. The work described in this paper was co-supported in part by Zhejiang Public Technology Application Project (2011C23027) and 2010 NingBo Science Popular Project.

References

1. Foley, J.D., Dam, A., Feiner, S.K., Hughes, J.F.: Computer graphics: principles and practices. Addison Wesley, Reading (1995)
2. Hearn, D., Baker, M.P.: Computer graphics with OpenGl, 3rd edn. Prentice-Hall, Englewood Cliffs (2004)
3. Watt, A., Watt, M.: Advanced Animation and Rendering Techniques. Addison Wesley, Reading (1992)
4. Maestri, G.: Digital Character Animation 2: Essential Techniques, vol. I. New Riders Publishing (1999)
5. Maestri, G.: Digital Character Animation 2: Advanced Techniques, vol. II. New Riders Publishing (2001)
6. Fernandez, I.: Macromedia Flash Animation and Cartooning: A Creative Guide. McGraw-Hill Companies (2002)
7. Rhodes G.: Macromedia Flash MX 2004 Game Development, Charles River Media Released (2004)
8. Perlin, K., Goldberg, A.: Improv: a system for scripting interactive actors in virtual worlds. In: Proc. SIGGRAPH 1996 Conf., New Orleans, Louisiana, USA, pp. 205–216 (1996)
9. Lewin, K.: Principles of topological psychology. McGraw-Hill (1936)
10. Gibson, J.J.: The ecological approach to visual perception. Lawrence Erlbaum Associates, Inc., Hillsdale (1986)
11. Knapp, M.L., Hall, J.A.: Norverbal communication in human interaction. Wadsworth Publishing, Belmont (2002)
12. Zeedyk, M.: Behavioural Observations of AdultChild Pairs at Pedenstrian Crossings. Accident Analysis and Prevention 35, 771–776 (2003)

The Evaluation on Ideological and Political Teachers' Competency

Yongfang Liu

School of Humanities, Economics and Law, Northwestern Polytechnical University, Xi'an 710129, China
liuyongfang125@163.com

Abstract. This study on a questionnaire survey in Hubei province, the first five using ability scale development the university thought politics theory class teaching effectiveness by our company SEEQ scale. Six of the result being selected, ability factors prediction variables and regression analysis teaching effectiveness scale has been SPSS17.0 total score. The results show that six dimensions have significant prediction function ability, teaching effectiveness scales. At the same time, the ability to score is selected for the six factors prediction variables and regression analysis method has taken all scale.

Keywords: ideological and political teachers, colleges, competency.

1 Introduction

The teaching effect of ideological and political course related the cultivation of university students, especially to improve the ideological and ethical standards. The teacher's teaching effectiveness evaluation was also called the teaching quality evaluation and teachers' teaching level evaluation. Foreign theory and demonstration research, on the longer and more mature. It started from the 1970 s and 1990 s fast growth. In the United States, the development of teachers' evaluation is put in the first place, regarded as the main principle and model. This SEEQ has been published in the British journal of anesthesiology 1982 years' education psychology start. In the public domain, SEEQ questionnaire were tested, extensively. There are more than 50000 of the course and 1000, graduated in 000, college students use questionnaire [1].

2 Research Methods and Design

Relationship between competency and teaching effectiveness of Ideological and Political Theory teachers.

Random distribution of 6,000 copies of 'Competency Scale' and 'Teaching Effectiveness Scale' has been done in Hebei University of Technology, Yanshan University, Hebei Normal University, Hebei Agricultural University, and Hebei University. 5317 copies were recovered and 5310 questionnaires were available (see Table 1). Competency and teaching effectiveness of 76 teachers have been tested (see Table 2).

Regression analysis and path analysis have been made by SPSS17.0 and structural equation model.

A. Xie & X. Huang (Eds.): Advances in Computer Science and Education, AISC 140, pp. 409–414.
springerlink.com

Table 1. Statistics of Surveyed Students

University Name	Total Number	Male	Female	Freshman	Sophomore	Junior	Graduate	Liberalarts	Engineering	Economic Management	Arts
Hebei Agricultural University	1118	464	654	408	710	0	0	183	542	310	76
Hebei University	1586	631	955	733	709	142	50	85	995	506	0
Hebei University of Technology	1423	774	656	585	321	320	0	437	703	157	51
Yanshan University	638	406	232	316	322	0	50	395	188	0	55
Hebei Normal University	525	128	410	120	43	28	253	384	133	28	0
Summary	5310	2403	2907	2162	2105	690	353	1504	2623	1001	182

Table 2. Statistics of Evaluated Teachers

University Name	Total No.	Male	Female	Prof.	Assoc.	Lecturer	Assistant	PhD	Master	Underg-graduate	Morality and legal basis	China History	Marxism Philosophy	Socialism Theory
Hebei Agricultural University	14	2	12	4	8	6	4	1	10	3	6	0	4	10
Hebei University	19	9	10	5	6	13	0	2	10	8	3	1	3	11
Hebei University of Technology	19	7	7	3	3	10	0	2	26	4	6	5	7	11
Yanshan University	12	5	6	0	2	3	0	2	3	1	1	2	3	4
Hebei Normal University	12	6	12	0	4	4	1	0	3	1	2	0	2	5
Summary	76	29	47	12	23	36	5	7	52	17	18	8	19	41

3 Data Analysis and Results

Table 3 shows that the average score of teaching effectiveness is 4.27 points. It indicates that teaching effectiveness has achieved a higher level in Hebei Province. Dimensions of the highest scores are 'teacher-student harmony' and 'course difficulty and homework quantity'. They were scored 4.3 points or more. The dimension of lower score is 'teaching course', scoring 4.1 points.

Table 4 shows that the average score of competency is 4.29 points, indicating that the competency level of Ideological and Political Theory teachers is higher in our province. Dimensions which get relatively higher scores are the first dimension (professional knowledge), the third dimension (professional ethics) and the sixth dimension (teacher-student harmony). The lower score dimensions are the second dimension (professional skills) and the fifth dimension (personal characteristics).

Table 3. Descriptive Statistics of Teachers' Teaching Effectiveness

Variable	Sample Size	Minimum	Max	Mean	Standard Deviation
ejg	5310	1.00	5.00	4.2593	.76842
etd	5310	1.00	5.00	4.2049	.74134
ehz	5310	1.00	5.00	4.2923	.76360
ehx	5310	1.00	5.00	4.3477	.80656
exl	5310	1.00	5.00	4.2641	.78387
egc	5310	1.00	5.00	4.1755	.67045

Table 4. Descriptive Statistics of Teachers' Competency Level

	Sample Size	Minimum	Max	Mean	Standard Deviation
czs	5310	1.00	5.00	4.3622	.65059
cjn	5310	1.00	5.00	4.2058	.68981
cdd	5310	1.00	5.00	4.3689	.63566
csy	5310	1.00	5.00	4.2446	.75899
ctz	5310	1.00	5.00	4.2185	.70783
chx	5310	1.00	5.00	4.3386	.69569

Table 5. Related Analyses of Six Competency Factors and Seven Teaching Effectiveness Factors

Variable	Mean	S.D.	1	2	3	4	5	6	7	8	9	10	11	12	13
1CZS	4.36	.65	1												
2CJN	4.21	.69	.35**	1											
3CDD	4.37	.64	.33**	.37**	1										
4CSY	4.24	.76	.29*	.33*	.34*	1									
			*	*	*										
5CTZ	4.22	.71	.32**	.38**	.37**	.35**	1								
6CHX	4.34	.69	.32**	.36**	.37**	.33**	.39**	1							
7EJN	4.26	.77	.24**	.25**	.28**	.24**	.26**	.25**	1						
8ETD	4.21	.74	.27**	.30**	.31**	.28**	.29**	.28**	.39**	1					
9EHZ	4.29	.76	.27**	.29**	.28**	.25**	.28**	.27**	.35**	.39**	1				
10EHX	4.35	.81	.24**	.25**	.27**	.23**	.26**	.25**	.33**	.35**	.36**	1			
11EXL	4.26	.78	.26**	.29**	.31**	.27**	.28**	.28**	.37**	.39**	.36**	.32**	1		
12EGC	4.18	.67	.33**	.37**	.36**	.37**	.36**	.35**	.35**	.39**	.38**	.34**	.38**	1	
13EKL	4.32	.68	.34**	.37**	.38**	.37**	.35**	.35**	.27**	.36**	.29**	.27**	.31**	.37**	1

As can be seen from the Table 5, the six competency factors and seven teaching effectiveness factors are significantly related, indicating that competency can predict teaching effectiveness.

Six of the result being selected, ability factors prediction variables and regression analysis method to teachers' teaching efficiency scale the total score is carried away captive. Results table 6, professional knowledge, professional skills, professional moral, political quality, personal characteristics, teachers and students of the harmonious prediction effect significantly the teaching effect. Among them, DuoYuan correlation coefficient is 0.800, 64% of the total variance in performance could be explained by ability factors.

1) Regression on competency about learning values and sense of achievement (ejg): The scores of six competency factors are selected as predicting variables and regression analysis of subscale score about 'learning values and sense of achievement' has been made. The result shows that professional knowledge, professional skills, professional ethics, political qualities, personal characteristics, teacher-student harmony have significant predictive effects to 'learning values and sense of achievement'. Among them, the multiple correlation coefficient was 0.598, and 35.8% of total variance in teaching effectiveness can be explained by all competency factors.

Table 6. Regression Analysis of Teaching Effectiveness with Competency

Variable	Regression Coefficient B	Standard Error	Standardized Regression Beta (β)	t Value
Constant	16.932	1.276		13.271
czs	4.562	.392	.146	11.651a
cjn	5.286	.448	.179	11.794a
cdd	8.705	.470	.271	18.522a
csy	3.125	.338	.116	9.235a
ctz	3.016	.451	.105	6.685a
chx	3.009	.431	.103	6.977a
R=.800 R2=.640 Adjusted R2=.640 F=1561.069a				

Shows that significance level P<0.05

2) Regression on competency about teaching enthusiasm and attitude (etd): The scores of six competency factors are selected as predicting variables and regression analysis of subscale score about 'teaching enthusiasm and attitude' has been made. The result shows that professional knowledge, professional skills, professional ethics, political qualities, personal characteristics, teacher-student harmony have significant predictive effects to 'teaching enthusiasm and attitude'. Among them, the multiple correlation coefficient was 0.677, and 45.9% of total variance in teaching effectiveness can be explained by all competency factors.

3) Regression on competency about teacher-student interaction (ehd): The scores of six competency factors are selected as predicting variables and regression analysis of subscale score about 'teacher-student interaction' has been made. The result shows that professional knowledge, professional skills, professional ethics, political qualities, personal characteristics, teacher-student harmony have significant predictive effects to 'teacher-student interaction'. Among them, the multiple correlation coefficient was 0.642, and 41.2% of total variance in teaching effectiveness can be explained by all competency factors.

4) Regression on competency about teacher-student harmony (ehx): The scores of six competency factors are selected as predicting variables and regression analysis of subscale score about 'teacher-student harmony' has been made. The result shows that professional knowledge, professional ethics, political qualities, personal characteristics, teacher-student harmony have significant predictive effects to 'teacher-student harmony'. Only professional skills have no significant predictive effect to it. Among them, the multiple correlation coefficient was 0.589, and 34.7% of total variance in teaching effectiveness can be explained by all competency factors.

5) Regression on competency about knowledge width and information amount (exl): Six of the result being selected, ability factors prediction variables and regression analysis method to the width of the mass fraction of knowledge and information of amount. The results show that, professional knowledge, professional skills, professional moral, political quality, the teachers and students have significant prediction effect of harmony of the width of the knowledge and information quantity ". Only personal characteristics, no significant prediction function. Among them, some correlation coefficient is 0.660, 43.5% of the total variance in the teaching effect can explain the factors for all ability.

6) Regression on competency about teaching course (egc): Six of the result being selected, ability factors prediction variables and regression analysis method to the mass fraction of teaching course ". The results show, professional knowledge, professional skills, professional moral, political quality, personal characteristics, a significant prediction function the teaching process of "harmony between teachers and students. Among them, some correlation coefficient is 0.808, 65.3% of the total variance in the teaching effect can explain the factors for all ability.

7) Regression on competency about teaching course (egc): Six of the result being selected, ability factors prediction variables and regression analysis method to the mass fraction of social etiquette course to the difficulty and homework quantity ". The results show that professional knowledge, professional skills, professional moral, political quality, personal characteristics, a significant prediction function of the course and harmonious teachers and students work difficulty quantity ". Among them, DuoYuan correlation coefficient is 0.826, 68.2% of the total variance in the teaching effect can explain the factors for all ability.

4 Conclusion

The results show that the prediction of scoring ability have significant factors such as six effect seven subscales-learning the values and the sense of achievement, teaching enthusiasm and attitude, the teachers and students interact, the teachers and students, the width of the harmonious knowledge and information quantity, the teaching process, the difficulty of the class and homework quantity.

References

1. Marsh, H.W.: Student's Evaluations of University Teaching: A Multimensional Perspective. Higher Education 64(1) (January 1993)
2. McClelland, D.C.: Identifying competencies with Behavioral Event Interview. Psychological Science 9, 331 (1998)
3. Chen, H.Y.: Evaluation of the Teachers' Competency in Ideological and Political Theory Courses Based on the CompetencyModel in Hebei Province. Journal of Heibei University of Technology 39(5), 84–91 (2010)
4. Zhang, Y.: The Forefront of the ideological and political education. People's Publishing House, Beijing (2006)
5. Yu, X.: Inter-subjectivity: A new perspective to understandteacher-student relationship. Contemporary Education, Science (2004)
6. Zhang, T.: Contact to the interactive practice subjectivity of education. Education Science Press, Beijing (2005)

The Application of Set Pair Analysis in College Ideological and Political Education

Yongfang Liu and Zhikai Yun

School of Humanities, Economics and Law, Northwestern Polytechnical University, Xi'an 710129, China
liuyongfang125@163.com

Abstract. The method is not only effective evaluation and comparison as a whole, and at the same time, can the actual situation of each college evaluation. For the general evaluation, also pointed out the shortcomings of the effective evaluation and finishing work. This method is simple and accurate, and the College of Political referenced in the fiber, the college thought politics fiber good methods to evaluate.

Keywords: evaluation, college ideological and political education, set pair analysis.

1 Introduction

Students must inherit and develop the party's ideological and political work of the glorious traditions, colleges should adapt to efforts to create a new situation of ideological and political work of the effective way and method, ideological and political work in the whole teaching system, permeability and college students' socialist initiative and creativity and loyalty, there is high quality of college students power to agglomerate of building the socialism with Chinese characteristics, and ensure that the big enterprise up China's socialist modernization construction four healthy development and direction [2]. The quality of ideological and political problem has a large number of researchers, college students' ideological and political quality of the training strategy, but no scholar's ideological and political quality evaluation, puts forward the model analysis set of college students' ideological and political quality, ideological and political education evaluation, the level of university. Also in the ideological and political quality education of university system provides a foundation construction.

2 The Meaning of Ideological and Political Quality

Ideological and political education in China is to promote cultural and spiritual civilization construction of the most important content, and also one of resolving social contradictions and problems of the main way. Ideological and political education is very important, and they are quite difficult to do. Especially in the market economy condition, our country's ideological and political work is relatively wear

A. Xie & X. Huang (Eds.): Advances in Computer Science and Education, AISC 140, pp. 415–420.
springerlink.com

conditions, does not meet the requirements of the development of modern society is very good. Make the thought political work strength for many reasons, but not a important reason why we overlook the talent education and improve college students for a long time [3].

Ideological and political education is a guiding people to form the correct thought behavior science; it formed a man's thoughts and behavior, and realizes the change rule of ideological and political education principles as his object of study.

Through the research widespread contact, fascinating social factors and education object outlook on life, the view of the world of the formation and development of relations, the paper discusses how to adjust the enlightenment, the social environment of the 3 d object of education, as well as to the education alternative action of the object in the standard guide received time environmental impact will move function.

3 Set Pair Analysis of Model Presentation and the Ideological and Political Quality of Evaluation Index System

Set pair analysis (SPA), is the collection and relation as the basic concept, clear and uncertainty and conversion of the rules a system analysis technology in the description and research system widely exist [4]. The collection to refer to have certain relation two set composition antithetical couplet, the Set Pair Analysis is to the collection to some characteristic relation and the opposition description, the available relationǴexpresses the collection to dialectical relations.

$$\mu = a + bi + cj \tag{1}$$

In this formula, a, b, c are separately called the set pair's same, difference and opposition. i Iandjare the coefficient of difference and opposition, and $i \in [-1, 1]$, $j = -1$, $a + b + c = 1$ .In connection degree μ, a and c attribute the thing separately definite identical and oppose two sides, indefinite item of bi has manifested in system's definite uncertainty dialectically, can reflect the relatively complete information, the portray system's security problem.

The difference is absolute, outstanding is relative. Bad and outstanding is the condition which two kinds oppose, therefore, may the college student come to understand personally takes a collection with Ideological and political education system these two sets to be right. Therefore, takes the college student ideological and political quality level group of collections to be right, analyzes outstandingly with SPA to the theory or not, with the connection degree expression is

$$\mu = bi + cj \tag{2}$$

Type (2) reflects the college student ideological and political level and the college Ideological and political education set pair to connection degree. In this set pair, no absolute safe, any small dangerous factor will cause an accident. So, in this formula, there are only bi and cj items. And cj manifests system's critical condition; bi manifests between the differences, the outstanding two conditions the intermediary transition risk, is may to the difference or the outstanding condition transformation.

The SPA to analyzes the connection degree is appraises the factor to have one kind of situation probability. Data which grades according to the expert, may divide into the college student ideological and political quality level five ranks, as shown in Table 1:

Table 1. level

I Level	II Level	III Level	IV Level	V Level
100	90-90	80-80	70-70	60-60

In order to cause the connection degree metrication, the structure connection degree function is as follows: (Value is bigger superiorly) regarding target U, the college student ideological and political quality of evaluation index is in level (Σ) when:

$$\mu_{11}=\begin{cases} 1 \quad x\in[U_1,U_0) \\ \dfrac{x-U_2}{U_1-U_2} \quad x\in[U_2,U_1) \\ 0 \quad x\in[0,U_2\ ) \end{cases} \tag{3}$$

The evaluation index will be in level (II) when:

$$\mu_{12}=\begin{cases} \dfrac{U_0-x}{U_0-U_1} \quad x\in[U_1,U_0) \\ 1 \quad x\in[U_2,U_1) \\ \dfrac{x-U_3}{U_2-U_3} \quad x\in[U_3,U_2) \\ 0 \quad x\in[0,U_3\ ) \end{cases} \tag{4}$$

The evaluation index will be in level (III) when:

$$\mu_{13}=\begin{cases} 0 \quad x\in(0,U_4\ ]\cup[U_1,U_0) \\ \dfrac{U_1-x}{U_1-U_2} \quad x\in[U_2,U_1) \\ 1 \quad x\in[U_3,U_2) \\ \dfrac{x-U_4}{U_3-U_4} x\in[U_4,U_3) \end{cases} \tag{5}$$

The evaluation index will be in level (IV) when:

$$\mu_{14}=\begin{cases} 0 \quad x\in(0,U_5\,]\cup[U_2,0) \\ \dfrac{U_2-x}{U_2-U_3} \quad x\in[U_3,U_2) \\ 1 \quad x\in[U_4,U_3) \\ \dfrac{x-U_5}{U_4-U_5} x\in[U_5,U_4) \end{cases} \tag{6}$$

The evaluation index will be in level (V) when:

$$\mu_{15}=\begin{cases} 0 \quad x\in[0,U_5)\cup[U_3,0) \\ \dfrac{U_3-x}{U_3-U_4} \quad x\in[U_4,U_3) \\ 1 \quad x\in(U_5,U_2\,] \end{cases} \tag{7}$$

In type (3)~(7), U_0-U_5 respectively are college student ideological and political quality of evaluation index rank limiting value;x is the expert grades actual value; i is each different target arrangement serial number. Likewise, May also determine that the value is smaller a more superior indexU is corresponding connection degree function.

Extracts the i index after the different appraisal rank relation μ_{ij} (j=1,2,3,4,5), determines the total connection

$$\text{degree } \mu_j=\sum_{i=1}^{n}\mu_{ij}\omega_j \tag{7}$$

In this formula, μ_{ij} is each index regarding j rank's total relation. ωj is the weight each index occupies in the system of index. The rank which has the biggest relation is considered the final rank.

As follows according to the college student ideological and political quality evaluating indicator system:

Table 2. College Student Ideological and Political Quality Evaluating Indicator System

Element layer	Index layer
College Ideological and political education system principl	Seeking truth from facts principl Exemplary education principle Treating people equally principle Grading education principle Positive guidance principle Stressing practical results princip Direction goal principle Multiplicity principle
The college student came to understand personally	psychological disposition living habits value concept world outlook and outlook on lif moral consciousness
Teacher quality	scientific research ability managerial ability Personal glamour teaching ability knowledge level

4 Result Analyses

Ideological and political education of university students is based on students' ideological, views, with attitude, political position and thought consciousness of students. Ideological and political education of university is one of the important part of higher education, and train high quality powerful guarantee of college students' [1]. The ideological and political work must obey and service, the main task of the party and should have the bright party spirit, its practicability and mass character. It with Marxism-Leninism, MAO Zedong thought as a guide, with communist ideology, for college students to cultivate college students', set up the correct standpoint, viewpoint and the correct throughway and working methods. Ideological and political work is a science is based on the theory of dialectical materialism and historical materialism, the combination system basis; Marxist theory of psychology, pedagogy, sociology, and ethics is a body comprehensive applied science. It has its internal work rules and its characteristic, and has the work principle and scientific methods of work, should check repeated.

References

1. Lai, P.: By college graduate's Ideological and political education. Success (education) (2) (2009)
2. Li, Z.: New time college graduate Ideological and political education work study. Education and Occupation (3) (2009)

3. Ding, J., She, S.: How does the institute complete the thought politics theory class the education teaching profession - - take the Wuhan College politics and the public administration institute as an example. Thought theoretical education (3) (2008)
4. Li, S.: Project fuzzy mathematics and application. Harbin Industry College Publishing, Harbin (2004)
5. Yu, L.: College campus media Ideological and political education function research. Central China Normal college graduate student graduation thesis (2008)

The Improvement of Enterprise-Oriented Scale-Free Network Model

Yongxian Chen[1], Dichong Wu[2], Canlin Mo[1], Dongfang Zhao[1,2], and Renwang Li[1]

[1] Department of Industrial Engineering Zhejiang Sci-Tech University
310018 Hangzhou, China
{chenyongxian-2,zdf115}@163.com, arenwangli@foxmail.com
[2] Business Administration College, Zhejiang Institute of Economy and Finance
310018 Hangzhou, China

Abstract. This article firstly discussed a special kind of business cooperation network, scale-free network of business cooperation, and some of its features are mentioned. Then based on scale-free network and the actual process of enterprise improvement, an improved model of business scale-free network is put forward. There are two areas for improvement, one is in the business cooperation network, the business as node's in addition to increased, as well as reduced, the other is in order to better describe the nodes information, the concept of nodes age is introduced. Computer simulation results show that , even taking into account the probability of 70% of new businesses close down, business degree is still showing power-law distribution, but the advantage of initial business node becoming the key one is not obvious.

Keywords: scale-free network, enterprises cooperation, computer simulation.

1 Introduction

With continuous development of economic globalization, enterprises are exposed to a fierce global competition when obtaining global market. It's difficult for Individual enterprise to survive the increasingly intense competition. Through the cooperation between enterprises, to obtain the developmental needs of their own funds, technology, upstream and downstream enterprises are having increasingly close business contacts, which becomes the main themed of today's world.

Linkages between enterprises form a business cooperation network. Cooperation network is the concern of economics, principles of management engineering, etc. Related theories keep emerging. Among these, R. Albert and AL Barabas physicists proposed scale-free network in 1999, and became a very popular tool.[1] Enterprise is abstracted as a node and the linkage between enterprises is abstracted as a side in the model. The linkage contains supply chain (material flow), economy co-operation (funds flow) and Internet (information flow)[2]

A. Xie & X. Huang (Eds.): Advances in Computer Science and Education, AISC 140, pp. 421–427.
springerlink.com

2 The Features of Enterprise-Oriented Scale-Free Network

Scale-free model is a new model to study large and complex network structure in recent years, by Barabas in University of Notre Dame, in Indiana, whose characteristic is non-homogeneous (inhomogeneous). A scale-free network includes the basic parameters of the average path length, network clustering coefficient and network degree distribution. This paper focuses on the degree distribution. The degree of a node in the network means the number of a node connected to the others. The degree distribution-p(k), refers to the degree the probability of k nodes, and it reflects the relation between nodes and edges in the complexity network. Another way of describing the degree distribution is the cumulative distribution function: node is greater than or equal to the probability of k. Cumulative degree distribution has an advantage of all the original exam data, and controls of the tail noise. The results show that: the Internet, movie actor collaboration networks, scientific collaboration network, the network of human sexual contact, nervous system and the language network, etc. have scale-free properties.

Overall, business-oriented cooperation scale-free network's scaling, small world, growth, robustness and selection of the best selectivity, according to the need for simplified discussion of previous studies, often differs with reality. Below, the characteristics of the scale-free organization network are summarized, and some improvements will be pointed out.

2.1 No Scaling

No scaling refers to a corporate network, node degree (each node connected to the number of other nodes) of the distribution is scale-free, showing the characteristic of power law distribution, and a small number of core nodes, has a large number of connection, while other large number of nodes need only a small number of connections.

2.2 Growth

Growth means in the real world, when a new company set up with the existing enterprise network connection, it will be added to the network. However, the growth of the enterprises has only considered the case of establishment, not a member of business bankruptcies— the process of nodes disappearance. As a result, the network will expand without limitation, which is clearly different from the actual situation

2.3 Choose the Best

The positions of different network nodes are not the same, and a small number of points are keys. So, when a new point intervened, it will give priority to the core point. This preference will lead to "the rich get richer" situation.

2.4 Coexistence of Robustness and Fragility

The so-called robustness refers to the network is little impact by the external. Network has a large number of nodes in non-critical positions; they appear and disappear almost no effect on the entire network. Minor enterprise's creation and collapse is little impacted to the entire industry. the impact is negligible. While, a enterprise in a key position in the network important implications. Once these key points under attacked by a purpose to make paralyzed, then the whole network may collapse [3,4,5].

3 The Establishment of Cooperation-Oriented Scale-Free Network Model

Previous studies on the problem of scale-free usually focus on the increase of the network under ideal state. The fact is more complicated. Thc success rate of founding an enterprise in China is less than thirty percent by some related research. Therefore, the demise of enterprises is an unavoidable problem in the process of the network increase. In the published literature, there is no relevant in-depth study.

In order to take into account the company's demise, the author presents the basic parameters of a new node age. The age of the node Y is defined as: the time of nodes after generating using each node create time as unit 1. For example, after a node K joined in the network, there are three nodes join the network, then the node K recorded the age of 3. As follows, the author, explore the new node has seventy percent increase in the extinction of the situation through computer simulation, companies are able to form a scale-free network.

Hypothesize there were companies Ei (i = 1,2, ... n) at the scale-free network,. The initial form of the network as Figure 1, there are three nodes, all of them are fully connected structure, namely each point and there is a connection to the others.

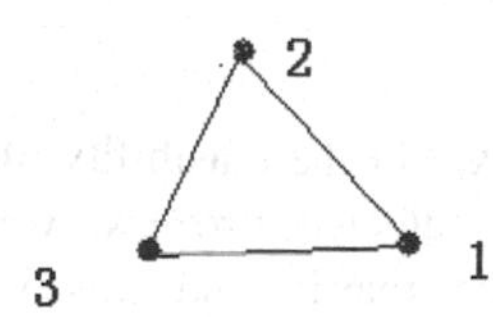

Fig. 1

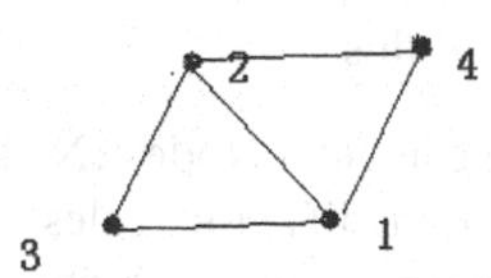

Fig. 2

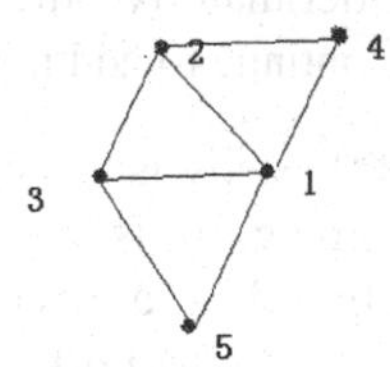

Fig. 3

According to certain rules to add and delete nodes, the entire network will develop into a network with many nodes. The network is scale-free networks. In order to better characterize the network, we introduce the matrix to describe network. Fig.1, Fig.2 and Fig.3 are described by

$$A = (a_{ij})_{3\times4}, B = (b_{ij})_{5\times4}, C = (c_{ij})_{7\times4},$$

$$A=\begin{pmatrix}1&1&2&1\\2&1&3&1\\3&2&3&1\end{pmatrix}\quad B=\begin{pmatrix}1&1&2&2\\2&1&3&2\\3&2&3&2\\4&1&4&1\\5&2&4&1\end{pmatrix}\quad C=\begin{pmatrix}1&1&2&3\\2&1&3&3\\3&2&3&3\\4&1&4&2\\5&2&4&2\\6&1&5&1\\7&3&5&1\end{pmatrix}$$

The Matrix has four columns and each row represents a connection between two points. Column 1 is the number of the connection side. Column 2 and Column 3 represent linked two points. Column 4 indicates that the node age. The increase and delete rules of nodes as follows

(1) The initial number of nodes in the network is m_0.
(2) Add a new node and create connection to any two existing nodes. The selected node is proportional to the node's degree.

$$p(k_i)=\frac{N(k_i)}{\sum_{j=1}^{n} N(k_j)} \tag{1}$$

(3) If the new node's age is less than five years old, it will have seventy percent probability of extinction. It also will cancel the connection relationship. If the node's age is elder than five, the node will survive forever.
(4) Continue to add new nodes as this.

After 4000 steps as this cycle, we can get a Nodes (N, 4) matrix. For the feasibility of programming, it is necessary to modify the rules. In the selected process, we randomly select a node at column 2 and column 3 in the matrix and this is equivalence to the Eq.1.

4 The Analysis of Enterprise-Oriented Scale-Free Network's Property

4.1 The Analysis of Node's Degree

After 4000 simulation steps, getting a No1 (5230, 4) of the matrix. Transform this matrix to get a No2 (N, 2) matrix; Transform No2 (N, 3) to get No3 (N, 3) matrix.

$$No1=\begin{pmatrix} 24 & 12 & 14 & 3990 \\ 25 & 6 & 14 & 3990 \\ 28 & 15 & 16 & 3988 \\ 29 & 11 & 16 & 3988 \\ \vdots & \vdots & \vdots & \vdots \\ 8002 & 2468 & 4003 & 1 \\ 8003 & 316 & 4003 & 1 \end{pmatrix} \quad No2=\begin{pmatrix} 2 & 3 \\ 3 & 20 \\ 6 & 22 \\ 11 & 6 \\ \vdots & \vdots \\ 4002 & 2 \\ 4003 & 2 \end{pmatrix} \quad No3=\begin{pmatrix} 1 & 154 & 0.0502 \\ 2 & 1495 & 0.4873 \\ 3 & 608 & 0.1982 \\ 4 & 277 & 0.0903 \\ \vdots & \vdots & \vdots \\ 46 & 1 & 0.0003 \\ 75 & 1 & 0.0003 \end{pmatrix}$$

The column 1 of matrix No1 (5230, 4) are node's number. Column 2 and column 3 are connected two points and column 4 is node pair's age, meanwhile, column 4 is the age of node of column 3.The column 1of matrix No2 is the node's number and the column 2 is the node's degree. The column 1 of matrix No3 is the array of the node's degree of matrix No2,the column 2 is the number of such ones and the column 3 is the proportional of these nodes its degree is describe at column 1.The diagram draw according to matrix No3,using column 2 , the node's degree, as abscissa and using column 3, the node distribution, as ordinate.

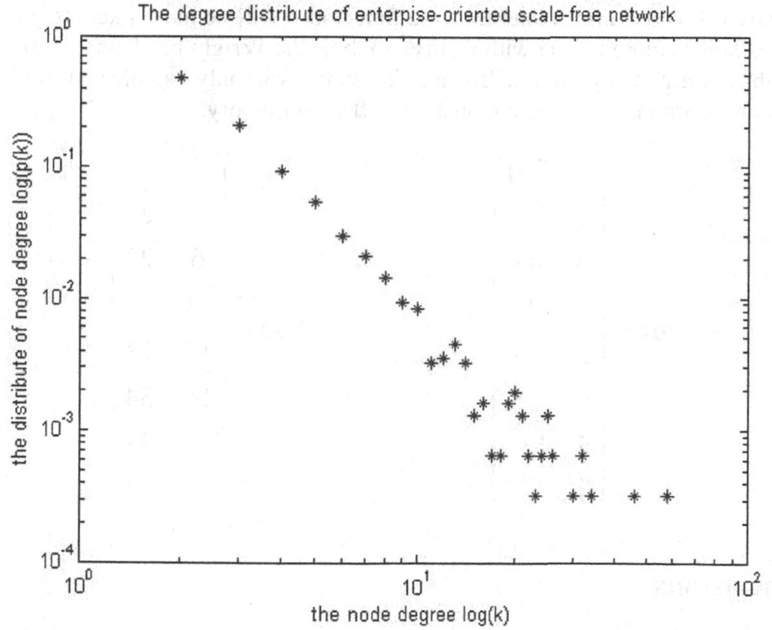

Fig. 4. The degree distribute of enterprise-oriented scale-free network

In accordance with the theory, if the node degree distribution power law distribution program, then there are

$$p(k) = lk^{-r}$$
$$\log p(k) = -r\log(lk) \tag{2}$$
$$\log p(k) = -r\log(l) - r\log(k)$$

From the diagram, we know that, even taking into account the Seventy percent of nodes die, then the degree of the nodes in the network will still show power-law distribution.

4.2 Analysis of the Robustness of the Network

In accordance with the original model, not considering the demise of the node, the initially node will grow to key node , namely the node '1' '2' '3' would a large number of connection points, the matrix No4 is the original model simulate by 4000 steps, the resulting node degree. However, if you take into account the phenomenon that the newly established enterprises will be a great probability of extinction; the original node will not have a clear advantage. Or even disappear, such as Node 1.Matrix No5 is the initially node degree distribution of new model simulation by 4000 steps, This matrix shows the node arisen later than the first nodes will be more likely to grow as a core node. This is more in line with reality. Such as the Wright brothers in 1909 set up the Wright Aircraft Corporation, Boeing Company was only established until 1917. Now Boeing is among the largest global aircraft manufactory.

$$No4 = \begin{pmatrix} 1 & 201 \\ 2 & 119 \\ 3 & 47 \\ 4 & 96 \\ 5 & 18 \\ 6 & 78 \\ 7 & 137 \\ \vdots & \vdots \end{pmatrix} \qquad No5 = \begin{pmatrix} 2 & 3 \\ 3 & 20 \\ 6 & 22 \\ 11 & 6 \\ 12 & 13 \\ 14 & 34 \\ 15 & 38 \\ \vdots & \vdots \end{pmatrix}$$

5 Conclusions

The innovation of this paper lies in improving the existing enterprise-oriented model of scale-free network on the base of 70% probability of new enterprises bankruptcy. For a more accurate description of the node, I introduce a new concept, the node's age, to imitate the age of business growth. At the same time, it explores whether the first nodes in a network is certain to become a key node in the final or not, which provides a theoretical basis for governmental support for new-born business. This article only discusses the problem of degree distribution, without concern of the other network parameters such as average path and the network's clustering coefficient. Those problems should be discussed in other articles.

Acknowledgement. The work is supported by the Zhejiang Province Natural Science Foundation (Grant No. R6080403, No. Z6090572, No.Y6110568), Zhejiang Province Key Science and Technology Innovation Team (Grant No. 2011R09015)

References

1. Wang, K., Wang, J.: Scale-free organization enterprise network characteristics and evolution model. Micro Computer Information 22, 210–212 (2006) (in China)
2. Zhang, L.: Enterprise cooperation complex network of research. Shanxi university (2008) (in China)
3. Luo, Y.: Scale-free network model of the research and application. Donghua university (2009) (in China)
4. Zhang, Z.: The complex network of evolution model research. Dalian university of technology (2006) (in China)
5. Wang, L., Dai, G., Hu, H.: A new topology parameters of Scale-free network. The System Engineering Theory and Practice 26(6), 49–53 (2006) (in China)

Research on Model of Ontology-Based Semantic Information Retrieval

Yu Cheng[1] and Ying Xiong[2]

School of Computer, Hubei University of Technology, Wuhan, China
[1]958161176@qq.com,[2]1654149247@qq.com

Abstract. This paper studies mainly how to apply the ontology into information retrieval system, so as to achieve semantic retrieval. Starting from the introduction of status quo of traditional information retrieval , and analyses of its main Problems, the paper describes the key technologies involved in the semantic retrieval, including ontology building a database, ontology reasoning, semantic search tools Jena and OWL language of ontology, which have been combined to complete the design of semantic retrieval model.

Keywords: Ontology, Information Retrieval, Jena, Semantic reasoning.

1 Introduction

With the development of information society, information rapid growth on the Internet has begun to show its complexity and diversity. The traditional keyword-based information retrieval technology can't meet people's needs of information search. The problem exits mainly in that the traditional information retrieval system can't understand the inner meaning of information resources and their relationship, i.e. the lack of semantic understanding. The information people want to obtain just can be matched to knowledge in the professional field. However, the results retuned by matching with keyword are usually only the literal information, getting far away from information people really want to get. The semantic search technology is the most promising methods to solve this problem, which has become one of the hot fields of information retrieval.

2 Key Technologies of Semantic Retrieval

2.1 Ontology Description Language

Ontology description language is mainly used to describe the ontology which can be used to write a clear and formal concept description for a domain mode. Therefore, Ontology description language should meet the requirements of good semantic, supporting effective semantic reasoning and full expression. Nowadays, a variety of ontology description languages exist, that including RDF, RDFS, OWL, DAML, KIF, etc. The paper chooses OWL as the ontology constructing language. OWL is W3C

A. Xie & X. Huang (Eds.): Advances in Computer Science and Education, AISC 140, pp. 429–434.
springerlink.com

recommendation standard based on description logic, and absorbing Web resource as its describing object, XML language as described base. Description Logic (abbreviated DL) can reason conditionally ontology classes, instances, and its attributes to rich ontology semantics. Therefore, OWL ontology based on description logic is superior to other description languages in the concept expression and reasoning ability as well as the integration of Web resources' knowledge and logical testing when ontology library building. Relationship of ontology language and XML, RDF / RDFS, OWL is shown as the Figure 1.

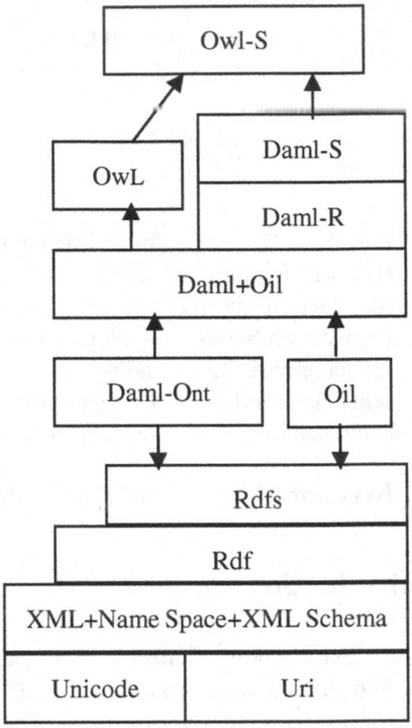

Fig. 1. Semantic Description Language Derivative Sketch Map based on XML

2.2 Ontology Constructing Tool

There are many techniques and ways for building ontology, the paper selects Protégé as the ontology constructing tool. Protégé had been developed by the research team of School of Medicine, Stanford University. The thought of Protégé object is very similar to the principles of classification of ontology, which is easy to expand for developers to use a variety of plug-ins to increase the Protégé function. For example, supporting graphical ontology editing mode, Protégé can express the relationship among concepts through graphic representation; supporting database storage mode, built ontology can be recorded into the database for user-friendly operation; supporting group development, OWL files is easily extended, read and use the other ontology. Therefore, a large ontology can be put into several smaller parts, which is responsible by different developers for; supporting the logical testing, in order to avoid errors existing in persons' collaboration, the final merged ontology needs conceptual consistency testing and conflict testing, so as to detect and correct the contradictions in the concept of ontology and the error in case-attribute.

2.3 Semantic Reasoning

Compared with traditional information retrieval, the biggest feature of semantic information retrieval is its introducing into the retrieval processing of the resource object. Data, information and knowledge are three concepts people daily exposed to, which is the object the user retrieve. The purpose of user's information retrieval is to obtain valuable knowledge. Domain ontology describes the relationship among the concepts, providing the logical rules semantic reasoning required, and the object of reasoning of the metadata information stored as XML / RDF or OWL, i.e. semantic reasoning allows computers to recognize and understand the structure of domain ontology and metadata information so that it can seek the closure of the existing information base in accordance with relevant logical rules

2.4 Semantic Search Tools

The paper makes currently more popular choice of the Jena Semantic Development Kit. Jena is toolkit of Java programmer development, which is open source and supporting RDF, OWL and semantic reasoning. Jena framework consists of three parts: the Graph, Enhanced-Graph and Mode layer. The Model layer is the entrance programmer operating RDF, OWL data. The Enhanced-Graph layer can provide a variety of different views for Graph layer which is responsible for data persistence, and display in the appropriate manner.

3 Ontology-Based Semantic Information Retrieval Model

3.1 Method of Semantic Retrieval

Semantic information retrieval is a combination of semantic retrieval, natural language, artificial intelligence technology. It analyzes search requests and information resource object from the perspective of the semantic understanding, which is a matching mechanism based on the relation of concepts. The key lies in representation of information resources and reasoning among concepts. Currently there are two type of semantic retrieval, which is based on ontology and concepts, the former achieve semantic search based on ontology constructing space, while the latter is based on relational database or conceptual dictionary to build the conceptual space. In summary, relationship of ontology, conceptual space and semantic retrieval can be shown as Figure 2:

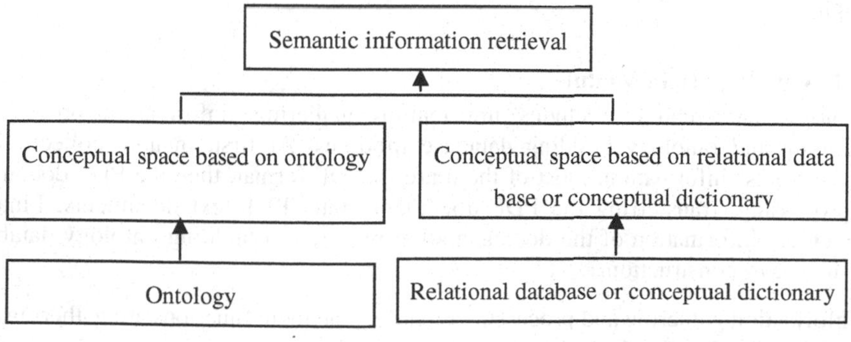

Fig. 2. The Method of Semantic Retrieval (the Relationship among Ontology, Space of Concept and Semantic Retrieval)

3.2 Design of Information Retrieval Model

Ontology-based semantic information retrieval model includes four modules of input and output of user, Lucene index, semantic query module and ontological data. Figure 3 shows the relationship of four modules.

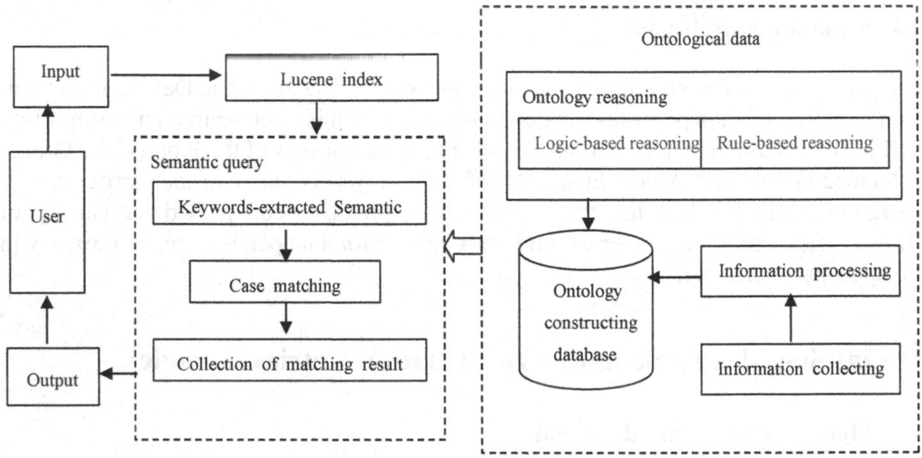

Fig. 3. Ontology-based semantic information retrieval model

The model carries on systematic semantic function, through the ontology database and ontology reasoning technology. System uses Protégé to build ontology structure, as well as the OWL ontology language to describe the ontology, and then fill in the content to the ontology database through Jena, open-source toolkit based on Java. At the same time, semantic reasoning has been carried on through Jena and the rule base for the ontology. Also Lucene, Java-based open-source toolkit, has been used to detach the word. Finally, information can be queried in the database through using the Jena toolkit.

3.2.1 Ontology Data Module

Ontology data module includes information gathering, information processing, reasoning and ontology building database modules. At first, manual collection of related fields' information, most of them are in PDF format, then the PDF document preprocessing, transferring the PDF documents into TXT text documents. Finally, extract key information of the documentation required in building ontology database for the use of construction.

①Information gathering and processing module: The main functions are gathering and processing the data for the preparation of next work.

②Ontology reasoning module: includes logic-based ontology reasoning and rule-based ontology reasoning. Logic-based ontology reasoning is mainly used for testing the logical correctness of concepts, cases and its attributes; rule-based ontology reasoning is for exploring the tacit knowledge and richening ontology database, which can be realized through the method that building the ontology database, calling Racer inference engine in Jena package, logic testing and modifying or deleting an entry by the use of the ontology inference engine based on Tableaux algorithm for the easy calls on ontology database, meanwhile, the preparing a reasoning rules, reading the OWL ontology file using Jena, then rule-based reasoning to ontology database based on the use of Jena embedded inference engine.

③Ontology building module: by using the ontology building language, OWL, to construct the ontology-database structure, then store information extracted from text documents into the corresponding classes and attributes of the ontology structure. Its implementing steps is to determine the domain ontology building process based on the skeleton rule and seven-step method according to the specific content of ontology, as shown in Figure 4.

3.2.2 Semantic Query Module

The key information of the user retrieving and extracted, should be matched with instances of ontology database to obtain the matching-results, which is assembly of instances combining with the critical information of the user. Any semantic query to one of the assembly of instances can obtain results, which is tri-assembly of instances. This can be realized through the method, that is to get natural language keywords processed through the systematic front-end module, to get the class instance corresponding to keywords by using listSubClasses way of ExtendedIterator class of Jena package, to traverse tri-assembly of instances corresponding to keywords through using listProperties methods of StmtIterator class, and the next Statement of Statement class, and the last to get the matching tri-assembly.

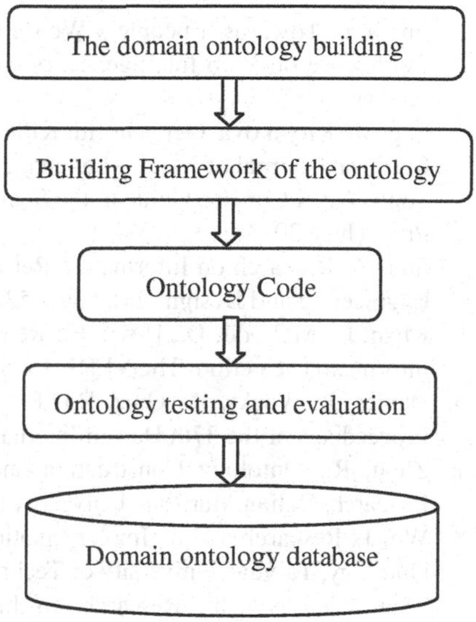

Fig. 4. Domain ontology building process

3.2.3 Lucene Indexing Module

Through Lucene indexing, the module, on the one hand, sends the results obtained by inputing information word to the semantic query module, for the semantic matching. On the other hand, it can meet the user's needs to traditional search through full-text index function by the use of Lucene.

3.2.4 User Input and Output Module

User input module is a simple search entrance. User output module is the graphical visualization of the result of semantic query module. Visualize graphic description of tri-assembly can be carried on by using owl2prefuse engineering package.

4 Inclusion

With the massive growth of information on the Internet, current information retrieval systems can't meet the user's information requirements. How to improve the current information retrieval systems to meet the growing information needs of users has become a very important issue. Ontology as a shared conceptual model has a better conceptual level and semantic description ability. The paper try to introduce the ontology into information retrieval systems, so as to semantic the traditional information retrieval systems, to realize clearly the real information needs of user, and understand real semantics of the information resources, to match information resources and user's information needs in the semantic level, to improve precision rate and recall rate of information retrieval system.

References

1. Im, M.I.: Towards a people's Web: metalog. In: Proc. of IEEE/WIC/ACM International Conference on Web Intelligence, pp. 320–328. IEEE Computer Society, Washington DC (2004)
2. Popovb, Kiryakova, Ognyanoffd: KIM: a Semantic Platform for Information Extraction and Retrieval. Journal of Natural Language Engineering 10(3), 375–392 (2004)
3. Song, W.: A Concise Guide to the Semantic Web, 1st edn., pp. 117–118. Higher Education Press (June 2004)
4. Sun, Z.: Research on Information Retrieval System based on Ontology of Grid. Computer Engineering and Design (23), 5392–5399 (2009)
5. Khan, L., McLeod, D., Hovy, E.: Retrieval effectiveness of an ontology-based model for information selection. The VLDB Journal 13(1), 71–85 (2004)
6. Davies, J., Weeks, R., Quiz, R.D.F.: Search Technology for the Semantic Web [C]. In: Proceedings of the 37th Hawaii International Conference on System Sciences (2004)
7. Zhou, R.: Ontology Construction and Its Application in Book Information Retrieval Research. Dalian Maritime University (6), 4–5 (2009)
8. Wu, J.: Research and its Implementation of Semantic Retrieval System Based on Domain Ontology. Taiyuan University of Technology (May 2010)
9. Zhang, Y., Nankai: Research on Information Retrieval Model Based on Ontology. Computer Application Research, 2240–2245 (August 2008)
10. Tang, W.: Research and Its Application of Semantic Retrieval System Based on OWL. Wuhan University of Technology (June 2010)
11. Shen, Y., Tian, A.: Research and Its Implementation of Information Retrieval System Based on Semantic Grid. Shandong University of Technology (January 2011)

Research of ICE-Based Unified Identity Authentication Model

Yue Zhang and BingBing Xia

Department of Information Engineering, Shandong Jiaotong University
lvlv1919@163.com, jennifer_xiababy@yahoo.com.cn

Abstract. In the application integration, the distributed authentication applications have become urgent and important issues. First, it describes the characteristics of a Unified Identity Authentication. Unified Identity Authentication is a platform that provides certification services for specialized enterprise in various application subsystems. It changes the original distributed certification into focus certification, which greatly improves the credibility and reliability. Secondly, it describes characteristics of Unified Identity Authentication and restrictions in application. Thirdly, in the combined with the latest ICE (Internet Communication Engine) technology it presents a distributed application for authentication interaction model. Lastly, it describes particularly the realization of Unified Identity Authentication.

Keywords: Unified Identity Authentication, ICE, authentication interaction model.

1 Introduction

With digital information developing rapidly, enterprises run more and more application subsystems. Now every application system establishes an independent identity module, authenticating with a separate authentication mechanism in their respective file or database [1]. It is inconvenience for that the most obviously is it must create user accounts for each application system. It increases the difficulty of management and inconvenient to the user. Therefore a Unified Identity Authentication System can be used in a business or a school. Unified Identity Authentication System establish uniform format for authentication data easy to expansion and modification. Unified Identity Information database can avoid the various application systems database data updating and increasing data security.

2 Unified Identity Authentication Introduction

Unified Identity Authentication is a platform that provides certification services for specialized enterprise in various application subsystems. It changes the original distributed certification into focus certification, which greatly improves the credibility and reliability.

Common authentication interactive modes are the following:

(1) Component-based Authentication Interactive Mode.
(2) Unified Authentication-based Interactive Mode.
(3) Trust Agent-based Interactive Mode.

The above three models are generated by interactive user login session token to achieve. It can only provide the user's logon session token, but the other user authentication information is difficult to provide. This can't have advantages for unified authentication center management unified identity.

(4) Web-based services interactive mode: Authentications certified by enter a user name, password through the client to call the web service of web authentication service center.

WEB-based authentication interactive mode can provide more of authentication information, but the reason of network bandwidth and CPU load of the bottleneck of WEB services technologies used in the Simple Object Access Protocol (SOAP) and SOAP low security makes this interactive mode in practical applications has been greatly restricted.

Therefore, unified authentication deployment application requires new technologies to achieve a more secure and a deeper level of authentication information interaction model.

3 Internet Communications Engine (ICE)

ICE is a new object-oriented intermediate platform. It provides for the establishment of object-oriented client server application tools required, API and library. ICE enables distributed client server application becomes easy to establish. ICE applications can be deployed in a variety of environments: client and server can use different programming languages to be achieved; it can run on different operating system platforms and different machine architecture; communication can be with a range of network technology. ICE application source coded are portable available by no matter what kind of deployment environments [2].

4 ICE Authentication-Based Interaction Model

4.1 Interaction Model Framework

Certified Interaction Model Framework Chart:

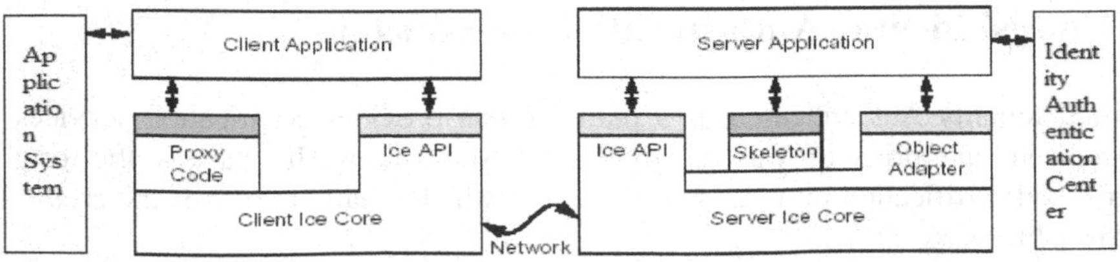

In this model contains the application code, library code, and a Slice definition generated code [3]:

Ice Core code section contains the client and server run-time support for remote access, mainly to solve the network, thread, byte stream-related issues such as network communications.

Ice API is used to access the Ice Core common part of the code set. Ice is the completion of the main run-time initialization and cancellation of operations. This part of the code of client and server are very important.

Proxy code is code generated by the Slice definition to specify the objects defined in the Slice or data type. It consists mainly of a client call down the interface and data structure definition code.

Skeleton code is equivalent to the client's Proxy code, except that it contains the server-side call up the interface.

Object adapter is part of Ice API, except that is only applications on the server side. It completion client requests corresponds to programming language object in a particular method.

Client Application is implemented in different languages. It calls the interfaces or data types defined in the Proxy code. Applications get certified information through Client Application.

Server Application is program code implemented in different languages. It used to implement the interface functions or data structures defined in Skeleton code. It can use the API authentication access authentication platform to complete the certification operation [4].

4.2 Interaction Model Implementation

In the authentication model, Ice Core, Object Adapter and the Ice API and other parts of the code are provided by the ICE of the library, so the implementation process in the model need to achieve Proxy Code, Skeleton Code as well as the part code of client and server side applications, which Proxy Code and Skeleton Code generated by the ICE's Slice. Slice (Ice Specification Language) is descriptive defined for applications interfaces, operations and interaction data type of client and server.

A simple Slice definition:

```
module idstar{
        struct Group{
         string id;
         string name;
        };
```

```
            sequence<string> StringSeq;
            dictionary<string, StringSeq> attrMap;
            class Attribute {
              string name;
              StringSeq values;
            };
            interface IdentityManager{
            bool checkPassword(string id, string password);
            };
    };
```

After the definition of Slice of the interface, it compile these definitions into the desired language can be applied to achieve the API, then you can model the client and server applications.

For achieving authentication interaction model of client, the main thing is to establish the connection to the server, call the Slice definition of interfaces, provide the appropriate variables to the server and send an authentication request, and finally get the server response. For example, the client program can call the interface *idstar checkPassword* to verify a user password.

For achieving authentication interaction model of server, first it must define interfaces in Slice. Authentication information is achieved by API access through unified authentication platform; build server, monitor port, accept the request to respond. For example, the server must implement *checkPassword* defined by the interface *idstar* operation. The data access authentication platform determines whether the user password is valid, then return the results to the client.

5 Conclusion

Now, with Unified Identity Authentication in the enterprise wide application, the requirements of distributed authentication interactive mode will be higher. During ICE design, it gives full consideration to security, providing SSL communication. Based on ICE technology as a simple, seamless and secure interactive mode, it is bound to be more and more attention and widespread application. Meanwhile, security, scalability and other issues of ICE is studied further.

References

1. Wu, X., Zhang, Y.: LDAP-based authentication system unified campus network design. Huazhong University of Science and Technology in (October 2003)
2. Henning, M., et al.: A New Approach to Object-Oriented Middleware (EB / OL) IEEE Internet Computing 1,2 (April 2004)

3. Gamma, E., et al.: Design Patterns: Elements of Reusable Object-Oriented Software. Addison-Wesley (1999, April 2004)
4. Henning, M., et al.: Distributed Programming with Ice, ZeroC (EB / OL) (2003), `http://www.zeroc.com/Ice-Manual.pdf` (April 2004)

Research of Key Technologies in Development in Thesis Management System

Huadong Wang

Department of Computer Science
Zhoukou Normal University
Zhoukou, Henan
wanghuadong@zknu.edu.cn

Abstract. To achieve comprehensive and effective management of university thesis, using development of technology ASP.NET2.0, an university thesis management system based on B / S mode is designed and implemented. In this paper, several key technical in system development such as controlling of user access rights, using of master pages, production of Word format document, editing and publishing online information are focused on. These technologies have certain versatility in the Web application system. The results of the development and operation show that application of these key technologies improved the development efficiency and practicality of university thesis management system significantly.

Keywords: ASP.net, thesis management system, B/S, editing and publishing information online.

1 Introduction

In recent years, with the expansion of college enrollment, workload of thesis guidance and administrative increased exponentially, the traditional thesis management methods cannot meet the actual requirements, a new way to efficiently manage thesis process is urgent needed. With the popularity of computer networks in universities, the necessary hardware infrastructure and operating platform of office management by the campus network are provided. University graduation thesis management system achieve business logic of the graduate design process through the network, build a software platform between the teachers , students and managers that enables teachers and students can use the system to interact, to complete their course in the graduate design needs to be done , and then achieve the thesis network automation and management purposes.

In this paper, with design and development of university thesis management system, several key technical in system development such as controlling of user access rights, using of master pages, production of Word format document, editing and publishing online information are focused on.

A. Xie & X. Huang (Eds.): Advances in Computer Science and Education, AISC 140, pp. 441–446.
springerlink.com

2 Design of University Thesis Management System

2.1 System Architecture and Development Environment

System is based on ASP.NET three layers structure: presentation layer, business layer and data layer, as shown in fig 1. The application layer provides the user interface: the client IE browser; users access the system with IE browser. Business layer provide business functions, it is the core of the system. This layer provides function calls for the presentation layer, but it also calls functionality provides by the data layer to access the database. Data layer is in the bottom, using ADO.NET as the interface, dealing mainly with data request of the business layer, insert, modify and delete the data stored in the database.

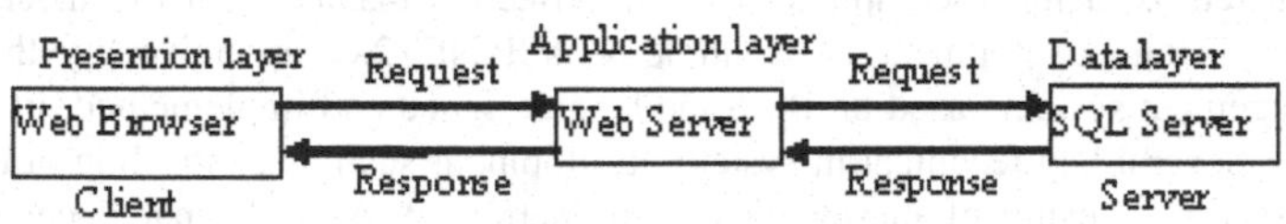

Fig. 1. Three layers structure of the system

System use Microsoft's Visual Studio2008 and Visual Developer2008 as a development platform, SQL Server 2005 as back-end database, based on B / S development model, build applications using ASP. NET technology, combined with C # and JavaScript.

2.2 System Function Design

System's main functions are: system management, teacher-tasking function, students-tasking function, defense management, forms printing, and similar paper detection. Systems management capabilities offer a simple system management to system administrators; teacher-tasking function help teachers to complete the subject of submission, submit the mission statement, opening report reviews, assessments, and other major work; students-tasking function help students to complete opening report submission, paper submission, etc.; respondent management features provide the respondent during the scoring function; form printing capabilities available to the teachers and students to print the relevant forms; identical paper detection can detect similar paper to reduce the phenomenon of student thesis.

According to the papers management processes and different user roles, establish four systems function "student management", "Teacher Management", "Department of Management" and "faculty management" and the corresponding sub-functions. Basic structure of the system is shown in Figure 2. Implementation of the system function, using some of the key technologies can not only improve system availability, but also has some versatility.

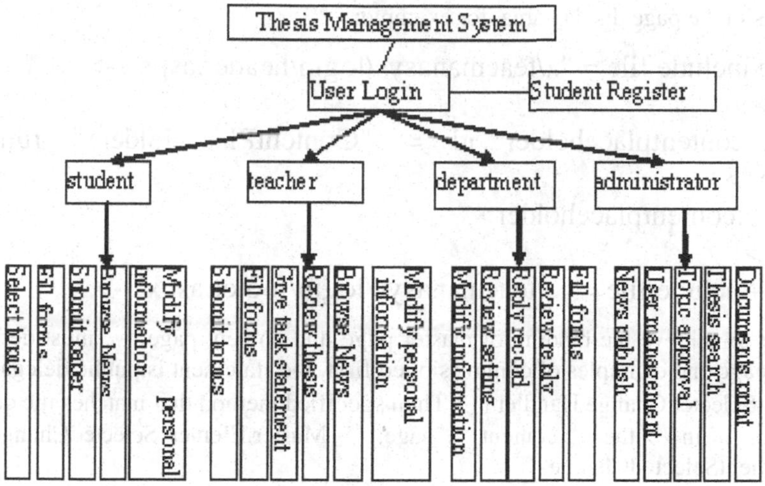

Fig. 2. System Functions

3 Usage of ASP. NET2.0 Master Pages

All pages of thesis management system have some common elements, such as copyright notices, navigation bar, Session judge and help Tips functions. We use ASP.NET2.0 master pages to create a unified user interface and styles, and to define the common elements of the system page.

Master pages are an ASP.NET file with extension of .Master, which can contain static text, HTML elements and ASP.NET server controls and other Web elements. Master page identificative by the special @ Master directive, the directive replaced the ordinary .Aspx page's @ Page directive, which written statements<%@Master Language = "C#"%> in master page code file. We design and define the master page file named Thesis Page. master. In this paper, use HTML table tags for page layout, set the system LOGO by an image element, display the copyright notice by a static text, and achieve navigation with the server control SiteMapPath. Add other controls in the master page needs to increase the common functions of system page is allowed.

In the master page, CSS cascading style sheets are used to achieve the reunification of the page style. the CSS cascading style sheets file set up by us named face.css, reference in the master page through <link> mark elements as follows:

```
<link href = "../css/face.css" rel = "stylesheet" type = "text/css">
```

Master page must be accessed by users with support by content page (ie, after binding). Content page is actually common .Aspx file. In college thesis management system, the common elements of the function code is written in two content pages, where the navigation bar, Session judge and other elements in the page header function code written in header.aspx file, and footer.aspx files are used to achieve function such as copyright notice, help tips and other elements in the bottom of the page. In the master page, ContentPlaceHolder controls of ASP. NET2.0 are called to identify common elements in the page display area, for example:

```
  <!--#include file = "../teacmanasys/login/header.aspx"-->
 <div>
  <asp:contentplaceholder id = "ContentPlaceHolder1" runat =
"server">
  </asp:contentplaceholder>
 </div>
  <!--#include file = "../teacmanasys/login/footer.aspx"-->
```

In order to achieve the binding of master page and content page, we must define the master page and examples of commissioned first, the statement is public delegate void ElementSelectedChange Handler () ; Then specified method that matches the delegate signature in the content page: Master.ElementSelectedChange = his.ElementSelectedChange

The method to load the master page in each feature page is add the sentence asterPageFile = "~/ facuinfomana/default/ ThesisPage.master in the Page directive of the page code.

4 Classification Rights Management Based on RBAC

Many existing information management systems use a centralized access management, which all system maintenance and user permissions, etc. done by the system administrator to be responsible. However, in the actual information management work, often require the use of decentralized multi-level access control. For example, in the thesis management system, to meet the Department / Teaching And Research Section level management system requirements, business director can set the function operation range and data access permissions of the related Department Director, and the function operation range and data access permissions of other teachers access should set by the Department Director. Access control systems can effectively manage all information access requests and decide whether to allow users to access based on the system safe. Therefore, the study and design of the hierarchical rights management scheme can meet the management needs and solve the contradictions between hierarchical nature and the concentration of data management, became a key issues need to address.

In teachers integrated management system, based on role-based access control (RBAC) freedom customization of the role and operation rights are achieved. As required the appropriate role can be create for different jobs, and assign responsibilities based on user roles, user obtain the corresponding function and operating authority by assigned roles. An authorized user can have multiple roles, a role can form by multiple users; each role can perform different operations, each operation can also perform by different roles. Specifically, the allocations of user rights implement Teaching And Research Section and department two level management; centralized data storage, and data input / import implement the Department, Teaching And Research Section, teachers and students four level management, that all levels of legitimate users within the limits of their authority distributed data management system; the data submitted by next level of user should be reviewed by the superior users; access permissions of users of the system functions and data depends on their respective roles, to facilitate the implementation of role-based user security policies.

5 Editing and Publishing Information Online

Editing and publishing information online is a commonly used Web application component. Use university thesis management system as an example, online editing and publishing of related information of thesis are needed to achieve, so FCKeditor online text editor control is use to achieve the functionality. FCKeditor is compatible with most major browsers, including Internet Explorer, Mozilla Firefox, etc. In ASP. NET using FCKeditor control to achieve online text editor needs the following steps:

1) Download FCKeditor library archive from FCKeditor's official website (http://www.fckeditor. net) and copy FredCK.FCKeditorV2.dll to the bin directory of the current project;
2) Define FCKeditor setting item In ASP.NETWeb application configuration information Web.config:

```
<appSettings>
<add key = "FCKeditor:BasePath" value = "~/fckeditor/"/>
<add key = "FCKeditor:UserFilesPath" value = "/Files/"/>
</appSettings>
```

FCKeditor: BasePath is the default directory of the editor kernel, after set it can be used without specified BasePath by the FCKeditor instance attributes. In addition the control supports file upload, upload file type image, text, etc, FCKeditor: UserFilesPath is used to specify the directory where all uploaded files are.

3) Directive of add registered FCKeditor control in a Web Forms page used to publish information is: where TagPrefix attribute values determined the alias of only namespace in FCKeditor control; Namespace attribute defines a namespace associated; Assembly property defines the associated assembly, which is the dynamic link library files FredCK.FCKeditorV2.dll.

```
<%@Register TagPrefix = "FCKeditorV2" Namespace = "Fred-
CK.FCKeditorV2" Assembly = "FredCK.FCKeditorV2"%>,
```

4) the code for add the FCKeditor control to a Release information Web Forms page is <Fckeditorv2:fckeditor id = "FCKeditor1" runat = "server" DefaultLanguage= "zh-cn">

</Fckeditorv2:fckeditor>, Which Fckeditorv2 is the alias of the control Identified in the registration instructions; FCKeditor1 is designated as the control identifier, runat = "server" means that the control runs on the server side; DefaultLanguage= "zh-cn" indicates that the default language is Chinese.

5) In .cs file writing event code for FCKeditor control, define variable text_value. Users use FCKeditor control to edit the information to release and format processing them such as setting the text font, color, paragraph, etc. the information after editing is stored in the control object FCKeditor1.Value properties in the form of a string, the stored information is the content and the

HTML code to display the format of the page. In the page using the control achieve publishing information, click "Save" button, triggered event is assign the value of FCKeditor1.Value variable to text_value variable, and store the value of the variable in a database table

6 Conclusion

The implementation method based on key technologies such as RBAC hierarchical rights management, Word format document production, editing and publishing online information describes in this article, has been in-depth application development in the university thesis management system, improved the development efficiency and practicality of system. These techniques have universal applicability in Web application development, and have reference value in the development of other Web application

References

1. Chen, X.: Design and Implementation of Journal Manuscript Management System Based on B/S. Journal of HangZhou Normal University (Natural Science) 5(1), 37–41 (2007)
2. Chen, W.: Solution of Manuscript Management System Based on Embedded Word Technology. Journal of NingBo Vocational and Technical College 12(5), 87–90 (2009)
3. Liao, W.: Thesis Management System Based on ASP.NET and XML Technology. Journal of HuNan Science and Technology College 29(8), 89–91 (2009)
4. Yu, W., Zhang, W., Liao, F.F.: Discussion of Three-tier Development Framework Application Based on ASP.NET. Software Guide 7(8) (2009)
5. Sun, B.X., He, Y.S., Wu, Z.X.: Implementation of Printing Industry's Online Quotation System Based on.NET Three-tier Architecture. Computer Knowledge and Technology 3(4), 599–600 (2009)
6. Xiong, Z.B., Chen, W.Y.: Design and Implementation of Universal Data Access Component Based on.NET. Software Guide 7(8), 95–97 (2009)

The Localization Algorithm for Wireless Sensor Networks Based on Distance Clustering

Xiaohui Chen[1,*], Jinpeng Chen[1], and Bangjun Lei[2]

[1] College of Computer and Information Technology, China Three Gorges University, 443002 Yichang, Hubei, P.R. China
[2] Institute of Intelligent Vision and Image Information, China Three Gorges University, 443002 Yichang, Hubei, P.R. China
chui@ctgu.edu.cn

Abstract. To improve the localization precision of nodes in Wireless sensor networks (WSN_s), this paper analyses the shortage of least square localization algorithm (LSL) in WSN_s, and then proposes two improved least square algorithms. It is the distance clustering in LSL(LSL-DC) and the distance clustering and minimum related distance in LSL(LSL-DCR). In the two improved algorithms, we give a new method to choose the anchor nodes with the clustering. The simulation results indicate that the proposed algorithms can improve the localization precision.

Keywords: WSN, localization precision, clustering, least square method.

1 Introduction

The Wireless sensor networks (WSN_s) consist of a collection of wireless networked low-power sensor devices, with each node integrating an embedded microprocessor, radio, and a limited amount of storage[1]. It comprehensively uses the microelectronics technology, embedded computation technique, wireless communication technology, distributed information processing technology. Its aim is to perceive, collect and process the information of the perceived objects by the distributed sensors in the region, and sends the information to the observers. With the recent advances in WSN_s research, applications employing WSN_s have become quite popular[2]. Today, WSN_s are widely used in industrial automation, traffic control, environmental monitoring, and military reconnaissance and so on. To make the information obtained from the sensor nodes effectively, we should also precisely know the location of sensor nodes. Therefore, the basis and key technology is the precise localization of the nodes in WSN_s, and the information which lacks precise localization is meaningless. The localization of nodes in WSN_s can be achieved by the perceived information and information processing of nodes.

To achieve more accurate localization, the localization algorithms of the WSN_s need to be improved. Savarese proposed two localization algorithms: cycle accuracy-Cooperative

* Corresponding author.

A. Xie & X. Huang (Eds.): Advances in Computer Science and Education, AISC 140, pp. 447–452.
springerlink.com

ranging[3] and Two-Phase localization[4] that can decrease the influence of distance error to localization; Avvides[5] proposed n-hop multilateration primitive localization algorithm, and he used Kalman filtering technique to calculate the accurate coordinates circularly, it reduced the accumulation error; The existent localization refinement algorithms are based on the circulatory method[6]-[10]. At present, the refinement methods that already existed can reduce localization errors in different degree, but there are still many deficiencies:

1) The existent refinement methods focus on the improvement of algorithms, and obtain high precision with the method of circulation, but its data operations are excessive, and it increases the energy of the nodes.

2) The measures of the refinement algorithm are multiple, and the information exchange between the nodes is frequent, so it increases the cost of the networks' energy.

Based on the above issues, this paper aims at the influence of the anchor nodes in the least square algorithm, and then optimizes the selection of the anchor nodes. So the goal to improve the localization accuracy can be achieved.

2 The Model of Nodes' Localization

The trilateration is a classical deterministic localization method. We adopt the trilateration to calculate coordinate of the unknown node. Set the coordinate of unknown node D (x, y), and the coordinates of A, B, C 3-point are known as (x_1, y_1), (x_2, y_2), (x_3, y_3). The distance from unknown node D to the known nodes A, B, C is d_1, d_2, d_3 in order. Then we can get the following equations:

$$\begin{cases} (x - x_1)^2 + (y - y_1)^2 = d_1^{\,2} \\ (x - x_2)^2 + (y - y_2)^2 = d_2^{\,2} \\ (x - x_3)^2 + (y - y_3)^2 = d_3^{\,2} \end{cases} \quad (1)$$

With the simultaneous equations, we can calculate the coordinate of the unknown node D.

3 Localization Refinement Algorithm

3.1 Least-Square Localization Algorithm (LSL) for Localization Refinement

The LSL algorithm is a mathematical optimization technique. It looks for the best function matching of the data in the condition of minimizing the square sum of errors.

Supposing the number of anchor nodes is n, then

$$\begin{cases}(x_1 - x)^2 + (y_1 - y)^2 = d_1^2 + e_1 \\ (x_2 - x)^2 + (y_2 - y)^2 = d_2^2 + e_2 \\ \vdots \\ (x_{n-1} - x)^2 + (y_{n-1} - y)^2 = d_{n-1}^2 + e_{n-1} \\ (x_n - x)^2 + (y_n - y)^2 = d_n^2 + e_n\end{cases} \tag{2}$$

Let the first, second… (n-1)th equation subtracts the nth equation, then we can calculate the accurate location of the unknown node.

3.2 The Distance Clustering in LSL (LSL-DC)

3.2.1 Clustering

The clustering method depends on a matrix of independent variables to classify different individuals. The individuals that have similar properties are classified together, while the individuals that have high differences properties are classified different classes. This paper adopts the hierarchical clustering method to cluster the nodes, and the similarity of the nodes is measured by the distance of the nodes.

3.2.2 The Distance Clustering in LSL (LSL-DC)

As some location information are lost during the process of elimination and reduced-order, so it increases some extra errors. Based on the method of LSL, we calculate the coordinate (x, y) of the unknown node in the condition of minimum $\sum_{i=1}^{n-1}(e_i - e_n)^2$, but $\sum_{i=1}^{n}(e_i)^2$ can not get minimum at the same time. So we attempt to seek the condition that $\sum_{i=1}^{n-1}(e_i - e_n)^2$ and $\sum_{i=1}^{n}(e_i)^2$ get minimum at the same time. This paper introduces the thought of clustering and takes the distance between the anchor node and the unknown node as the parameter to cluster. We select the cluster that has the maximum and more than 3 nodes as the corrected anchor nodes, and the distance between each corrected anchor node around the unknown node and unknown node is approximate after clustering, as well as the distance error. Then the LSL algorithm is adopted to calculate accurate location of the unknown node. Its steps are shown as follows:

a) Aiming at the unknown node, calculate the distance between the all anchor nodes around the unknown node and the unknown node, and then form the distance set.
b) Cluster the distance set and select the set that has the maximum and more than 3 nodes as the corrected anchor nodes to calculate the coordinate of the unknown node.
c) The least square algorithm is adopted to calculate accurate location of the unknown nodes.

3.3 The Minimum Related Distance in LSL-DC(LSL- DCR)

The method of LSL-DC can further improve the localization accuracy, however, it still exists the problem that how to further select the corrected anchor nodes reasonably. Based on the method of LSL-DC, this paper further selects the corrected anchor nodes to get better localization accuracy, and proposes the method of LSL-DCR. That is to say, selects the central point from $\{e_1, e_2, \cdots, e_n\}$as the benchmark anchor node, and then the LSL-DC algorithm is adopted to improve the localization accuracy.

The specific steps to select the benchmark anchor node are shown as follows:

1) Select the position(line number) of the distance of maximum polymerization degree in the distance matrix.
2) Take a distance as a number.
3) Each distance subtracts other distance, and then calculates the summation of the differences.

$$DClus_i = \sum_{j=1\&j\neq i}^{n}(BeaD_i - BeaD_j) \tag{3}$$

4) Take the minimum i to Dmin, and take the maximum to Dmax, and then take the corresponding anchor node of Dmin as the benchmark anchor node.

The steps of LSL-DCR are shown as follows:

a) Aiming at the unknown node, calculate the distance between the all anchor nodes around the unknown node and the unknown node, and then form the distance set.
b) Cluster the distance set and select the set that has the maximum and more than 3 nodes as the corrected anchor nodes to calculate the coordinate of the unknown node.
c) Aiming at the clustering that has been selected, we calculated the central point from distance set, and take the corresponding anchor node of the central point as the benchmark anchor node in the process of reduced-order.
d) The least square algorithm is adopted to calculate accurate location of the unknown node.

4 The Simulation Results

The performance of the proposed localization algorithms can be evaluated by a series of simulations.100 unknown nodes are randomly deployed, and there are 7 random anchor nodes around each unknown node. The error is a random distribution and is proportional to the distance. The accurate distance between the unknown node and the anchor node is dc^*. Set the measurement error of the distance is $dc^* \times er$, in which er is the random noise of [-0.3, 0.3] and its mean value is 0.

The localization error of each unknown node is the distance between the calculated coordinate and standard coordinate. In consideration of the whole situation, we adopt the mean square deviation of the all nodes' localization error.

We compare the estimated locations of a number of nodes with their actual locations. The comparison of corrected effect among the three algorithms is shown as follows:

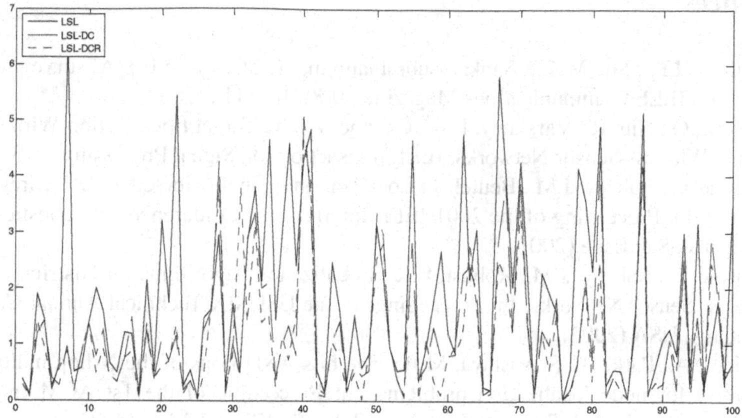

Fig. 1. Corrected effect of the three algorithms

The comparison of errors among the three algorithms is shown as table1.

Table 1. The comparison of average errors among the three algorithms

algorithm	LSL	LSL-DC	LSL-DCR
Average error	1.7998	1.1976	1.0939

Based on the thought of clustering, we can conclude that the localization accuracy of the most nodes has greater improvement than LSL, and the phenomenon of big errors is eliminated partly. As shown in fig1 and table1, it is clear that the LSL-DCR is better than others.

5 Conclusion

This paper analyzes the affecting factors of localization algorithms in WSN_s, and the key novelty of the proposed algorithms is the use of the method of clustering. We have also compared the localization accuracy of the unknown nodes among the three proposed algorithms and the simulations have been conducted to show the effectiveness of the localization algorithms based on distance clustering. This paper mainly aims at the nodes of simulation in 2d layout environment. For future study, we will do more researches on the 3d physical model of the nodes.

Acknowledgments. Project supported by the National Natural Science Foundation of China (Grant No.60972162), 2010.1-2012.12.

References

1. Akyildiz, I.F., Su, W.L., Sankarasubramaniam, Y., Cayirci, E.: A survey on sensor networks. IEEE Communications Magazine 40(8), 102–114 (2002)
2. Ozdemir, O., Niu, R., Varshney, P.K.: Channel Aware Target Localization With Quantized Data in Wireless Sensor Networks. IEEE transactions on Signal Processing 57(3) (2009)
3. Savarese, C., Rabaey, J.M., Beutel, J.: Localizationing in distributed ad-hoc wireless sensor network. In: Proceeding of the 2001 IEEE International Conference on Acoustics, Speech, and Signal, Salt Lake (2001)
4. Savarese, C., Rabaey, J.M., Robust, L.K.: Localization Algorithms for Distributed Ad Hoc Wireless Sensor Networks. In: Proceedings of the USENIX Technical Annual Conference, Monterey, USA (2002)
5. Avvides, A., Park, H., Srivastava, M.B.: The bits AND flops of the N-hop multilateration primitive for node localization problems. In: Proceeding of the 1st ACM International Workshop on Wireless Sensor Networks and Applications, Atlanta (2002)
6. Bergamo, P., Mazzini, G.: Localization in sensor networks with fading and mobility. In: Processing of the 13th IEEE International Symposium on Personal, Indoor and Mobile Radio Communications, NJ, USA (2002)
7. Guha, S., Murty, R.N., Sirer, E.G.: A unified node and event localization framework using non-convex constraints. In: Proceeding of the 6th ACM International Symposium on Mobile Ad Hoc Networking and Computing, IL, USA (2005)
8. Cao, M., Anderson, B.D.O., Morse, A.S.: Sensor network localization with imprecise distances. Systems and Control Letters 55, 887–893 (2006)
9. Srirangarajan, S., Tewfik, A.H., Luo, Z.Q.: Distributed sensor network localization with inaccurate anchor positions and noisy distance information. In: IEEE International Conference on Acoustics, Speech and Signal Processing, HI, USA (2007)
10. Guo, Y., et al.: Hop-based Refinement Algorithm for Localization in Wireless Sensor Networks. Computer Engineering (3), 145–147, 151 (2009)

About the Author(s)

Xiaohui Chen, he received his bachelor's degree from Wuhan University of Technology (WHUT) in 1988 and the master's degree from Huazhong University of Science and Technology (HUST) in 1996, He is currently an associate professor in the college of computer and information technology of China Three Gorges University. His research interests include wireless sensor networks, data mining, and intelligent control.

Jinpeng Chen, he received his bachelor's degree from Huazhong University of Science and Technology (HUST) in 2010, he is seeking for his master's degree in the college of computer and information technology of China Three Gorges University now. His research fields include wireless sensor networks, intelligent control, and embedded system.

Research and Analysis of the Social Access to College Sports Venues in Wuhan

Han Jiang

Wuhan Institute of Physical Education, Wuhan, Hubei, China 430079

Abstract. Sports venues, not only is a major component of college teaching resources, but also undertakes the task of holding sports and games for the surrounding communities, enterprises and government institutions. This paper, adopting the methods of documentation, interview, questionnaire, mathematics statistics and logical analysis, conducts a comprehensive research on the social access to the college sports venues in Wuhan as well as explores an effective way for social access to the college sports venues.

Keywords: Colleges in Wuhan, Sports venues, Social access.

1 Introduction

Along with the rapid development of society and economy as well as the advancing of the people's living standard and the implement of the nationwide fitness campaign, people's desire for sports participation for the sake of health-keeping has never been stronger. Therefore, the demand for the basic material guarantee-- sports venues-- has been on the rise, which leads to the expose of the obvious problem of lacking sports venues. In view of the fact that social sports venues in existence could not meet people's needs and college sports venues not being fully utilized, it is a matter of great urgency to research on how to open and utilize the college sports venues. This paper aims at probing how to realize full social access to the college sports venues through investigating the present status of the distribution of college sports venues and social access to them. Meanwhile, it also tries to analyze and study the advantages of social access to college sports venues as well as the influencing factors, so as to explore effective methods appropriate to cope with social access to college sports venues in Wuhan.

2 Research Object and Methods

2.1 Research Object

The object of this paper is the present status of distribution of college sports venues and social access to them. A general survey has been conducted in the 37 chosen colleges and universities in Wuhan.

A. Xie & X. Huang (Eds.): Advances in Computer Science and Education, AISC 140, pp. 453–458.
springerlink.com

2.2 Research Methods

This paper adopts the methods of documentation, interview, questionnaire, mathematical statistics and logical analysis. Questionnaire is used to investigate the present condition, management, advantages and difficulties of social access to college sports venues as well as the influencing factors. 37 questionnaires are administrated, of which 30 are effective (see table 1).

Table 1. Administration of questionnaires

NO. of questionnaire distributed	NO. of questionnaire collected	Percentage of collection	Effective questionnaire	Percentage of effective questionnaire
37	30	81.1%	27	90.0%

3 Research Result and Analysis

3.1 Present Status of Social Access to College Sports Venues

Paid social access to college sports venues should be on the basis of maintaining the regular sports work inside colleges and for the purpose of addressing the problem of sports venues shortage in society, meeting the public's demands for the gymnasiums fulfilling their self-value as well as covering part of sports expenses. In the chosen 37 colleges and universities, 32 of them open their sports venues to society, accounting for 81.5%. 44.4% of these colleges adopt the model of combining paid and free social access together. 5 of 37 colleges and universities do not open their sports venues to society, accounting for 18.5%. These statistics show that some of the colleges are conscious of opening to the public and are able to develop and make full use of their sports venues.

Table 2. Social access to college sports venues in Wuhan

Types of social access	Combination of paid and free social access	Free social access only	No social access	Total
No. of colleges	15	6	5	27
Percentage	44.4%	37.1%	18.5%	100%

3.2 The Content of Social Access to College Sports Venues in Wuhan

83.5 % of the colleges, which bring into effect paid social access, mainly conduct sports-related business, 16.5% of them engage themselves in comprehensive business including non-sports services. However, sports-related business still accounts for 83% and non-sports service only 17 %(see table 3). Our survey shows that the content of paid social access is mainly mass sports activities.

Table 3. Order of the content of paid social access

Content	Times of selection	Percentage	Order
Mass sports	14	93.4%	1
Training course	10	66.7%	2
Social gathering	9	60.0%	3
Exhibition hall	3	20%	4

Table 4 clearly shows that there are three levels of the content of the paid social access to college sports venues. Badminton and table-tennis is in the first level; football, tennis and swimming constitute the second level and last level includes basketball, volleyball, and aerobics. Consumers are willing to spend money on their beloved activities. Therefore, in order to attract more people to participate sports activities, the development of social access to college sports venues should be oriented towards the needs of the masses and conduct businesses in light of the market demand.

Table 4. Order of sports service in paid social access

Items	Times of selection	Percentage	Order
Badminton	14	93.3%	1
Table-tennis	14	93.3%	2
Football	12	80%	3
Tennis	10	66.7%	4
Swimming	7	46.7%	5
Basketball	5	33.3%	6
Aerobics	4	26.7%	7
Volleyball	3	20.0%	8

3.3 Charging Method for Social Access to College Sports Venues

Colleges and universities in Wuhan adopt a single mode of charging. Most colleges (about 86.7%) charge all exercisers a single fee for sports venues used, i.e. consumers pay for the sports venue you use every single time. Some colleges (about 46.7%) charge the morning exercisers per month. 66.5% colleges choose to collect field rentals for external units to hold sports in the college sports venues. 13.5% of colleges charge external units by day. Take the South-Central University for Nationalities as an example, it charges the external units and individual for the access to the swimming pool by using a combination charging method of individual charge, group chartering and annual card.

Table 5. Charging method for the paid social access to college sports venues

Charging method	All exercisers	Fitness card for different time period	Membership System	Field rentals for sports	Group charter charges
NO. of colleges	13	7	3	10	10
Percentage	86.7%	46.7%	20%	66.7%	66.7%

4 Conclusions and Proposals

4.1 Conclusions

- The survey shows that there are plenty of outdoor sports venues but relatively fewer indoor ones. The distribution of the college sports venues indicates: how many sports venues a college possessing has little to do with the types, affiliation and geographical location of the college but is closely related to the scale of the college. A college of comparatively larger scale tends to boast more sport revenues.
- The college sports venues in Wuhan enjoy less social access compared with those in the front-line and coastal cities in that the portion of paid social access is much lower.
- Those college sports revenues which enjoy paid social access mainly open at noon and in the evening. 54.3% of the surveyed college sports venues choose to open all th e year round; while those for free access tends to open only on workdays.
- As for the types of the college sports revenues which enjoy social access, swimming pool boasts the highest social access rate. Then are badminton halls, table-tennis rooms, football fields, tennis courts, basketball gyms and gymnasiums for body building. Last are the free outdoor sports venues, such as open and idol fields for basketball, volleyball and tracking.
- The factors affecting social access to college sports venues in Wuhan mainly includes merits of the college itself, the conditions of the sports facilities in sports venues, the opening time, the content of service, financial input of the college, college leaders' mindset, the publicity and publicity effort, the management system of the sports venues, security situation of the facilities and in colleges as well as the financial state, consciousness and behaviors of physical exercises of the surrounding neighborhood.

4.2 Proposals

- Social access to college sports venues should be based on that the needs of teaching and training shall be met firstly. Then can each college decides whether to conducat the social access or not according to their own conditions. Paid social access shall be in coordination with teaching and training. On top of that colleges shall make a reasonable time schedule for social access to the sports venues and try to conduct more paid social access.
- Colleges which have social access to their sports venue should realize that government plays a leading role in this respect. Government of all levels shall support the colleges in policy making and financial input, etc. Meanwhile, the colleges shall seek support, understanding and cooperation from all government departments, other colleges and institutions, community committee as well as the public.
- A scientific management system shall be established, workable measures shall be taken, security, inspection and maintenance of the sports venues and facilities shall be strengthened to ensure orderly social access to college sports venues.
- Colleges shall choose the type of social access in line with their own situations. The social access to the college sports venues should take society as the platform and be oriented to market needs so as to set up charging standard and the time for social access.
- Colleges should try to realize social access through various channels, including holding high-level activities like sports games and performances; organizing college fitness clubs and sports training courses; cooperating with local community committee to carry out public fitness activities; strengthening the relationship with local department in charge of teaching and physical education to set up activity base for physical education.

References

1. B. X: Research on the Management problem of China University City Gymnasium Resource——Taking Guangzhou University City as the Example. Journal of Guangzhou Physical Education Institute (05) (2007)
2. F. Y: Investigation on the Operation of Some College Sports Facilities in China. Sports Science Research (2) (2005)
3. G. M. W: Feasibility of Social Access to College Sports Venues and Facilities– Taking Nanjing as an Example. Social Sciences in Nanjing (9), 81–84 (2003)
4. J. C. Y: Comparative Research of Foreign and Chinese Sport Ground. Journal of Shenyang Sport University (4), 25–27 (2006)
5. L. W.: The Opening of Sports Stadiums and Grounds in Institutes and Universities of Hubei Province to the Mass Sports. Journal of Wuhan Institute of Physical Education (6), 170–172 (2003)
6. L. L. Z.: Research on Influential Factors of The benefit of stadium resources in Colleges and Universities. Sports & Science (6) (2005)
7. Q. X. H: The current state of and measures for the paid opening of sports venues in Guangzhou University City. Journal of Physical Education (7) (2010)

8. T. L. Z.: Sports Management and Administration, vol. 4, p. 144. Fudan University Press, Shanghai (2004)
9. X. C.: Exploration on the Construction of Chinese Sports Venues and their Management in the 21st Century. Journal of Xi'an Physical Education University (4) (2002)

Prediction of Zenith Tropospheric Delay Based on BP Neural Network

Yong Wang[1,*], Lihui Zhang[2], and Jing Yang[2]

[1] Hebei Researching Center of Earthquake Engineering, Tangshan, Hebei, 063009, China
[2] College of Mining Engineering, Hebei United University, Tangshan, Hebei, 063009, China
wangyongjz@126.com

Abstract. This paper used improved Levenberg-Marquart BP neural network technology to predict GPS tropospheric delay with the data of Beijing GPS data and meteorological data, which the input parameters are either space position (longitude, latitude, elevation) or space position, temperature, pressure. The testing results showed that the two predicted results are mostly closer; meanwhile the prediction results considering space position, temperature and pressure are slightly superior to the other method. The deviations between the predicted values of most GPS Stations and the actual atmospheric delay values are about 5mm. The accuracy of tropospheric delay predicted by BP neural network amounted to millimeters.

Keywords: BP Neural Network, Zenith Tropospheric Delay, GPS.

1 Introduction

In recent years, Interferometric Synthetic Aperture Radar (InSAR) is a new technology of detecting the surface of the earth. It shows excellent application in the 3-D reconstruction of the terrain, the deformation field detection of the earth surface and the land use classification. The most important factors impacting the accuracy of InSAR are the variable atmospheric conditions and different temporal characters which the effects amount to a single pixel. The variable atmospheric conditions lead to different phase delay, the inconformity performance not only on time scale but also on spatial scale. The serious atmospheric delay even covers other interesting signals. How to improve atmospheric effect is a key problem of further study of InSAR data process. The atmospheric delay correction of InSAR using GPS is a better method of the atmospheric delay correction of InSAR. For the spatial resolution of the atmospheric delay with GPS inferior to the resolution of SAR image, it need interpolate the GPS atmospheric delay in order to correct the atmospheric delay of InSAR.

As the distances of the GPS stations are tens of kilometers commonly, in order to use GPS tropospheric delay to reach the requirements of SAR application, this paper used improved Levenberg-Marquart BP neural network technology to predict GPS

---

* Corresponding author.

A. Xie & X. Huang (Eds.): Advances in Computer Science and Education, AISC 140, pp. 459–465.
springerlink.com

tropospheric delay with the data of Beijing GPS data and meteorological data. According to the various input parameters, the used BP Network applied two methods which are space position or space position, temperature, pressure.

2 BP Neural Network

Artificial Neural Network is a complicated network system consist of a large number of neurons that those are connected each other, it has perception, memory, learn and associate function, which can imitate the physical process that the human brain process information. It has parallel simulation processing in large scale, distributed store information, nonlinear dynamics and global network function, as well as omecharacteristics, such as powerful adaptation, self-learning, fault-tolerant ability, etc, so it has become a powerful nonlinear means of information processing. BP neural network is the core part of feed forward neural network, present the most essential parts of Artificial Neural Network, widely used in function approximation, pattern recognition/classification, data compression etc.

BP neural network consist of input layer, inter layer, output layer, its learning process include two stages: the first stage is that inputting the known learning samples array, by means of setting up the network structure and the last iterative power and threshold, from input layer by inter layer to output layer, working out outputs of each layer's nodes, the second stage is that modification of power and threshold, according as the output nodes' error between the actual value and the anticipant value(error more than required precision), transmitting the error signal along original connected channels in reverse way, minishing the error by doing modification of power and threshold of each layer. Repeat the process above, until the error of network outputs minish to allowable presion or reach the scheduled learning time.

3 Prediction of Zenith Tropospheric Delay Based on BP Neural Network

Beijing is surrounded with mountains in the west, north and northeast, its southeast is a great plain gently towards the Bohai. The plain elevation is from 20 to 60 meters, the country elevation is from 1000 to 1500 meters commonly in Beijing area. The ground of Beijing area is 16410 km^2, there into, the ground of plain is 6338 km2, account for 38.6%; the ground of coteau is 10072 km^2, account for 61.4%; the ground of Beijing city is 87.1 km2 .According to table 1, the elevation of Beijing GPS stations are different obviously, it is in plain and coteau, as a mountain area with lesser hypsography only considering the stations' elevation.

The data used in this paper consist of the GPS data of some Beijing GPS stations and temperature, pressure data of these stations, the date of the data are. The GPS tropospheric delay are worked out used GAMIT by hours, the GPS tropospheric delay picked up in 1 hour, 5 hour, 9 hour, 13 hour, 17 hour, 21 hour every day regard as study data of this paper.

Table 1. Heights of GPS stations in Beijing area

GPS network	*Station name*	**Height (m)**
	BJFS	46.6
	DAXN	37.6
	SHIJ	65.6
	CHPN	76.2
Beijing GPS network	CHAO	35.3
	PING	28.1
	THKO	331.6
	MYUN	71.8
	YANQ	487.9
	ZHAI	440.3

The GPS troposphere delay of some Beijing GPS stations used in this paper is extracted, combined with longitude, latitude, elevation, temperature and pressure of the stations, used BP neural network technology improved by Levenberg-Marquart theory for training and prediction. The data of this paper are the GPS troposphere delay of five days which consist of different stations. The same date of data is as a set data, which is from 1 hour to 21hour at intervals of 3hour every day, 60 sets of data in all. For the stations with atmosphere data are minority, so adopt the following practices: regard one station of a set data as the prediction station, other stations as training stations, predicted the prediction station, up to each station of the set stations. This paper used BP neural network is a three-layer network, the network type is feed-forward back prop, the transfer functions are tansig and logsig, the training function is trainlm, the adaption learning function is learngdm, the performance function is msereg. For the mapping relation is nonlinear, the initial value is very important for the learn is whether or not to local minimum, whether or not convergence and the training time is long or short, so the preprocessing of sample data is necessary in the beginning. This paper designs the BP neural network with Neural Network Toolbox of MATLAB7.0. When the GPS troposphere delay is predicted, regarding two aspects of longitude, latitude, elevation and longitude, latitude, elevation, temperature, pressure as input factors, the GPS troposphere delay as output factor, by means of training and prediction, then the GPS troposphere delay of prediction sample is got.

For practical application of the predicted value of the GPS troposphere delay, the predicted results are checked out necessarily. The comparisons of Beijing GPS stations between the actual value and the two prediction results of the GPS troposphere delay with BP neural network are shown up in Fig 1~10.In the figures, the horizontal axis mean the date of the data, the vertical axis mean zenith troposphere delay, units is cm ,the black solid lines mean the actual value, the black dotted lines mean the predicted value 1, predicted by longitude, latitude, elevation, temperature and pressure, the black solid lines with triangle marks mean the predicted value 2, predicted by longitude, latitude, elevation.

According to the figures from 1 to 10, the two predicted results of the GPS troposphere delay are mostly close, the results predicted by longitude, latitude, elevation, temperature, pressure are slightly better than those predicted by longitude, latitude, elevation. BJFS、DAXN、CHAO、CHPN、PING 、SHIJ、THKO、MYUN, the actual value of the eight stations accord with the two predicted value, except the deviations between the predicted value of several date and the actual value are slightly much, are about 1cm,the other deviations are mostly about 5mm; the deviations of several date at YANQ station and ZHAI station are larger, the deviations between the predicted value and the actual value of YANQ on December 3, 2007 and ZHAI on August 4, 2008 are obviously different, the predicted results of other date at THKO and ZHAI and the predicted results of other stations at the same time are very good, the most deviations are about 5mm, the minority are about 1cm.

The troposphere delay consist of dry delay and wet delay, the dry delay can be worked out by meteorological data, the wet delay means that the troposphere delay removes the dry delay, the simple conversion relations between wet delay and PWV is 0.15, we can see that 1 cm error of the atmosphere delay is equal to 1.5 mm error of water vapor, 2 cm error of the atmosphere delay is equal to 3.0 mm error of water vapor.

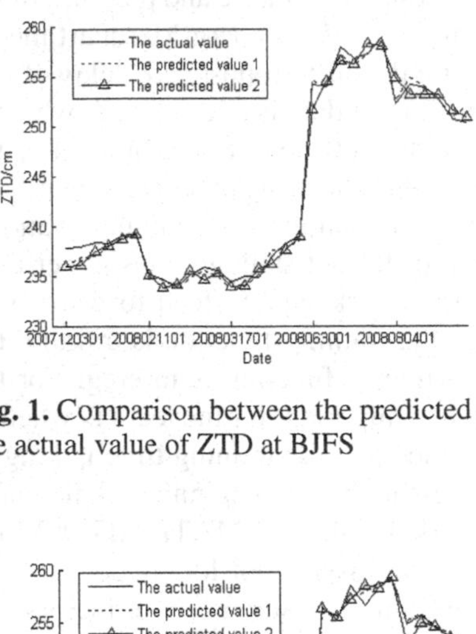

Fig. 1. Comparison between the predicted and the actual value of ZTD at BJFS

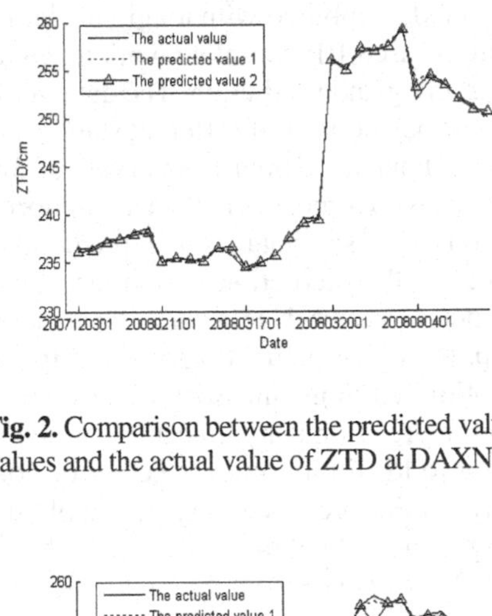

Fig. 2. Comparison between the predicted values values and the actual value of ZTD at DAXN

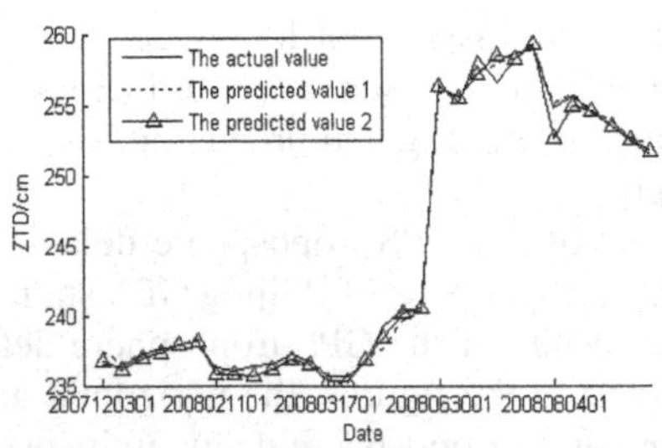

Fig. 3. Comparison between the predicted and the actual value of ZTD at CHAO

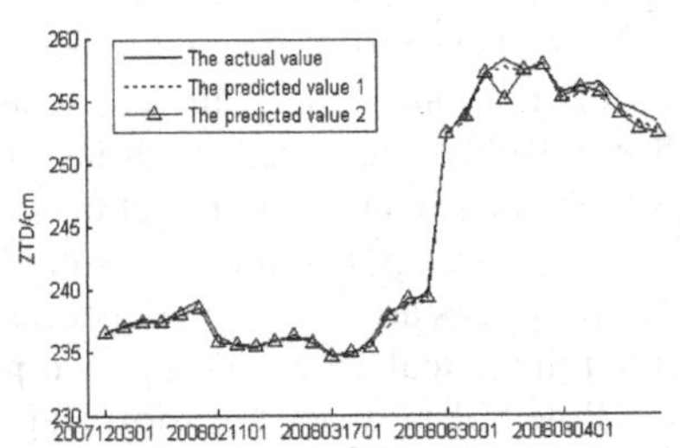

Fig. 4. Comparison between the predicted values values and the actual value of ZTD at CHPN

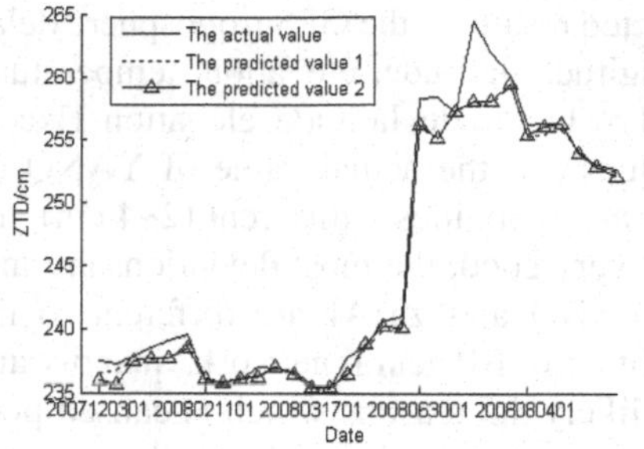

Fig. 5. Comparison between the predicted and the actual value of ZTD at PING

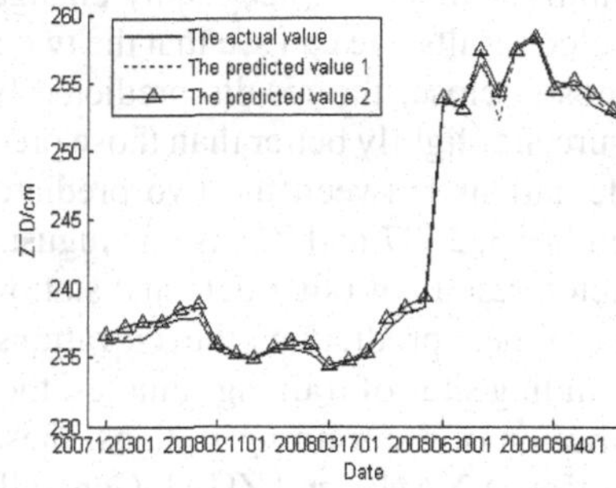

Fig. 6. Comparison between the predicted values values and the actual value of ZTD at SHIJ

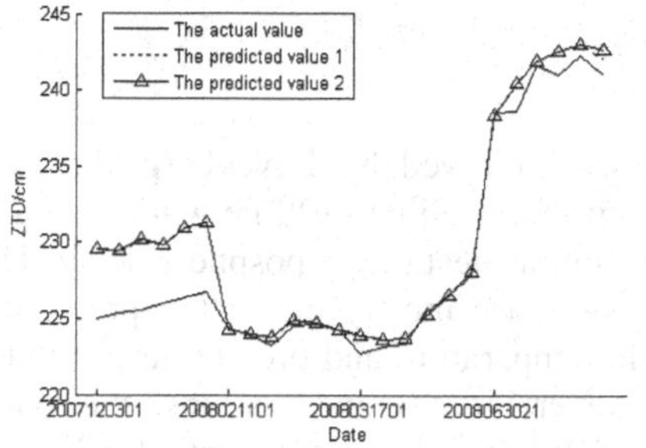

Fig. 7. Comparison between the predicted and the actual value of ZTD at YANQ

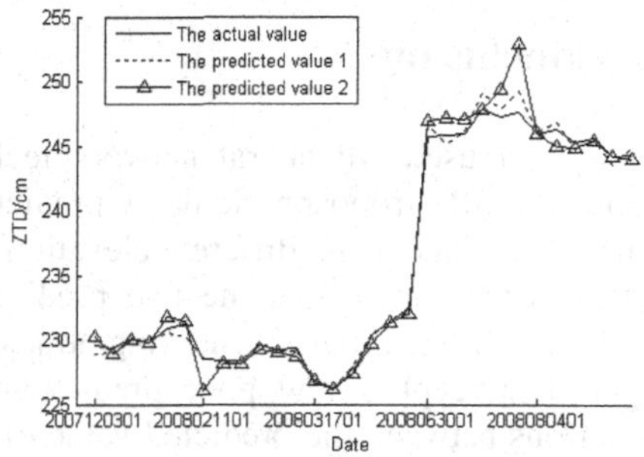

Fig. 8. Comparison between the predicted values values and the actual value of ZTD at THKO

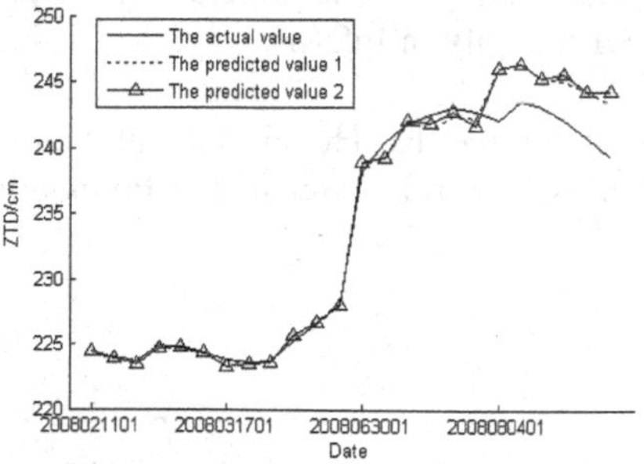

Fig. 9. Comparison between the predicted and the actual value of ZTD at ZHAI

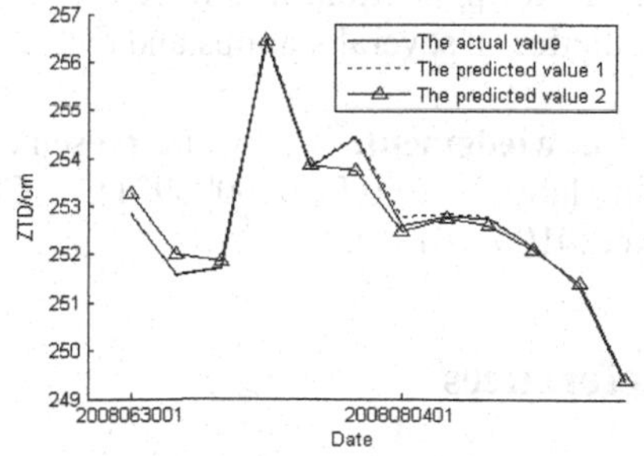

Fig. 10. Comparison between the predicted values values and the actual value of ZTD at MYUN

From the above we can see that the elevations of Beijing GPS stations are from 28 to 490 meters, the elevations of seven stations among the stations used in the study are less than 80 meters, the elevations of the other three stations are respectively 331.6, 440.1 and 487.9 meters, the elevations changes greatly, the area of the stations regards as the mountain area with hypsography changes commonly. From the overall effect of the predicted results, we can see that the two predicted results of the GPS troposphere delay are mostly close, the results predicted by longitude, latitude, elevation, temperature, pressure are slightly better than those predicted by longitude, latitude, elevation. Except the deviations between the two predicted values and the actual value of YANQ on December 3, 2007 and ZHAI on August 4, 2008 are obviously different (2~4 cm), the predicted results of other date and stations are very good, the most deviations are mm levels; When predicting, the elevations of YANQ and ZHAI are different to the surrounding sites of training samples, the condition of BP neural network, quantity and distribution of the training stations, these are likely the reasons which is caused poor prediction at YANQ and ZHAI. Generally, the predicted results are better than general spatial interpolation method, the predicted results considering longitude, latitude, elevation, temperature, pressure have better effects, if there are other factors relatively, likely to get better results or the reason for poor prediction at several stations and dates.

4 Conclusions

This paper used BP neural network technology improved by Levenberg-Marquart theory for GPS tropospheric delay prediction with the Beijing GPS network data and atmosphere data, using different elevation GPS stations test the troposphere delay. The testing results show that the two predicted results are mostly close; the prediction results of GPS stations considering atmospheric temperature and pressure are slightly superior. Except a few poor predictions at several dates and stations, the other deviations between the predicted value of most GPS Stations and the actual value are about 5mm and a few stations are about 1cm. In general speak, the precision of BP neural network achieved inferior centimeter level, most of the sites amounted to millimeters. Restricted in the data in the study, the condition of BP neural network, it is likely to get poor predictions of several dates and stations, if considering more relative factors for prediction, it is likely to get more ideal results or the conclusion for poor prediction at several stations and dates, it's better to apply in InSAR.

Acknowledgment. The work presented was supported by Hebei Natural Science Foundation (No.D2010000921), China Postdoctoral Science Foundation (No.20100470144).

References

1. Liu, G.: Application examples of InSAR and its limitation analysis. Surveying and Mapping of Sichuan 28(3), 139–143 (2005)
2. Shan, X., Liu, H.: Detecting digital elevation model by InSAR technique. Remote Sensing for Land & Resources (2), 43–47 (2001)
3. You, X.: Synthetic aperature radar interferometry and its application. Journal of Geodesy and Geodynamics (3), 109–116 (2002)
4. Lu, X.: Experiment on land subsidence monitoring by INSAR. Journal of Geodesy and Geodynamics (4), 66–70 (2002)

5. You, X.: Quantitative estimation of effect of atmospheric refraction on InSAR. Journal of Geodesy and Geodynamics (2), 81–87 (2003)
6. Qiao, X.: Acquisition of DEM of Three Gorges area by INSAR. Journal of Geodesy and Geodynamics 23(2), 122–126 (2003)
7. Liang, W., Wang, Q.: Application of INSAR to monitoring and studying volcano. Journal of Geodesy and Geodynamics 23(4), 120–124 (2003)
8. Li, Z., Liu, J., Xu, C.: Error analysis in InSAR data processing. Geomatics and Information Science of Wuhan University 29(1), 72–76 (2004)
9. Xu, J.: Spatial interpolation methods for correcting atmospheric effects using interferometric SAR. Journal of Electronics & Information Technology (4), 911–915 (2008)
10. Wang, Y., Jiao, J., Zeng, Q.: Study on Zenith Tropospheric Delay Interpolation Methods for different terrain. Journal of Geodesy and Geodynamics 30(3), 132–136 (2010)

Advance in Scence Change-Based H.264 Rate Control Methods

Feifei Lu[1] and Xiao Chen[1,2]

[1] School of Electronic and Information Engineering,
Nanjing University of Information Science and Technology,
Ningliu Road 219, 210044 Nanjing, China
[2] Jiangsu Key Laboratory of Meteorological Observation
and Information Processing
chenxiao@nuist.edu.cn

Abstract. Scene change occurs frequently in the video, which has an important impact on the video quality. The JVT-G012 rate control algorithm for H.264/AVC uses fixed-length group of picture structure, which can not effectively deal with the video sequence scene change, resulting in the quality decline of frames after scene change. We reviewed the scene change-based H.264 rate control methods in this paper. They are classified into two categories-scene change detection-based rate control method and scene adaptive rate control method. A comparative study of the interaction and discussion about the various algorithms and their advantages and disadvantages are presented. Based on these research results, some improvedment are discussed.

Keywords: H.264, video coding, rate control, Scene change.

1 Introduction

In the past decade, the multimedia technologies become more sophisticated with the rapid development of Internet. The video communications market more widely. To achieve video communication has become a variety of ways, such as video telephony, instant messaging, video chat, IPTV and remote monitoring, telemedicine. Unlike the picture and voice information, digital video data is large. It is is unrealistic simple to use expanded memory capacity, increased communications lines and the transmission rate of approach. The data compression technology is a proven solution. Through data compression, we can the amount of pressure down the information and data, stored in compressed form, transmitted, which saves storage space, but also improves the efficiency of the transmission lines of communication. What's more, it enables the computer to real-time processing of audio, video information, to ensure high-quality video playback, audio programs. Therefore, the video communication has a high compression requirements.

Video coding rate control is the key to the process. Its role is to ensure the rational allocation of coding bits to accommodate the network, or media storage needs. The effect of rate control not only affects the video stream bit rate stability, but also affect the image quality of video sequences.

A. Xie & X. Huang (Eds.): Advances in Computer Science and Education, AISC 140, pp. 467–471.
springerlink.com

Video sequences are bound to be a scene change. The time correlation of scene change frames greatly weakened, and will affect the video encoding results. The JVT-G012 rate chontril method in the H.264/AVC encoder uses fixed-length group of picture structure, which can not effectively deal with the video sequence scene change, resulting in the quality decline of frames after scene change. On the other hand, the method is mainly based on the linear model to determine the allocation of coding bits and quantization parameters. When there is scene change, the predicted MAD deviation larger, leading to frame up the scene change frame encoding a serious decline in quality. Here are some recent methods to deal with scence change in the video sequence.

2 Types of Scene Change

The abrupt change and gradual change are the two types of scene change. This transformation occurs in accordance with the editing features of different decisions. In the abrupt change, the image content on the performance takes an inconsistent and sudden changes. The abrupt change is directly connected to two different scenarios, without any conversion frame. The gradual change is by some way to gradually transition to the previous scene after scene. The methods include the following situations: fade in, which refers to the picture slowly emerging from the darkness gradually clear in the case; fade out, which means a clear picture gradually hidden by the darkness into the case; dissolving, which refers to the screen there is a scene fade into the darkness and then gradually clear the scene of another show.

Scene change impacts on video coding decoding image quality and is generally reflected in a sharp decline near the scene change. If there are frequent video sequence scene changes, loss of time between adjacent image correlation, the rate control algorithm will lead to inappropriate allocation of resources to be encoded, the encoding quality down, and the validity of the model for rate control parameters update reduced, thereby affecting more follow-up frame.

3 Methods Based on Scene Change

Consider the scene change rate control, there are two common ways: First, scene change detection. When scene change is detected, the termination of the current GOP is taken and we start a new GOP. This avoids the emergence of two within a GOP had spent bits number of cases. Tt can solve the problem of scene change. The second is adaptively adjust the quantization parameter based on the degree of scene change. We can make the JVT-G012 method does not take into account the complexity of the frame and the bit evenly.

3.1 Rate Control Methods Based on Scene Change Detection

Scene change processing is usually considered from two aspects: how to detect scene changes and the treatment after scene change is detected.

The scene change detection methods are the pixel comparison method [1], histogram-based detection method [2,3], the gray value-based detection method [4,5], the macroblock encoding-based detection method [6,7], DCT coefficient method [8], et al.

In the pixel comparison method, the scene change detection is by calculating the two images correspond to points of all components of the difference between the brightness or color. Then we compared with thresholds to determine whether there is scene change. The easiest way is to calculate the absolute difference of pixels and then compared with the threshold. Pixel comparison algorithm is based on a comparison between the pixels, so this method is difficult to distinguish between small area or large area of big change in small change [1].

Histogram-based Detection Method. In this method, we use the nature of brightness and color histogram changes before and after the scene change to detect scene change. Let the n-frame histogram of H (n), then the histogram of the absolute frame difference HD (n) is $HD(n) = |H(n) - H(n-1)|$. We obtained the adjacent HD HDSum: $HDSum(n) = HD(n-1) + HD(n) + HD(n+1)$. After HDSum (n) and set the threshold comparison to determine whether scene change occurs [2].

Brightness histogram value is not sensitive to the movement, which can prevent the movement of the interference. However, there is no record of this eigenvalue pixel position information, there is a completely different location for the pixels of the two may have similar histogram distribution. Then use the histogram values may cause leakage detection or error detection.

Gray Value-based Detection Method. In the scene change frame position, the absolute difference image pixel gray frame image is larger than the previous one order of magnitude or more. Therefore, by determining the difference between adjacent frames absolute size of pixel gray scale, can easily detect a scene change frame position [4].

Macroblock Encoding-based Detection Method. This method takes advantage of B frames in macroblock coding information. Specific approach is the B frame macroblock is divided into bi-directional prediction and two types of non-bi-directional prediction, which predicted macroblocks when the ratio of two-way below a certain threshold to determine when the scene switches to occur [6].

This method is highly sensitive to the movement of objects, when there is more movement of the object detection accuracy dropped immediately severe.

DCT Coefficient Method. At present, image and video compression standards are international DCT-based. Frequency domain transform coefficients with the pixel domain are closely related. Therefore, DCT coefficients of compressed video sequences can be used for scene boundary detection. If a change in the number of blocks exceeds a certain threshold, said the video sequence scene change occurred. However, only direct access to I-frame DCT coefficients, B and P frame can not be directly obtained DCT coefficients. Motion compensation must be calculated in order to calculate the B and P frame DCT coefficients. Not computationally intensive, and likely to cause miscarriage of justice [8].

Post-treatment after Scence change Detected. H.264/AVC standard rate control algorithm can not effectively deal with scene change. When the scene change frame is detected, we must give the new adjustment method to compensate for the scene change frame and adjacent frames decoded image visual quality decline, and to ensure that the cache is neither underflow nor overflow. The appropriate approach are fixed-length GOP algorithm [9,10] and variable length GOP algorithm [11].

In the fixed-length GOP algorithm, we does not change the type of scene change frame encoding, but by more for the scene change frame target bit allocation or adjust the scene change frame QP (Quantization Parameter) to improve the quality of the video scene change impact. There are obvious flaws in this algorithm. If scene change frame allocated more coding resources, it will inevitably increase the pressure on the encoder buffer, the subsequent P frame encoding quality impact. If the scene change frame can not be sufficient coding resources, it will lead to the quality of the video scene change at the dump.

In the variable length GOP algorithm, the new scene will be transformed in the first P frame to I frame as the beginning of the new GOP. That there is a new GOP partitioning problem. In reference 12, when the scene change is detected, the current GOP is divided into two GOP, the current frame from the P frames into I-frame, that frame this as a new starting point for the GOP, the consequences of doing so is leading to increased I-frame, resulting in waste of resources, rate, making the overall image coding quality degradation [12]. Literature 13 suggested that the switch frame and subsequent frames in the first P frame into I-frame, start a new GOP; then extended to the original follow-up frame GOP structure, therefore, GOP is too long will lead to later in GOP frame coding performance [13].

For scene change detection algorithm based on application needs in the detection accuracy and a compromise between the complexity of the algorithm. In general, real-time is not critical you can use a higher computational complexity algorithms, such as the first pre-coded to determine the location of scene change, then two coding switch on the scene to make appropriate treatment; and real-time of demanding applications such as video communications, require low complexity of the scene change detection method.

3.2 Scene Adaptive Methods

In scene adaptive methods, the detection of scene change is not required. We consider the scene between adjacent frames relative change. Unlike the methods based on scene change detection, there is not necessary to change the GOP structure, to explicitly determine whether scene change occurs, thus avoiding the missed and false.

In the H.264/AVC standard rate control JVT-G012 algorithm, because the average number of bits for frame allocation and MAD linear prediction inaccuracies can not switch on the scene to make the right treatment. Therefore, at present there are two types of scenarios adaptive method: adaptive adjustments MAD value [14] and adaptive distribution of bits [15].

In the literature 14, the authors proposed sliding window based on image histogram and improved method of adaptive technology. According to the image complexity and scene changes, a sliding window through the image histogram and the introduction of the ratio factor. Adaptive MAD prediction of adjustment makes it closer to the real MAD, effectively improving the rate control performance.

In the literature 15, The proposed algorithm compensates the degradation of quality by adjusting the target bit numbers dynamically.

4 Conclusion

When the video sequence is detected in the scene change, we need to be handled the scene switches to improve the image quality of subsequent frames at the same time to ensure that buffer overflow does not occur. Although there have been many effective scene change rate control methods, the image quality for video coding requirements are increasing, making the rate control method to be further developed and improved.

Acknowledgments. This work was supported by by Qing Lan Project, the National Natural Science Foundation of China (No. 10904073) and the Priority Academic Program Development of Jiangsu Higher Education Institutions.

References

1. Wang, B., Huang, Y.: Retrieval of video shot boundary detection. Infrared and Laser Engineering 29, 32–35 (2000)
2. Fernando, W.A.C., Canagarajah, C.N., Bull, D.R.: Automatic Detection of Fade-in and Fadeout in Video Sequence. ISCAS 5, 255–258 (1999)
3. Fernando, W.A.C., Canagarajah, C.N., Bull, D.R.: Fade-in and Fade-out Detection in video sequence using Histograms. In: ISCAS 2000-IEEE International Symposium on Circuit and Systems, pp. 259–262 (2000)
4. Luo, L.J., Zhou, C.R., He, Z.Y.: A New Algorithm on MPEG-2 Target Bit-Number Allocation at Scene Changes. IEEE Trans. Circuit. Syst. Video Technol. 7, 136–147 (1997)
5. Yu, Y., Zhou, J., Wang, Y.L.: A Fast Effective Scene Change Detection and Adaptive Rate Control Algorithm. Image Processing 2, 379–382 (1998)
6. Lelescu, D., Schonfeld, D.: Real-time scene change detection on compressed multimedia bitstream based on statistical sequential analysis. Multimedia and Expo 2, 1141–1144 (2000)
7. Lelescu, D., Schonfeld, D.: Statistical sequential analysis for real-time video scene change detection on compressed multimedia bitstream. IEEE Transactions on media 5, 106–117 (2003)
8. Aiman, F., Hsu, A., Chiu, M.-Y.: Image processing on compressed data for large video databases. In: Proceedings of First ACM International Conference on Multimedia, pp. 267–272 (1993)
9. Jiang, M., Ling, N.: On enhancing H.264 /AVC video rate control by PSNR-based frame complexity estimation. IEEE Trans on Consumer Electronics 1, 281–286 (2005)
10. Lee, C., Lee, S.: Real-time H.264 rate control for scene-changed video at low bit rate. In: ICCE 2007, pp. 1–2 (2007)
11. Lee, J., Shin, I., Park, H.: Adaptive intra-frame assignment and bit-rate estimation for variable GOP length in H.264. IEEE Trans. on Circuits and Systems for Video Technology 10, 1271–1279 (2006)
12. Yu, Y., Zhou, J., Wang, Y.: A fast effective scene change detection and adaptive rate control algorithm. Journal of China Institute of Communications 20, 50–55 (1999)
13. Wang, L.: Rate control for MPEG video coding. In: Proc. of SPIE, vol. 2501, pp. 53–64 (1995)
14. Duan, D.: An Improved H.264 Rate Control Algorithm. Microelectronics & Computer 25, 90–94 (2008)
15. Bai, X.: Research on scene adaptive rate control algorithm. Microcomputer Information 25, 110–114 (2005)

Optimizing the Network Topology in Gnutella P2P Networks

Jin Bo

Science and Technology Department, Shunde Polytechnic,
Foshan 528333, P.R. China
wlmingyue@163.com

Abstract. We first introduce the searching mechanism in Gnutella peer-to-peer (P2P) networks, and then analyze the poor scalability in such networks. To improve the search performance of Gnutella networks, we propose a new method which is used to dynamically optimize the network topology. The simulation results show that the policy can effectively reduce the network resources consumption and optimize the load-balance between nodes, and then improve the scalability of Gnutella network and the search efficiency.

Keywords: Gnutella, search efficiency, network topology, load-balance.

1 Introduction

The peer-to-peer (P2P) resource-sharing revolution started with the emerging of Napster [1] in 1999. Napster was the first system to recognize that requests for popular content need not be sent to a central server but instead could be handled by the many hosts, or nodes, that already have the content. Such serverless P2P systems can achieve astounding aggregate download capacities without requiring any additional expenditure for bandwidth or server farms. Moreover, such P2P resource-sharing systems are self-scaling in that as more nodes join the system to look for resources, they add to the aggregate download capability as well.

However, to take advantage of this self-scaling behavior, a node looking for resources has to find the nodes that have the desired content. Napster utilized a centralized search method based on resource lists provided by each node. By centralizing search (which does not require much bandwidth) while distributing download (which does), Napster achieved a highly functional hybrid design.

These centralized systems have been replaced by new decentralized resource sharing systems such as Gnutella [2] which distribute both the download and search capabilities. These systems establish an overlay network of nodes. Queries are not sent to a central server, but are instead distributed among the nodes. Gnutella, the first of such systems, uses an unstructured overlay network in that the topology of the overlay network and placement of resources within it is largely unconstrained. It floods each query across this overlay with a limited scope. Upon receiving a query, each node sends a list of all content matching the query to the query generator.

A. Xie & X. Huang (Eds.): Advances in Computer Science and Education, AISC 140, pp. 473–477.
springerlink.com

Because the load on each node grows linearly with the total number of queries, which in turn grows with system size, this approach is clearly not scalable.

2 Search Efficiency in Gnutella Network

2.1 Search Scheme in Gnutella Network

Gnutella is a decentralized P2P resource-sharing model developed in the early 2000 by Justin Frankel's Nullsoft [3]. Gnutella's development was halted shortly after its results were made public, and the actual protocol was reverse engineered using the code that was downloaded from the Nullsoft's web site just before its closure. Today there are numerous applications (referred to as Gnutella clients) that employ the Gnutella protocol in their own individual way and that allow their users to access the Gnutella network.

To share resources, a node starts a Gnutella client A on its local networked computer. This client will then connect to an already-existing Gnutella client B, finding its address through some out-of-band means. Now, client B will announce to all of the clients it knows (its neighbors) that a new client has joined the network. This process occurs recursively out into the network, until the announcement message travels a certain distance.

Similarly, when querying for a resource, client A will send out a query message telling its neighbors that it is looking for a certain resource. As other clients receive this message, they check their locally stored resources to see if any of them match. If a match is found, a Query_Hit message is returned to the sender along the path taken by the query. Subsequent to checking for local matches, the client repeats the broadcasting of the query message to all of its neighbors. The amount of messages, and hence bandwidth, required for a query is clearly exponential in the breadth and depth of the broadcast; moreover, if a resource exists in the network, it is not guaranteed to be found if the query message does not reach a client that is sharing the resource.

2.2 Scalability of Search Scheme in Gnutella Network

Flooding in Gnutella network has several merits: short response time, large coverage, and high reliability. A measurement-based study conducted in 2000 and 2001 have shown that 95% of the nodes in a Gnutella system could be reached within seven hops by flooding [4]. This is because more and more nodes join to route the message in parallel while the flooding is going on, and the number of nodes reached could be increased exponentially. The departure or failure of individual nodes can hardly have a disruptive impact on the system ability to transmit messages in flooding, because all possible routes in the specified neighborhood are utilized simultaneously. For these merits, flooding is widely used in various unstructured P2P systems.

However, with the increasing popularity of P2P systems and the rapidly expanding system scales, we are meeting a serious problem of flooding. Excessive traffic overhead is caused by a large number of redundant message forwarding, particularly in a system with a high connectivity topology. When multiple messages with the same message ID are sent to a node by its multiple neighbors, all, except for the first message, are considered as redundant messages. These redundant messages are pure overhead: they increase the network transfer and node processing burden without enlarging the propagation scope. As a result, some researchers think that fully decentralized Gnutella-style systems with flooding do not scale [5-6]. It was estimated that the total traffic on a Gnutella system of 50,000 nodes, where flooding search is used, accounted for about 1.7% of the total traffic over the US Internet backbone in December 2000 [4]. Considering that this volume is only the amount of messages for resource search and does not include resource transfer traffic, which is out of the Gnutella overlay, flooding search becomes a bottleneck for the scalability of unstructured P2P systems.

3 Optimizing Network Topology in Gnutella Network

Considering the dynamicity and the node heterogeneity in Gnutella network, we propose a new topology optimization algorithm. In the proposed network, network topology can be dynamically evolved according to the resource distribution among nodes so that queries in the system can be routed to the nodes with amount of resources.

(1) Node N checks if the number Number_Neighbor_n of its neighbors reach its maximum number res_n
(2) if Number_Neighbor_n $\geqq$ res_n then node n select the node m with lowest neighbor number, and do the following things:
Node n check if m is over workload, if res_m - $Number\_Neighbor_m$<0, then node n disconnect the connection to node m, and also delete the neighbor information in Neigh_Infor table. Or else, node n will do the following things:
1) Node n looks for the neighbor k (except node m) which has maximum neighbors and also its res_k-$Number\_Neighbor_k$>0.
2) If k exists, node k sends its information to node m, and then disconnects its neighbor relation with node m. Node m builds the neighbor relation with node k, and then adds the necessary information to their Neigh_Infor table.
3) If node k does not exist, then node n disconnect its neighbor relationship with node n.

Fig. 1. Topology Optimization Algorithm in Gnutella Network

To make sure that the network topology can be evolved to be better, every node in the system can keep the neighbors with high degree, and change the neighbor relationship from the nodes with low degree to the nodes with high degree. Since the nodes with high degree means that they has built the neighbor relationship with other nodes, and the queries in such nodes can rout more nodes to find the target than other nodes, and those the search efficiency can be kept high. Moreover, the method that the nodes with low degree build the neighbor relationship with the nodes which have amount of resources can reduce workload of the query hotspot nodes.

In order to optimize network topology automatically, the nodes in the network need to maintain a Neigh_Infor table to record the necessary information of its neighbors. In the system runtime, the above table will be continually updated by the exchanging messages among the neighbor nodes and receiving Node_ADD message from the nodes which newly join the system. The above messages include the maximum number of neighbors the node can maintain and the number of current neighbors the node is maintaining. The network topology optimization process is shown in Fig. 1.

4 Performance Evaluation

We use NS-2 as our simulation tool to simulate the query routing process in Gnutella network. In the simulated network, there are five different types of nodes: T1 line access node, special line access node, ADSL access node, LAN access node, and dial up access node, and also the number of nodes in each type of simulated network is: 150, 300, 450, 600, 750, and 900. We simulate the network which has topology optimization and no topology optimization respectively, and also use the average query response time as the metric to evaluate the search efficiency of the proposed topology. The simulation results are shown in Fig. 2.

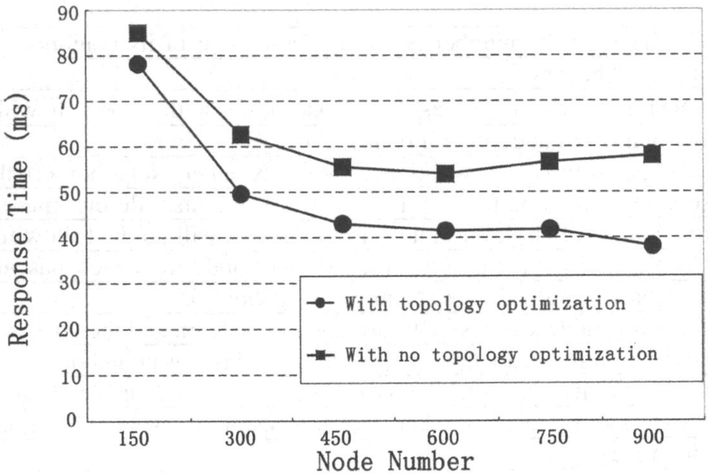

Fig. 2. Average Response Time Comparison

From Fig. 2, we can find that, with increase of nodes in the network, the average response time becomes shorter when nodes do not utilize topology optimization algorithm. However, when the node number reaches a certain level the response time becomes longer again. In comparison, when nodes utilize topology optimization algorithm, with increase of node number in the system, the response time becomes shorter and shorter.

5 Conclusions

In this paper, we propose a topology optimization algorithm based on the dynamic nature of nodes in the network to reduce the unbalanced workload among nodes and speed up the query routing process. To check the performance of the proposed network topology, we use simulation to simulate the Gnutella network with no topology optimization and topology optimization. The simulation results show that the suggested topology optimization can greatly improve the search efficiency in Gnutella network.

References

1. Napster Corporation, http://www.napster.com/pressroom/001013.html
2. Gnutella development forum. The Gnutella v0.6 Protocol (2001), http://groups.yahoo.com/group/the_gdf/files/
3. Gomes. Gnutella keeps growing and growing. WSJ Interactive Edition, http://www.zdnet.com/zdnn/stories/news/0,4586,2766234,00.html
4. Ripeanu, M., Foster, I.: Mapping Gnutella Network. IEEE Internet Computing, 50–57 (January/February 2002)
5. Scipanidkulchai, K., Maggs, B., Zhang, H.: Efficient Content Location Using Interest-Based Locality in Peer-to-Peer Systems. In: Proceedings of the IEEE INFOCOM (March 2003)
6. Lv, Q., Ratnasamy, S., Shenker, S.: Can Heterogeneity Make Gnutella Scalable? In: Druschel, P., Kaashoek, M.F., Rowstron, A. (eds.) IPTPS 2002. LNCS, vol. 2429, pp. 94–103. Springer, Heidelberg (2002)

Erratum: Inter-domain Communication Mechanism Design and Implementation for High Performance

Tan Jie and Huang Wei

School of Information Science and Technology, Jiu Jiang University, Jiu Jiang, China
1370188549@qq.com

A. Xie & X. Huang (Eds.): Advances in Computer Science and Education, AISC 140, pp. 385–390.
springerlink.com

DOI 10.1007/978-3-642-27945-4_76

The paper "Inter-domain Communication Mechanism Design and Implementation for High Performance" by Tan Jie, Huang Wei, DOI 10.1007/978-3-642-27945-4_61, appearing on pages 385-390 of this volume has been retracted due to a serious case of plagiarism. It is a plagiarized version of the paper "Inter-domain Communication Mechanism Design and Implementation for High Performance" Jianbao Ren, Yong, Qi, Yuehua Dai, Yu Xuan, DOI 10.1109/PAAP.2011.41.

The original online version for this chapter can be found at
http://dx.doi.org/10.1007/978-3-642-27945-4_61

[illegible] Inter-domain Communication Mechanism Design and Implementation for High Performance

[illegible]

Author Index